MILITARY BALLOONING

DURING THE EARLY CIVIL WAR

MILITARY BALLOONING
DURING THE EARLY CIVIL WAR

F. STANSBURY HAYDON

THE JOHNS HOPKINS UNIVERSITY PRESS
BALTIMORE AND LONDON

Originally published as *Aeronautics in the Union and Confederate Armies, with a Survey of Military Aeronautics prior to 1861*, vol. 1, by the Johns Hopkins Press, 1941
Johns Hopkins Paperbacks edition, 2000

The Johns Hopkins University Press
2715 North Charles Street
Baltimore, Maryland 21218-4363
www.press.jhu.edu

Library of Congress Cataloging-in-Publication Data

Haydon, Frederick Stansbury, 1908–
[Aeronautics in the Union and Confederate armies]
Military ballooning during the early Civil War / F. Stansbury Haydon—
Johns Hopkins paperbacks ed.
p. cm.
"Originally published as Aeronautics in the Union and Confederate armies, with a survey of military aeronautics prior to 1861, vol. 1, by the Johns Hopkins Press, 1941"—T.p. verso
Includes bibliographical references and index.
ISBN 0-8018-6442-9 (acid-free paper)
1. United States—History—Civil War, 1861–1865—Aerial operations.
2. United States—History—Civil War, 1861–1865—Technology.
3. Aeronautics, Military—United States—History—19th century.
4. United States. Army. Balloon Section—History—19th century.
5. Balloons—United States—History—19th century. I. Title

E492.7 .H3 2000
973.7'3—dc21   99-044060

A catalog record for this book is available from the British Library.

This work is dedicated to the author's

FATHER AND MOTHER

who made possible his residence in the Johns Hopkins University,

and to

RICHARD H. BAKER

the author's chaplain and constant advisor

CONTENTS

LIST OF ILLUSTRATIONS

PLATE

INTRODUCTION TO THE 2000 EDITION

THE MAKING OF A CLASSIC

By the spring of 1941 no one could doubt the immediate and terrifying impact of military aviation. The Luftwaffe returned to London on the night of April 15, 1941, raining almost 8,100 tons of high explosive bombs and more than 152,000 incendiaries down on the city in the heaviest raid of the war to date. Just three months before, test pilot Bill Thorn had taken the four-engine Avro Lancaster on its first flight. In the not-too-distant future, the Lancasters and Sterlings of RAF Bomber Command would be paying visits of their own to Berlin, Hamburg, Dresden, and virtually every other German city of any consequence.

Crete fell to the first airborne invasion in history on June 1 of the same year. Half a world away, Sergeant Floyd Beard became the first in a long line of American paratroopers to lose his life when his parachute failed to open during an exercise at Fort Benning, Georgia. Igor Sikorsky set a world endurance record for helicopters on May 6, when he kept his VS-300 in the air for 1 hour, 32 minutes, and 26 seconds. Just a week later, RAF Captain Frank Whittle, who had risked his career to develop the turbojet engine, watched as the Gloster E.28/39, the first English jet, circled the field at Cranwell. A month before, Fritz Schaeffer had test-flown the Heinkel He 280V–1, the world's first twin-engine jet aircraft.

That spring the Johns Hopkins Press published F. Stansbury Haydon's *Aeronautics in the Union and Confederate Armies*. With a total of 1,834 footnotes (an average of more than four per each of its 421 pages), the book was, as Lincoln biographer Paul Angle gently remarked, a poorly disguised doctoral dissertation "in the best academic tradition."[1] But Angle was much mistaken if he thought that the book would attract no more than a handful of specialist readers. Other reviewers were intrigued by the connection between

---

[1] Paul Angle, review of *Aeronautics in the Union and Confederate Armies*, by F. Stansbury Haydon, *Journal of Air Law and Commerce* 12 (April 1941): 201–202.

what seemed on the surface to be a very detailed, narrowly focused, and meticulously annotated account of a fairly arcane subject and the morning headlines describing the latest events in a global air war.

As J. Duane Squires noted in his review of the book for *Social Studies*: "The importance of military aeronautics at the present stage of world affairs is obvious to all who read the daily papers."[2] Francis Beirne agreed, remarking in his review for the *Maryland Historical Magazine* that "with aviation playing so large a part in the present war, Mr. Haydon's comprehensive study of the long-neglected subject of aeronautics in the Union and Confederate armies has a timely as well as a historical interest."[3]

The *Times Literary Supplement* (London) reviewed the volume on a page illustrated by a photograph of a Spitfire fighter plane, along with an account of a new book on the air battles of 1940. The *Times* reviewer suggested that Haydon's book would appeal not only to those interested in the fine points of the American Civil War but also to anyone interested in "the attitude of soldiers toward innovation."[4] The reviewer for the *Military Engineer* followed suit, recommending the book not only as an example of scholarship, "but as a lesson to all military men of the difficulties and misunderstandings which arise whenever a new means of conducting war is introduced into army circles."[5] A commentator in *Military Affairs* put the matter more colorfully, suggesting that the new volume offered much insight into "the snorting condemnation" of "army brass hats" toward a new technology.[6]

Over half a century after its publication, *Aeronautics in the Union and Confederate Armies* (retitled *Military Ballooning during the Early Civil War*) remains a fascinating case study of technical innovation in the context of a military bureaucracy. It is still the sin-

[2] J. Duane Squires, review of *Aeronautics in the Union and Confederate Armies*, by F. Stansbury Haydon, *Social Studies* 32 (October 1941): 282.

[3] Francis Beirne, review of *Aeronautics in the Union and Confederate Armies*, by F. Stansbury Haydon, *Maryland Historical Magazine* 36 (June 1941): 224–225.

[4] Review of *Aeronautics in the Union and Confederate Armies*, by F. Stansbury Haydon, *Times Literary Supplement*, 24 May 1941, 274.

[5] Review of *Aeronautics in the Union and Confederate Armies*, by F. Stansbury Haydon, *Military Engineer* 34 (July 1942): 370.

[6] Horace S. Mazet, review of *Aeronautics in the Union and Confederate Armies*, by F. Stansbury Haydon, *Military Affairs* 5 (winter 1941): 254–255.

gle best volume on the creation of the U.S. Balloon Corps during the American Civil War and is a model of historical scholarship.

The book has always been something of a puzzle, as well. It was clearly identified on the title page and in the preliminaries as the first volume of a two-volume work. Several generations of graduate students who have gone in search of the second volume can assure you that it was not published. What happened to volume two, and who was the author who contributed this one classic study to the literature of early aeronautics?

Frederick Stansbury Haydon was born at Great Neck, Long Island, New York, on September 1, 1908, and grew up in Towson, Maryland. As a boy, he was fascinated by natural science. "Ever since I can remember," he once explained, "I have had an intense interest in the study of insects."[7] He would retain his interest in natural history throughout his life. By 1935 he had amassed a collection of 5,000 insect specimens neatly labeled and arranged in trays. Following his return from World War II he placed an order with a scientific supply house for insect pins, killing jars, and spreading boards, describing himself as "a very ardent amateur" enthusiast who was anxious to resume his old entomological activities.[8]

Throughout his life, Haydon cherished warm memories of his youthful involvement in the activities of the Baltimore-based Maryland Academy of Sciences. While his professional career would move in a very different direction, it is interesting to note that the teenager who had arranged and identified his collections in such meticulous detail grew into a scholar who would pin his facts to the page with a footnote, as one reviewer noted, "for almost every sentence."[9]

Haydon was educated at the Boy's Latin School of Baltimore (1915–26), and earned a Bachelor of Laws degree from the University of Baltimore School of Law (1927–30). He entered the Johns Hopkins University College of Teachers in 1935 and two years later was admitted as a doctoral student to the history department,

[7] Frederick Haydon, "My First Night Away from Home" (an essay for English 2T [grade A-]) 21 January 1936, Papers of Frederick Stansbury Haydon, Milton S. Eisenhower Library, Baltimore, Md.

[8] Haydon to the Department of Entomology, Ward's Natural Science Establishment, 1 July 1946, Papers of Frederick Stansbury Haydon.

[9] Angle, review of *Aeronautics in the Union and Confederate Armies*, 201–202.

School for Advanced Studies. Haydon worked as a teaching assistant in 1938; he served in that capacity, and as an instructor and student advisor both, at Hopkins and nearby Goucher College, after he received his Ph.D. degree in 1940. In a short volume of poetry written near the end of his life, *Verses for the Johns Hopkins University and the Faculty Members Who Taught Me,* Haydon paid warm tribute to Eric Goldman, then just beginning his distinguished career; Kent Roberts Greenfield, chair of the history department; Frederick Chapin, under whose tutelage Haydon served as a teaching assistant and an instructor in modern European history; Sidney Painter, who taught his students "to appreciate the glories of Medieval France"; and John Calvin French, who taught English Literature and served as the university librarian. French, wrote Haydon, offered "gentle kindness and generous sympathy . . . during the ordeal of my doctoral oral examination."[10]

Haydon's primary focus in graduate school was on European history. Unable, however, either to afford a research trip to the continent or to obtain a travel grant, he was forced to find a topic that could be researched closer to home.[11] While a member of Professor William Stull Holt's Junior Seminar in American History late in 1936, Haydon had prepared a paper on ballooning in the Union Army during the Civil War. The topic interested him. More importantly, the subject was largely untouched, and the bulk of the research materials for a thorough study of the topic were available just an hour's drive south of Baltimore, at the National Archives, the Library of Congress, and the Smithsonian Institution.

"I began gathering material in the winter of 1936," Haydon explained in a 1939 grant application, "and continued without intermission until the fall of 1938." During this period, he became acquainted with leading Civil War scholars, as well as the descendants of T. S. C. Lowe, Gouverneur K. Warren, and other individuals whose lives and careers would figure in his text and who still guarded important manuscript collections. He also became acquainted with a small circle of individuals in Washington, D.C., and New York City who were pursuing and supporting research

[10] Frederick Stansbury Haydon, *Verses for the Johns Hopkins University and the Faculty Members Who Taught Me* (Baltimore, Md: By the Author, n.d.), Papers of Frederick Stansbury Haydon.

[11] Russell J. Parkinson, conversation with the author regarding his discussions with Professor Greenfield, and a meeting with Haydon in 1959.

into the history of flight, notably Paul Edward Garber, of the Smithsonian Institution; A. F. Zahm, N. H. Randers-Pherson, and Arthur Renstrom, of the Aeronautics Division, Library of Congress; and Lester Gardner, of the Institute for Aeronautical Sciences.

As the research drew to a close, Haydon realized that the entire study would far exceed the requirements for the degree. With the agreement of his advisors, he outlined the complete project, then began writing. He spun six solid scholarly articles out of his work in 1938–39 and completed the first half of the manuscript, which he submitted as his dissertation, "Aeronautics in the American Civil War, With a Survey of Military Aeronautics Prior to 1861, Part 1: From Bull Run to the Peninsular Campaign." Accepted for the degree granted in 1940, the dissertation drew the immediate attention of Civil War specialists and received the Mrs. Simon Baruch University Award of $1,000 "for the best university dissertation in Southern, Confederate, or Civil War History for the year."[12] The money went to the Johns Hopkins Press to support the cost of illustrating the published version of the study.

Haydon must have been especially pleased with the reviews his new book received—and also with the enthusiastic support of Lester Gardner, director of the Institute of Aeronautical Sciences, the primary technical society for American aeronautical professionals. Gardner informed Haydon in the spring of 1941 that he had sold fifteen copies of the book to friends in the aeronautical community in one week alone. He also passed along a special souvenir, a letter from Orville Wright, thanking him for the gift of a copy of *Aeronautics in the Union and Confederate Armies.*[13]

Then hoping to publish both volumes simultaneously, Haydon, in November 1939, applied for a John Simon Guggenheim Foundation grant to support completion of the second volume. When he failed to win the grant, university officials provided a monthly stipend of $100 to enable Haydon to continue work on the second half of *Aeronautics in the Union and Confederate Armies.* In addition, he was able to continue teaching at Hopkins and Goucher

---

12 Haydon, application for a Guggenheim Fellowship, "Appendix A," Papers of Frederick Stansbury Haydon.
13 Lester D. Gardner to Haydon, 18 February, 26 June 1941, Papers of Frederick Stansbury Haydon.

College. He seemed to be laying the foundation for a distinguished career, but as happened to so many others of his generation, fate intervened.

Having enlisted in the National Guard in 1927, Capt. F. S. Haydon commanded a battery in the 110th Field Artillery Regiment, 29th Infantry Division, in September 1940, when he received orders warning of the probable mobilization of his unit that winter. In December Haydon was called up, and volume two was very much on his mind. He deposited all of the "research materials, MSS, notes and other items relating to the unfinished second volume" with the aeronautics division of the Library of Congress. He authorized Dr. Ella Lonn, then professor of history at Goucher College and a specialist in Civil War history, to use the materials he had thus far collected and to complete the volume if he "did not survive the war."[14]

Haydon spent the first two years of his active military career in the United States. In addition to normal duties with artillery units, he received a series of assignments as an instructor in the Battery Officer's School, 54th Field Artillery Brigade; senior instructor in the Division of Chemical Warfare School; assistant director of schools, 29th Division; and instructor in the history of American military government in the War Department School of Military Government.

Shipped overseas in 1943, he saw combat in the North African, Sicilian, and Italian campaigns. In addition, he lectured on American staff functions and procedures in the Intelligence Officer's School, French Army, and as an instructor in comparative staff functions and procedures at the British Central Mediterranean Training Center. Haydon was featured in a 1943 *Chicago Sun* article on the training of U.S. Army officers to serve in occupied areas: "Typical of the 'faculty,' Major F. S. Haydon, who once taught history at the Johns Hopkins University, now teaches occupation officers to know the lands they may someday rule. He sketches the histories of the enemy nations, traces the rise of the dictators and explains how the average Axis citizen will probably react."[15]

[14] Haydon, application for a Guggenheim Fellowship, "Appendix A," Papers of Frederick Stansbury Haydon.

[15] "The Army of Occupation: They're Training U.S. Officers to Rule Conquered Lands," *Chicago Sun*, 18 April 1943.

Awarded the Order of the British Empire, he returned to the United States in May 1945. Initially assigned to the Pentagon as a lieutenant colonel in the Plans Section, Army Ground Forces, he accepted a request that he be transferred to the Army Industrial College. Writing to his friend Sidney Matthews in November 1945, Haydon noted that, while he and his wife, the former Margaret Kenway, were living comfortably in Alexandria, Virginia, an "assignment to the National Abortion called the Pentagon . . . would dampen the interest of the most ardent soldier." That being the case, he concluded, "I am most anxious to return to the old profession [of teaching], and hope to do so fairly soon."[16]

Haydon would find life in the academic world far more difficult than he remembered. After a good many letters to old professors, friends, and associates inquiring about teaching positions, he was promoted to full colonel, obtained his discharge, and accepted an offer to join the history faculty at the University of Maine beginning in the fall of 1946. While the source of the difficulty is not clear, things did not go well. "The standards of advanced education, especially in respect to MA degrees," he noted in a letter to one of his own graduate students in the spring of 1947, "have in latter years been so lowered that they scarcely remain as standards." "I do not mean to indicate any severe reflections on this present institution," he wrote to a colleague at the University of Rochester in March 1947, "but as you know, many state universities, especially in this present period of swollen enrollment, do not approximate anything like the standards which I naturally cling to after my training at Hopkins."[17] Given his assessment of the state of academic affairs, Haydon cannot have been too surprised when the department chair informed him that his "services and teaching for the past year have been unsatisfactory." He was told that he would remain on probation until the end of the term, when his case would be "considered." At any rate, he was assured, the university would give a year's notice.[18]

---

[16] Haydon to Sidney Matthews, 8 November 1945, Papers of Frederick Stansbury Haydon.

[17] Haydon to Dr. Printer, 31 March 1947, Papers of Frederick Stansbury Haydon.

[18] Haydon to Edward F. Dow (chair of History and Government at the University of Maine), 19 February 1947, Papers of Frederick Stansbury Haydon.

Kent Roberts Greenfield of Johns Hopkins came to the rescue. Now serving with the U.S. Army Historical Branch and preparing the multivolume history of the army in World War II, Greenfield arranged a job for Haydon at the Center for Military History in Washington. The records of the center indicate that he contributed to the general histories of the North African and Italian campaigns, although he is not listed as a contributing author of any official histories. When Haydon departed his post with Greenfield, he remarked that he was moving to a job in the intelligence community.

Recalled to active duty as a colonel of artillery during the Korean War, Haydon decided to remain in the army. After serving as commanding officer of the ROTC program at Yale, 1952–57, he moved on to serve as Chief for ROTC Affairs with the Department of Reserve Affairs at the Pentagon from 1957 until his retirement from the service in 1959.[19]

Recognizing that it would be difficult for someone of his age to find a college teaching position, Haydon applied for a post at a Massachusetts preparatory school, only to discover that the job would require certification. He enrolled in the Harvard Graduate School of Education, intending to earn an MA in teaching, but was once again appalled by the lack of standards. "Save for a schoolboy smattering," he complained, the professor who taught the teaching of history "knew no history."[20]

Haydon returned to the Baltimore area, where he taught at least occasional classes at the Boy's Latin School and became a familiar figure on the Johns Hopkins campus. A history department staff member remembered him as a "semi-elegant figure" during the closing years of his life. A Hopkins librarian described Haydon as "a character out of a Tennessee Williams play, dressed in a white suit and cadging cigarettes."[21] His losing battle to save an ancient tree threatened by the construction of a new building created a

[19] "Col. F. S. Haydon, 8th Army Veteran," obituary in the *Baltimore Sun,* 13 September 1992.

[20] F. S. Haydon, "Capital E Education: Bah Humbug," *Boxwooder,* no. 105, (1978).

[21] The comment on Haydon's "semi-elegant" appearance came from the administrative officer of the history department during a telephone conversation on 30 October 1999; the description of Haydon as resembling a character from a Tennessee Williams play is from a staff member of the Special Collections Division, Milton S. Eisenhower Library, on 12 August 1999.

minor stir on the campus. He commemorated the occasion with a poem lamenting the sight of "a solitary tragic stump that is all that is left of a once mighty Elm." Haydon expressed his lifelong love of the natural world and his affection for the Johns Hopkins Homewood campus in other poems, one of them "The Dirge to Trees" ("Farewell, defenseless victims, needlessly slain / To make way for the construction / Of harsh brick walls"). He wrote a eulogy for a squirrel found dead on the campus lawn and a tribute to a favorite pond on the university grounds.[22]

And what of volume two? Haydon attempted to get back to work on the project following his return from Europe. Failing to launch a traditional academic career, however, he must have found it difficult to devote serious time and energy to research and writing. When he died in his sleep at a nursing home in September, 1992, there was nothing in the few boxes of papers he left behind to indicate that he had made any progress on the old project—or preserved the results if he had.[23]

Fortunately, however, we do know how he intended to structure the second volume of *Aeronautics in the Union and Confederate Armies*. Before leaving uniform in 1946, he completed a second application for a Guggenheim fellowship, including a detailed outline for his new book. "This projected volume is planned to continue the tactical and technical operations of the Federal balloon service in the several campaigns in which the service participated," he explained, "including the Peninsula, Antietam, Fredericksburg and Chancellorsville."[24]

Haydon planned to demonstrate the value that balloon observations might have had at Fredericksburg and Chancellorsville "had they been acted on by the Federal commanders concerned." The circumstances surrounding the demise of the balloon corps after Chancellorsville would receive special attention. The central focus of the volume, however, would be on a careful assessment of the

[22] Haydon, *Verses for the Johns Hopkins University and the Faculty Members Who Taught Me.*

[23] Haydon, who was estranged from his family at the time of his death, was the father of three children: John Kenway Haydon, Peter S. Haydon, and Margaret Kenway Haydon.

[24] Haydon, application for a Guggenheim Fellowship, "Appendix C," p. 4, Papers of Frederick Stansbury Haydon.

"significance, value and importance of the balloon operations in the Army of the Potomac during the first three years of the War." There would also be chapters on the use of observation ballooning in the Confederate armies, and a discussion of "the numerous plans, proposals, and serious suggestions for aeronautical equipment, material, dirigible balloons, aerial bombing projects, etc., offered to the Federal and Confederate War Departments during the War."

In all, he projected fifteen chapters and four appendixes:

An Outline For: *Aeronautics in the Union and Confederate Armies,* vol. 2

Chapter I	From the Potomac Lines to Yorktown
Chapter II	The Siege of Yorktown
Chapter III	The Advance on Richmond: Seven Pines, Gaines' Mill, and Mechanicsville
Chapter IV	The Seven Days' Battles and the Retreat to the James
Chapter V	Interlude: Rest and Refit at Harrison's Landing
Chapter VI	Antietam and Operations in Maryland and Western Virginia
Chapter VII	The Fredericksburg Campaign
Chapter VIII	The Chancellorsville Campaign
Chapter IX	The Lowe-Comstock Controversy
Chapter X	The Final Operations of the Federal Balloon Service
Chapter XI	General Conclusions—Results of Three Years of Operations
Chapter XII	Confederate Aeronautical Operations: The Potomac Lines and the Peninsula
Chapter XIII	Confederate Aeronautical Operations: Charleston Harbor and Petersburg
Chapter XIV	Aeronautical Plans and Proposals, 1861–1865
Chapter XV	Foreign Comment and Influence on Foreign Development
Appendix I	Bibliographical Essay: A study, description, and commentary on all major sources relating to the subject, with critical treatment and historiographical evaluation.
Appendix II	Critical Bibliography
Appendix III	Roster and Biographical Notices of the Officers who Commanded the Balloon Service of the Army of the Potomac, 1861–1863
Appendix IV	Roster of Unites and Enlisted Personnel Serving in the Balloon Service of the Army of the Potomac [25]

[25] Ibid.

"The above tentative outline of the proposed volume," Haydon explained, "is naturally subject to revision which may be required as a result of findings from further research. . . . It is based on the applicant's present knowledge of the materials available which have been accumulated from previous research. About one-half of the research for this volume was completed before the applicant was called into Federal service, and from this a fair idea of the scope of the present work has been conceived. The above outline represents present plans based on the work accomplished thus far."[26]

Haydon remarked that considerable research remained to be done, particularly in the area of Confederate operations, where the material on hand was "scant." He attached a research plan and a list of collections that he hoped to visit. He estimated that the finished book would run to perhaps 550 pages and would require a year to fourteen months to complete.

The research plan suggests that Haydon wanted to remain focused on traditional archival sources. It should be noted, however, that in preparing volume one he had made good use of artifacts surviving in the collections of the Smithsonian Institution, including T. S. C. Lowe's binoculars, fragments of Civil War balloon fabric, and a balloon valve—it is an indication of his willingness to make imaginative use of nontraditional evidence. The second volume would have been heavily focused on the combat record of the balloon corps; given Haydon's love of the outdoors and his long years of experience as an artillery officer trained to understand and appreciate the importance of terrain, it is easy to believe that he would have tramped the battlefields of the Peninsula Campaign, Fredericksburg, and Chancellorsville in search of the exact locations where the balloons were inflated and flown. I myself have made that trek, and so have other historians of Civil War ballooning.[27] Surely Haydon would have done so as well, and learned a great deal in the process.

---

[26] Ibid., p. 5.

[27] For the pleasure of a day's search for probable balloon sites on the Peninsula battlefields, the author thanks Robert Krick of Richmond Battlefields National Park, and my fellow enthusiast, Dr. Jim Green of the NASA Goddard Research Center. The assistance of National Park Service personnel, who provided information about the sites of balloon operations at Fredericksburg and Chancellorsville, is also appreciated.

Going a step further, Haydon might have discovered for himself exactly what Thaddeus Lowe and his intrepid aeronauts saw from their perch hundreds of feet in the air. Until the emergence of modern propane-fired hot-air balloons in the 1960s, arranging for a balloon flight over one of the Virginia battlefields would not have been simple because gas balloons were relatively rare and expensive to inflate. Still, Haydon must have realized that the experiment of recreating Civil War balloon flight would have yielded considerable insight.

Unfortunately, we are left with only half the story that the author intended to tell—but that half is a very good book indeed. *Aeronautics in the Union and Confederate Armies* not only remains the basic account of the creation and early months of the federal balloon corps, it is recognized as something of a minor classic of historical scholarship. While reviewer Paul Angle feared that readers would find the level of detail and the sheer bulk of the documentation daunting, he also recognized that it was "a study quite likely to be definitive."

In fact it is Haydon's uncompromising scholarly rigor and his attention to the smallest detail that gives the book its extraordinary power. The author tells us how much fabric was used to manufacture every balloon that saw federal service, and he provides the formula for the varnish used to seal the envelopes. He explains the technical details of the mobile gas generators that Lowe designed to inflate his balloons in the field and provides the precise cost of the rubber hose used in their construction. And what color were those generators? Light blue. Haydon found the receipt for the paint.

As early reviewers of the book recognized, the author marshaled those details into a narrative that illuminates large and important questions. Haydon opened his account with a first-rate introduction to the history of military ballooning since the wars of the French Revolution, with special attention to discussions of military aeronautics in the United States since the time of the Seminole Wars. His detailed account of the complicated maneuvering among American balloonists who sought to aid the army prior to the Battle of Bull Run, and his description of the attitudes of various officers toward the balloons during the ensuing months of 1861–62, shed light on the forces that encourage and impede technological change in a military bureaucracy under pressure.

It would be a mistake to regard *Aeronautics in the Union and Confederate Armies* as incomplete. Rather, it simply covers less of the chronological ground than the author had originally intended. The book remains a major contribution to the study of the American Civil War and a classic account of interest to historians of military technology. Surely Frederick Stansbury Haydon would be pleased to know that his work has held up so well and that the Johns Hopkins University Press eventually brought his classic back into print in this new edition.

A Bibliography of Articles by F. Stansbury Haydon Related to *Aeronautics in the Union and Confederate Armies*

"A Proposed Gas Shell, 1862," *Journal of the American Military History Foundation* 2 (1938).

"First Attempts at Military Aeronautics in the United States," *Journal of the American Military History Foundation* 2 (1938).

"Confederate Railroad Battery Jacksonville, Florida,1863, Not the First Use of Railway Ordinance in the United States," *Journal of the American Military History Foundation* 2 (1938).

"An Identification Disc for the Army, 1862," *Journal of the American Military History Foundation* 2 (1938).

"Documents Relating to the First Military Balloon Corps Organized in South America," *Hispanic American Historical Review* 19, no. 4 (1939).

"Grant's Wooden Mortars and Some Incidents of the Siege of Vicksburg," *Journal of the American Military Institute* 4 (1940).

Tom D. Crouch, Senior Curator, Aeronautics Division
National Air and Space Museum
Smithsonian Institution

PREFACE TO THE ORIGINAL EDITION

The ever-increasing importance of the air as a major factor in modern warfare needs neither comment nor discussion. The development of the air weapon with respect to reconnaissance and offensive operations since its initial large-scale employment in the First World War, and the rôles that air forces are now playing in the present Second World War speak for themselves so far as the importance of military aircraft is concerned. Although the air force has become one of the most familiar as well as one of the most powerful and dreaded objects in the world today, the fact that the origin of military aeronautics as an adjunct to warfare dates from the Eighteenth Century is seldom recognized, even by individuals reasonably familiar with military history.

Like so many of Man's accomplishments in the realm of science, his first conquest of the air had been realized less than a week when public suggestion was made to utilize this new power for warlike purposes; and in a scant decade after the two French paper-makers at Annonay sent up their first balloon, the Republican armies of their country were using the result of their discovery over the battlefield. When we witness today with revulsion the results of the ultimate progress of the air weapon, which is daily being employed in the destruction of non-combatant lives and the irreplaceable monuments of civilization and culture, we may pause and wonder at this juggernaut which indirectly stems from the earliest of successful efforts to conquer the air. But it may give us even more pause when we stop and consider that this slaughter and destruction from the air is nothing recent in the minds of men. More than a century before the brothers Montgolfier sent aloft their first balloon in 1783, the Italian, Francesco Lana, had dreamed of such an accomplishment, and prominent among his considerations of the usefulness of his " aerial ship " was the idea of employing it as an engine of war for the bombing of enemy forces and cities from the air—an idea that he published to the world in the year 1670. Similarly Joseph Montgolfier, when the revolutionary wars ravished France, sought to persuade his government to subject the rebellious city of Toulon by bombing it from the air. On down the succeeding decades this idea has progressed, in the wars of the Napoleonic period, in the revolutionary wars of the mid-Nineteenth Century, in the War of the Crimea, and in the struggle which concerns this present study, the American Civil War. The plans and suggestions presented by Northern inventors and air-minded men for the destruction of Richmond and other Southern cities were numerous, and though the progress of aeronautical science had not approached the stage where such offensive tactics were possible, nevertheless the continuity of the idea from Lana to the War of Secession in America is easily apparent. In the present debacle of civilization its fulfilment seems achieved.

Notwithstanding the fact that this development of the air weapon in its present dreaded form has sprung from the invention of the heavier-than-air machine, the balloon is still significant in modern warfare. We may look at the wide-spread use of captive balloons in the First World War; to the recent developments by the American army at Fort Sill with small dirigible balloons for the observation of artillery fire; to the alleged ascension of large numbers of German observation balloons immediately on the Polish border just before the invasion of that unhappy country; and above all to the large-scale and effective use of the " balloon barrage " which has served with appreciable results in the defense of London in the present war. All these examples tend to show that the balloon is still a factor in the scheme of modern warfare.

Although the United States today ranks as a leading air power, little or nothing appears to be known of the inception of American military aeronautics, which took place during our Civil War of 1861-1865. During this struggle for the Union, in which many of the more modern aspects of warfare first made their appearance in this country, was initiated what may be called the first American air force. This novel service, consisting of the operation of observation balloons with American troops for the first time, was limited in scope, and did not attain a state of organization and efficiency commensurate with its possibilities or the abilities of its chief had he been properly supported and encouraged. For reasons not the fault of the aeronauts or of the service performed, the aeronautic establishment did not endure the full period of the war, but when all the facts are known, the balloon service of the Army of the Potomac operated with creditable success, and deserves added recognition because of its direct influence in leading to the later adoption of aeronautics in the British army. For these reasons this first air service should receive a high place in the annals of our military history.

Yet in those annals nothing adequate has been recorded of the work of the aeronauts and liberal-minded officers who supported their efforts in creating and operating the forerunner of one of our most potential armed services. No single epoch in our national existence has produced any greater array of literature, historical or otherwise, than has this sectional struggle of three-quarters of a century ago. But in the thousands of volumes and publications which have appeared relating to almost every phase of this war, not one has properly recounted the history of America's initial adaptation of aeronautics to warfare. It is to fill this gap in our Civil War history that this present work has been undertaken. If it fulfills this purpose, and recalls from oblivion the pioneer work of Thaddeus S. C. Lowe, creator of the first air force in America, and also the work of the other lesser military aeronauts of the same period, John Wise, James Allen, John La Mountain, and others, then the effort and labor expended in its preparation will not be considered in vain.

ACKNOWLEDGMENTS

This study owes its inception to an investigation of the subject as a seminar exercise in the Department of History of the Johns Hopkins University. The subject was suggested by Dr. William Stull Holt, Associate Professor of American History, who has since become head of the Department of History in the University of Washington at Seattle. Since the beginning of this seminar exercise, which developed into a doctoral dissertation and finally into this book, it has been the author's privilege and pleasure to have received the kindly guidance and assistance of Dr. Holt throughout the many stages of this undertaking. It is only fitting in this general expression of appreciation to all who have assisted in the task, that he should be first in the order of sincere acknowledgments. For his encouragement and advice, and general assistance in matters both technical and personal, the writer is infinitely grateful.

To Mrs. Henry M. Brownback, daughter of the chief figure in the present study, the writer wishes to extend his deep appreciation of her innumerable courtesies and assistance, without which the completion of this book could not have been possible. In making fully available the wealth of documents contained in the papers of her father, and in freely contributing her painstaking assistance in searching this large collection, Mrs. Brownback is largely responsible for making possible the writing of this study. In expressing his appreciation, the writer wishes particularly to include his gratitude for the charming hospitality extended to him over the several weeks of research in the Lowe Papers, and to assure Mrs. Brownback that her indispensable contribution in making available her father's papers has been equalled only by the graciousness with which she bestowed this privilege. To the members of her family, Mr. and Mrs. Henry M. Brownback, Jr., and Mr. and Mrs. Henry Lowe Brownback, the writer is likewise grateful for the many courtesies he enjoyed during his stay at Norristown.

Particularly helpful in rendering technical advice and assistance was Mr. Nils H. Randers-Pehrson, of the Division of Aeronautics, Library of Congress. His valuable criticism, expert knowledge, and assistance in locating published sources have been a most valuable contribution. His attendance at sessions of the Seminar in History of the Johns Hopkins University in which chapters of the book were discussed proved to be especially helpful. The writer wishes also to express his thanks for the many privileges extended to him by the Division of Aeronautics, Library of Congress, as a result of the interest shown by Dr. Albert F. Zahm, Chief of the division, by Mr. Randers-Pehrson, and other members of the staff, among whom were Mr. Arthur G. Renstrom and Miss Hollis Piatt.

Through the gracious permission of Miss Emily B. Warren, valuable material in the papers of her father, General Gouverneur K. Warren, was made available for study. Her courteous assistance in searching the private letters of General Warren among her family papers, and her interest in helping to locate other source collections among the descendants of her father's associates and brother officers, will always be remembered. The writer is also indebted to Miss Edna L. Jacobsen, Head of the MSS and History Section of the New York State Library, for her cooperation in searching the Warren Papers.

The writer is also grateful to Brigadier General E. T. Conley, Adjutant General of the United States Army (1937), for permission to search the records of the Old Records Section of his department, and for other assistance. Most cooperative and helpful in the examination of this material were Mr. Thomas E. Blades and Mr. Charles E. Gause, of the Organizational Records Section, Old Records Division of the Adjutant General's Office, whose courteous and willing assistance smoothed many difficulties and materially shortened the laborious task of searching the voluminous collections in the custody of their division. Similar appreciation is extended to the Chief of Engineers, United States Army, and members of his civilian staff, who granted permission to search and assisted in the examination of the old records of the Corps of Engineers and the Bureau of Topographical Engineers.

Valuable assistance was also rendered by members of the

staff of the National Archives, especially Dr. Dallas D. Irvine, Chief of the War Department Division; Mr. Frederick P. Todd, and Mr. Jesse N. Douglas, of the same Division; Dr. Nelson Blake, Chief of the Navy Department Division, and members of his staff; and Miss Edna Vosper, of the Division of Reference.

Most helpful was the cooperation of Mr. Paul E. Garber, Curator of the Section of Aeronautics, Smithsonian Institution, who painstakingly furnished photographs and photostatic copies of material in the Lowe Collection in the custody of the Smithsonian. Mr. Garber also gave valuable aid in suggesting and helping to locate references to a number of important sources. The kind cooperation of Mr. H. W. Dorsey, Administrative Assistant of the Smithsonian, in searching the records of the Institution and the extant records of Professor Joseph Henry, is also greatly appreciated.

To the members of the staff of the Library of Congress the writer is indebted for the many courtesies extended to him during the period of research and especially during the task of preparing the copy for the press. Particularly kind in this respect was Mr. David C. Mearns, Superintendent of the Reading Rooms, who graciously extended privileges of the stacks and made available study-room facilities, and Mr. Henry S. Parsons, Chief of the Division of Periodicals, who bestowed similar courtesies with regard to the files of Civil War newspapers. Equally helpful and cooperative were the members of the staff of the Library of the Peabody Institute, the Librarian of the Army War College, and Captain Dudley G. Knox, U. S. N. (Ret.), in charge of the Naval Records and Library.

Others who gave kindly assistance were Dr. Kent Roberts Greenfield, of the Johns Hopkins University, who translated several Italian sources; Miss Frieda C. Thies, of the library of the same university, who located a MS letter of General E. Porter Alexander and references to other material in the Mackall Collection; Dr. Douglas Southall Freeman, who graciously furnished notes on Civil War balloon operations from his own file of research material; Mr. Roy Knabenshue, a friend of the late Thaddeus S. C. Lowe, who furnished detailed information concerning Lowe's formula for balloon varnish;

Mr. J. T. O'Shaughnessy, of the Photostatic Laboratory of the War Department, who was most helpful in furnishing photostatic copies of documents in the War Department collections; the late Mr. John O'Callaghan, who faithfully gave time and labor in typing drafts of the early chapters of the present work; and Mr. Forrest H. Sweet, of Battle Creek, Michigan, who kindly furnished a photostatic copy of an original MS letter of General James Longstreet.

The writer wishes particularly to express his sincere gratitude also to Dr. and Mrs. St. Julien Ravenel Childs, his former associates in the Johns Hopkins University, and to Mr. and Mrs. J. Frank Williams, whose many kindnesses throughout the years of research in Washington have had a major share in making possible the preparation of this book.

To the Institute of the Aeronautical Sciences, and especially to its Executive Vice President, Major Lester D. Gardner, and Mr. George R. Forman the writer wishes to extend his grateful acknowledgments for their generous and timely assistance in making possible the illustrations and plates contained in this volume. On the eve of publication, when it was apparent that the volume would have to go to press without inclusion of any illustrative plates, Major Gardner and the Institute graciously undertook the support of this added item of publication. To this acknowledgment must also be added the writer's appreciation of the kindness and hospitality bestowed upon him during several visits to the Institute in New York.

F. S. H.

November 15, 1940.

ABBREVIATIONS AND SYMBOLS
CITED IN THE FOOTNOTES

Many of the collections of MSS in the War Department Division of the National Archives bear titles of unwieldy length, and have been so little used in historical publication that they have acquired no standard abbreviations through usage. Constant reference to documents in these collections has necessitated the arbitrary adoption of uniform abbreviations and symbols in order to avoid the repetition of voluminous titles. The writer has also found it expedient to indicate in the footnotes the titles of military rank, assignment, and office of individuals connected with the various armies and component administrative organizations, in order that these individuals may be readily identified and their significance in connection with the cited documents understood. This practice has also necessitated the inclusion of symbols and abbreviations which give this additional information without adding further voluminous titles to the references. The following glossary includes full information represented by all such abbreviations and symbols which appear throughout the notes.

AAAG	Acting Assistant Adjutant General
AAG	Assistant Adjutant General
AAQM	Acting Assistant Quartermaster General
ADC	Aide-de-Camp
AG	Adjutant General
AGO	Adjutant General's Office
AIG	Acting Inspector General
A and IG	Adjutant and Inspector General
AIO	Assistant Inspector of Ordnance
AQMG	Assistant Quartermaster General
AP	Army of the Potomac
BTE	Bureau of Topographical Engineers
C of A	Chief of Artillery
CAQM	Chief Assistant Quartermaster
CE	Chief Engineer
C of E	Corps of Engineers
CG	Commanding General
CQM	Chief Quartermaster
C of S	Chief of Staff

CSA	Confederate States Army
DQMG	Deputy Quartermaster General
HDQ	Headquarters
HQA	Headquarters of the Army
IG	Inspector General
LBAP	Letterbooks, Army of the Potomac
LBCE	Letterbooks, Corps of Engineers
LBHQA	Letterbooks, Headquarters of the Army
LBNV	Letterbooks, Department of Northeastern Virginia
LBSW	Letterbooks, Secretary of War
LBSC	Letterbooks, Signal Corps
LBTE	Letterbooks, Bureau of Topographical Engineers
LRAM	Letters Received, Army of the Mississippi
LRAP	Letters Received, Army of the Potomac
LRCO	Letters Received, Corps of Observation (Stone's Division)
LRDV	Letters Received, Department of Virginia
LRHQA	Letters Received, Headquarters of the Army
LRKB	Letters Received, Kearney's Brigade (Army of the Potomac)
LRNV	Letters Received, Department of Northeastern Virginia
LRPD	Letters Received, Porter's Division (Army of the Potomac)
LRSC	Letters Received, Signal Corps
LRSW	Letters Received, Secretary of War
LRTE	Letters Received, Bureau of Topographical Engineers
MSK	Military Storekeeper, Quartermaster Corps
NDNA	Navy Department Division, National Archives
OD	Officer of the Day
ORD	Old Records Division, Adjutant General's Office
PD	Porter's Division (Army of the Potomac)
PMG	Provost Marshal General
QM	Quartermaster
QMC	Quartermaster Corps
QMG	Quartermaster General
RCO	Records of the Corps of Observation (Stone's Division)
RIDM	Rhode Island Detached Militia
SOAOP	Special Orders, Army of the Potomac
USA	United States Army
USN	United States Navy
USNM	United States National Museum
TRNV	Telegrams Received, Department of Northeastern Virginia
TSAP	Telegrams Sent, Army of the Potomac
WDD NA	War Department Division, National Archives
WNY	Washington Navy Yard

A critical bibliography, including full descriptions of all sources used and cited, together with a bibliographical essay on the general subject of source material for this work, will appear in Volume II.

MILITARY BALLOONING
DURING THE EARLY CIVIL WAR

CHAPTER I

MILITARY AERONAUTICS PRIOR TO 1861

Although air observation as a branch of military science is generally associated with recent advancement in the field ot modern tactics, the use of balloons in warfare is not of recent origin. The earliest public suggestion of employing balloons with armies in the field seems to have been made in France shortly after the beginning of practical experiments in aerostation. The brothers Montgolfier demonstrated the first successful balloon at Annonay, near Lyons, on June 5, 1783, using heated air for ascensive power.[1] The experiment was closely followed by that of the distinguished physicist J. A. C. Charles and the brothers Robert, who sent aloft the first hydrogen balloon from the Champ de Mars in Paris on August 27 of the same year.[2] Some three weeks later Joseph Montgolfier repeated his demonstration at Versailles before the court of Louis XVI, sending up on this occasion the celebrated sheep, duck, and cock that have been known since as the first aerial passengers in history.[3] Thus far the balloon had been proved to be a practical reality, but no human being as yet had ventured aloft. On October 15, Jean François Pilâtre de Rozier, the first man to go up, made his initial ascension from the Faubourg Saint Antoine in Paris.[4] While repeating his experiment several times on the 17th, he was accompanied by André Giraud de Vilette,[5] who seems to have been the earliest recorded advocate

[1] Faujas de Saint Fond, *Description de la machine aérostatique de MM. Montgolfier* (Paris, 1783), I, 3-6. The use of rarefied air on this occasion led to the subsequent application of the name " montgolfier " to balloons operated by this form of lifting power.
[2] *Journal de Paris*, August 27 and 28, and September 13, 1783; Faujas de Saint Fond, I, 7-22. Saint Fond had charge of the arrangements for the ascension and raised a fund by subscription to defray the cost of the experiment.
[3] *Journal de Paris*, September 20, 1783. The ascension took place on September 19.
[4] *Ibid.*, October 22, 1783.   [5] *Ibid.*

1

of aerial reconnaissance in warfare. Vilette was apparently much impressed with his experience, and immediately foresaw the military possibilities of the new invention. On October 20 he wrote an account of his ascent to the *Journal de Paris*, in which he pointed out the advantages of a balloon to an army, and added that the apparatus might also be useful for operations at sea.[6] Thus the idea of employing the balloon as an implement of war was publicly expressed within four months after its invention, and within five days after the first ascent by a human being.

Vilette's suggestion was succeeded in less than a month by that of an Englishman, who published a pamphlet in the following November, in which he called attention to the military value of the newly invented " aerostatic globe." [7] He discussed the potential utility of the apparatus in warfare and advocated its adoption for reconnaissance in land and sea operations, as well as for long distance signalling.[8] The following year an author

[6] Vilette to the editors of the *Journal de Paris*, October 20, 1783, published October 26. After relating the details of his ascent, Vilette declared:

"I observed St. Cloud, Isty, Ivry, Charenton, and Choisy with ease, and perhaps Corbeil, which a light mist prevented me from distinguishing clearly; from this moment I was convinced that this apparatus, costing but little, could be made very useful to an army for discovering the positions of its enemy, his movements, his advances, and his dispositions, and that this information could be conveyed to the troops operating the machine. I believe that it is equally possible, with proper precautions, to make similar use of this apparatus at sea. There, Gentlemen, is an undeniable utility that time will perfect for us."

[7] William Cooke [?], *The Air Balloon: Or A Treatise on the Aerostatic Globe, Lately Invented by the Celebrated Mons. Montgolfier, of Paris* (London, 1783). The identity of the author of this tract is uncertain. It is generally attributed to " Will. Cooke," whose name thus appears handwritten on the title-page of a copy of the first edition formerly in the possession of Isaac Reed [1742-1807]. Cf. bibliographical note in J. H. Hodgson, *The History of Aeronautics in Great Britain* (London, 1924), p. 400. See also facsimile of title-page of Reed's copy, *ibid.*, p. 106.

[8] Cooke, pp. 24-26. A pertinent passage illustrates the author's views on the military possibilities of the aerostat:

" On the first report of a country being invaded, an *Air Balloon* would save the expenses of messengers, posts, &c., from the coasts to the main army, as at the height it ascends, with the assistance of glasses, the number of the enemy, together with their place of landing, might be communicated with great dispatch. . . . A general likewise in the day of battle would derive singular advantage by going up in one of these machines; he would have a bird's eye view of not only everything that was doing in his own, but in the enemy's army. . . . Observa-

of unknown identity published in Paris a remarkable tract also recommending the use of balloons in military operations. This writer went so far as to predict wide changes in the methods of warfare as a result of the adoption of such tactics. Indicating the advantages to be secured from such a new type of war equipment, he included in his program of proposed air operations, reconnaissance and observation, map making, and captive aerial scouts to protect a moving army from surprise.[9] Another volume from the pen of an English air enthusiast appeared in 1784, in which was proposed a scheme of aerial signalling not unlike that designed by T. S. C. Lowe for the Union army during the American Civil War. The author, Thomas Martyn, discussed the feasibility of employing pyrotechnics with balloons in accordance with prearranged codes, and asserted that such a system would be of great value in war, particularly with fleets operating at night.[10] In a lighter vein, as though in anticipation of Major Nicolas Lhomond's fantastic scheme twenty-four years later of invading England with a fleet of troop-carrying balloons, a broadsheet containing doggerel which foreboded such an event was hawked in London the same year that Martyn's suggestions appeared.[11] About the same time, the

tions may likewise be made at sea at a great distance . . . and during sieges they [balloons] may be rendered particularly useful by observing the works of the enemy."

[9] Anonymous, *L'art de la guerre changée par l'usage de machines aérostatiques* (Paris, 1784); also cited in Charles F. Snowden Gamble, *The Air Weapon* (London, 1931), I, 9.

[10] Thomas Martyn, *Hints of Important Uses to be Derived from Aerostatic Globes* (1784). A brief quotation in Gamble, I, 28, indicates Martyn's scheme:

" It has long occurred to me that one obvious purpose to which balloons may be made subservient, is the instant communication of events, by *night signals*. Balloons . . . can ascend to an elevation above the highest mountains; and raising them with *fireworks* of various colours, figures, and dimensions, might thus be made to describe with sufficient accuracy, a variety of interesting particulars. [The italics are Martyn's.]

Compare this idea with the system designed by Lowe for the Union army in 1862-1863, described in Chapter ix, *infra*, pp. 326-329.

[11] The author of the broadsheet is unknown, but his prophesy is amusing and interesting in view of Lhomond's scheme more than two decades later, and present day tactics practised in several European armies. The transport of invading German troops into Austria by aeroplane in March, 1938, the similar conveyance of large bodies of troops into Spain during the late civil war, and the use of parachute troops in the present Second World War, may be said to

4

American scientist Benjamin Franklin was discussing in a letter to a friend the immense tactical value of an army transported by air.[12] Likewise Philip Freneau, celebrated poet of the American Revolution, published a poem on the progress of balloons, in which he predicted the development of French military balloons to combat British naval power. He also prophesied the use of British "fighting balloons" against the United States in the event of another Anglo-American war.[13]

The foregoing examples serve to show that the military possibilities of the balloon were recognized at a very early date, and became the subject of much discussion and speculation. Doubtless many other suggestions of a similar nature were advanced by forward thinking men as well as by visionary

constitute to some extent a fulfilment of the idea. Two appropriate stanzas from the broadsheet are quoted in Gamble, I, 29:

> Should war 'gain break out,
> As is not a doubt
> With some that it may happen soon;
> The French will invade us
> Their troops will parade us
> Brought over in an Air Balloon.

> Their ships will appear,
> Not in water but in air,
> And come in a twinkling down;
> From Calais to Dover,
> How quickly they'll be over,
> Blown up in an Air Balloon.

[12] Franklin to John Ingenhausz, January 16, 1784, John Bigelow (ed.), *The Complete Works of Benjamin Franklin* (New York, 1888), VIII, 432-433.

The invention of the balloon, Franklin wrote, "appears, as you observe, to be a discovery of great importance. Convincing sovereigns of the folly of wars may perhaps be one effect of it, since it will be impossible for the most potent of them to guard his dominions. Five thousand balloons, capable of raising two men each, could not cost more than five ships of the line, and where is there a prince who could afford to cover his country with troops for its defense as that ten thousand men descending from the clouds might not in many places do an infinite amount of damage before a force could be brought together to repel them?"

In a sense, this statement by Franklin more than a century and a half ago may be quoted as the beginning of the modern controversy over the relative merits of aircraft and battleships. Franklin's comment on comparative costs is highly significant.

[13] Philip Freneau, "The Progress of Balloons," Fred Lewis Pattee (ed.), *The Poems of Philip Freneau* (Princeton, 1903), pp. 276-279. The verse was originally published in *Freeman's Journal*, December 22, 1784.

dreamers at this time, but it is apparent that they were unheeded by the military authorities in the countries in which they were offered. For a decade after the above cited examples, the science of aeronautics continued to be confined to civilian experimentation.

The outbreak of the French Revolution, followed by war on a large scale, eventually paved the way for a real trial of the balloon in military operations. In the spring and summer of 1793 the young French Republic faced its most crucial period. Confronted with foreign war against powerful adversaries as well as with internal disorder and rebellion, the revolutionary government was naturally willing to adopt any means that might in any degree contribute to victory. At such a period, proposals were considered seriously that probably would have been ignored by conservative war ministries in more tranquil times. During the crisis of 1793 the advocates of military balloons again came to the fore, this time to receive attention.

Early in that year Louis Bernard Guyton de Morveau [14] frequently pointed out to Lazare Carnot,[15] future director of the French armies, the "infinite usefulness" of balloons to the Republican forces.[16] About the same time, Joseph Montgolfier, co-inventor of the apparatus, urged that they be used for dropping explosives. He later presented a plan for the reduction of Toulon by dropping on the city bombs of immense size and weight.[17] Further influence was exerted by the celebrated mathematician, Gaspard Monge, who also suggested to the National Convention that captive balloons would be of value to the

[14] Sometimes spelled "Guyton-Morveau." He had been chancellor of the Academy of Dijon, and had conducted experiments with balloons and means of propelling them. In 1784 he published a summary of his work under the title of *Description de l'aérostat de Dijon.*

[15] Carnot was personally interested in the subject of aeronautics. In January, 1784, he read a paper on balloons before the Academie de Sciences, which he published the same year as *Mémoire sur les ballons.*

[16] Carnot to Antoine Buissart, February 15, 1793, Etienne Charavay, *Correspondence générale de Carnot* (Paris, 1892), I, 381. Carnot was at this time a deputy to the National Convention. In August of that year he was appointed to the Committee of Public Safety to direct military operations.

[17] R. Delachenal (ed.), *Un agent politique à l'armée des Alpes: Correspondence de Pierre Chépy avec le ministre des Affaires étrangères* (Grenoble, 1894), p. 367; also cited in Gamble, I, 10.

harassed armies of the Republic.[18] Additional proposals recommending the trial of dirigible balloons were made by individuals who claimed to have invented practical propelling machinery for navigating the air.[19] The Convention went so far as to appropriate funds for testing the proposed apparatus, and appointed commissioners to investigate the feasibility of the plans and to supervise the necessary experiments.[20] Likewise, the Committee of Public Safety, on the recommendation of Monge and de Morveau, made similar appropriations.[21] These projects having resulted in failure, the Committee thereupon turned its attention to the question of captive balloons for observation.[22]

On October 25, 1793, an act was passed ordering the construction of a balloon and attending equipment for service with the Army of the North, and the sum of 50,000 livres was

[18] Gaston Tissandier, *Histoire des ballons* (Paris, 1887), p. 133. Tissandier, an ardent student and prolific writer on aeronautics, collected during his lifetime an extensive and valuable library of rare books, pamphlets, journals, and documents relating to the subject. This collection is now in the Division of Aeronautics, Library of Congress. Tissandier was also an aeronaut of no small experience and reputation, having been one of the operators of the French balloons during the siege of Paris, 1870-1871.

[19] Camille Richard, *Le comité de salut public et les fabrications de guerre sous la terreur* (Paris, 1921), p. 619; Decree of the Committee of Public Safety, September 7, 1793, A. Aulard, *Recueil des actes de comité de salut public* (Paris, 1899, *et seq.*), VI, 326; Report of the deputy Moreau, October 10, 1793, *Archives parlementaires* [1st Series, 1789-1799], LXXVI, 309; *Le Moniteur*, October 12, 1793; Proposal of the deputy Seconds, October 16, 1793, *Archives parlementaires*, LXXVI, 631; *Le Moniteur*, October 18, 1793; Gabriel Vauthièr, "Un essai de ballon dirigeable en 1793," *Revue historique de la révolution française et l'empire*, VIII (1915), 305-311; MM. Marre and Desquiuemare to Minister of War Bouchotte, August 19, 1793, MS, *Tissandier Collection,* Division of Aeronautics, Library of Congress.

As early as 1791 a member of the Jacobin Club of Poitièrs named Alexandre had proposed building a dirigible balloon to be used for spreading Jacobin propaganda. Cf. Jean Falconetti, "Un aéronaute patriote en 1791," *La révolution française*, LVIII (1910), 270-271, in which is published Alexandre's MS letter from the Archives de l'Herault.

[20] Decree of the National Convention, October 10, 1793, *Archives parlementaires*, LXXVI, 309; *Le Moniteur*, October 12, 1793; Decree of the Convention, October 16, 1793, *Archives parlementaires*, LXXVI, 632; *Le Moniteur*, October 18, 1793.

[21] Decree of the Committee of Public Safety, September 7, 1793, Aulard, *Recueil des actes*, VI, 325.

[22] Richard, p. 620.

allotted for the purpose.[23] Jean Marie-Joseph Coutelle, a well-known chemist, the scientist Nicolas Conté, and an engineer, Nicolas Lhomond, were charged with the preparation of the apparatus.[24] The use of sulphuric acid for generating hydrogen was specifically forbidden, since every available supply of this substance was sorely needed for the manufacture of munitions. Coutelle accordingly substituted the water decomposition process with satisfactory results.[25] After arranging for his gas and equipment, he proceeded to the headquarters of General Jourdan, to make preparations in the field for the operation of the balloon.[26] He was received with indifference by Jourdan, who refused to take the project seriously,[27] despite a commendatory letter from Carnot.[28] Coutelle thereupon returned to Paris for further instructions, with the result that the Committee of Public Safety ordered additional tests and experiments.[29] These,

[23] Decree of the Committee of Public Safety, October 25, 1793, Aulard, *Recueil des actes*, VIII, 3; Order of the Committee to Nicolas Lhomond, October 25, 1793, Marc de Villiers de Terrage, "Les aérostiers militaires pendant la campagne d'Egypte," *L'Aéronaute*, XXXIV (1901), 90.

[24] Decree of October 25, Aulard, *Recueil des actes*, VIII, 3.

[25] Jean Marie-Joseph Coutelle, *Sur l'aérostat employée aux armées de Sambre-et-Meuse et du Rhin* (Paris, n. d.), pp. 3-4. The same process for generating hydrogen was attempted by John Wise for the Union army in 1861. His apparatus, however, proved to be too cumbersome and expensive for practical use. See Chapter iii, *infra*, pp. 76-77.

[26] Decree of November 1, 1793, Aulard, *Recueil des actes*, VIII, 161; Coutelle, p. 4.

[27] Tissandier, p. 134; Richard, p. 621; Coutelle, p. 4. Coutelle states that Jourdan simply told him to go back to Paris and report to the Committee that an Austrian attack was imminent. Duquesnoy, representative on mission with Jourdan's headquarters, also refused to consider the balloon operations favorably. He regarded Coutelle and his project with scorn and suspicion, and wrote to the Committee that he had no faith in Coutelle, adding that "a battalion is needed more at the front than a balloon." Dequesnoy to the Committee of Public Safety, November 1, 1793, Aulard, *Recueil des actes*, VIII, 166; Archives de la Guerre, Armée du Nord, quoted in Charavay, *Correspondence générale*, IV, 26, n. 2.

Jourdan later regarded balloons so favorably that the official correspondence forms for the Army of the North under his command bore an illustrated heading, showing a portion of his army in battle, over which floated a large observation balloon and its crew. Jourdan to General Ernouf, 23 Fructidor, An IV [September 9, 1796], MS, *Tissandier Collection*.

[28] Carnot to Jourdan, November 4, 1793, Charavay, *Correspondence générale*, IV, 25-26.

[29] Decree of November 24, 1793, Aulard, *Recueil des actes*, VIII, 672-674.

carried on over a period of several months, evidently convinced the Committee of the usefulness of the apparatus to the army, and on April 2, 1794 an act was passed creating a regular balloon company, later known as the *1ᵉʳ Compagnie d'Aérostiers*.[30] The first air corps in history thus came into existence; the ideas of Martyn, Villette, and others became reality.[31]

The organization of the newly created company is worthy of some attention. It consisted of a captain, a lieutenant, a sergeant-major, who served also as a quartermaster, a line sergeant, two corporals, and twenty privates. The personnel were to be selected because of special knowledge of chemistry, sketching, carpentry, and masonry. A distinctive uniform was prescribed for the unit, and each man was armed with two pistols and a saber.[32] The act creating the company required

[30] Decree of April 2, 1794, Aulard, *Recueil des actes,* XII, 349-350.

[31] The literature on the operations of the French military balloon service in the revolutionary wars includes a number of reliable primary and secondary accounts, of which the following give satisfactory details: Albert Joseph de Selle de Beauchamp, *Mémoires d'un officier des aérostiers aux armées de 1793 à 1799* (Paris, 1850); Coutelle, cited *supra*; Charles Dollfus and Henri Bouché, *Histoire l'aérostatique* (Paris, 1923); Paul Émile Faucart and Jules Finot, *La défense nationale dans le nord, de 1792 à 1802* (Lille, 1890); Gamble, cited *supra*; Gustav de Gaugler de Gempen, *Les compagnies d'aérostiers militaires sous la République de l'an II à l'an X* (Paris, 1857); Freiherr vom Hagen, " Geschichte der militärischen Aëronautik," *Zeitschrift des Deutschen Vereins zur Förderung der Luftschiffart,* I (Berlin, 1882); Lieutenant F. Letonné, " Les aérostiers militaires pendant les guerres de la révolution," *Revue de génie militaire,* XXV (Paris, 1903); Major Hermann W. L. Moedebeck, *Taschenbuch zum praktischen Gebrauch für Flugtechniker und Luftschiffer* (Berlin, 1904); *Moniteur de l'armée,* October 1, 1851; Ladislas d'Orcy, " Early History of the French Aeronautic Corps, 1794-1802," *Air Service Journal,* II (New York, 1918); Charles Oscar Paullin, " Early Use of Balloons In War," *United Service Magazine,* XXXIX, n. s. (London, 1909); Captain W. Peters, " Militärluftschiffart im Feldzug, 1794," *Mitteilungen des K. u. K. Kriegsarchivs,* VII (Vienna, 1907); Étienne Gaspard Robertson, *Mémoires récreatifs, scientifiques et anecdotiques du physicien-aéronaute* (Paris, 1833), II, 15-37; A. Sircos and Th. Pallier, *Histoire des ballons* (Paris 1876); Gaston Tissandier, *Histoire des ballons* (Paris, 1887); *idem.,* " Nouveaux documents sur les aérostats militaires de la première république," *La Nature,* V (Paris, 1877); R. S. Waters, " Ballooning in the French Army during the Revolutionary Wars," *Army Quarterly,* XXIII (London, 1932); " La Campagne de 1794 et l'armée du nord: Les aérostiers," *Revue d'histoire, rédigé à l'État-major de l'armée,* I (Paris, 1905).

The original MS journals of ascensions made with the Army of the North and the Army of the Sambre-et-Meuse are preserved in the archives of the French Corps of Engineers, Paris.

[32] The uniform consisted of coat, waistcoat, and breeches, blue, piped with red;

drill and maneuvers to be held at once, in order that the unit might attain a state of efficient training as soon as possible. The officers and men were to receive pay and allowances on the same basis as regular troops of the line.[33] In short, the Committee of Public Safety set up a well-organized military unit, subject to army regulations and training, with full status as regular troops.[34]

Equally interesting is the program of tactical objectives laid down for the training of the company. The object of the service was outlined under three sections, which included reconnaissance and observation, signalling between various friendly divisions in the field, and the spreading of propaganda pamphlets from the air.[35]

Coutelle, whose previous mission to Jourdan had been so discouraging, was commissioned captain to command the company, and was charged with the enrollment of the personnel.[36] The engineer Lhomond received the lieutenancy,[37] and his son was enrolled as an enlisted man in the ranks.[38] The unit was ordered to be mustered in and trained with the least possible delay and to report for active duty with the Army of the North.[39] By the end of May, 1794, the company with its new balloon

collar and cuffs black, with red facings; and buttons of the infantry pattern. A fatigue uniform of blue cotton slacks and jacket was also prescribed.

[33] Decree of April 2, 1794, Aulard, *Recueil des actes*, XII, 349-350.

[34] Compare this organization with that of the Balloon Corps of the Army of the Potomac, 1861-1863. The Federal War Department might have learned much from the earlier French example. See Chapter viii, *infra*.

[35] " Extract from Instructions Regarding the Service of a Company of Balloonists, Organized by the Committee of Public Safety," trans., by Captain H. A. Scholle [ACR], *United States Air Services*, XII (May, 1927), 41-42.

[36] Decree of April 2, 1794, Aulard, *Recueil des actes*, XII, 350.

[37] Coutelle, p. 13, n. 11; Hermann W. L. Moedebeck, *Pocketbook of Aeronautics* (trans. from the German by W. M. Varley, London, 1907), p. 230.

[38] Decree of April 10, 1794, Aulard, *Recueil des actes*, XII, 500. The Lhomonds, father and son, were the first of a long series of military balloonists who were closely related either in the degree of parent and child, or as brothers. In the Italian Campaign of 1859, the two Godard brothers served as military aeronauts with the army of Napoleon III. In the Federal armies during the Civil War, the balloonists included T. S. C. Lowe and his father, Clovis Lowe; John Wise and his son Charles; the brothers James and Ezra S. Allen; John B. Starkweather and his brother James. See p. 21 *infra*, and Chapters iii, and vii, *infra*.

[39] Decree of April 20, 1794, Aulard, *Recueil des actes*, XII, 705.

Entreprenant joined Jourdan's forces at Maubeuge, and on June 2 the first military reconnaissance from the air took place in the midst of a severe bombardment and the enthusiastic cheers of the French troops.[40] From June 23 to June 25 ascensions were made before Charleroi, in which a general officer went aloft to study the enemy positions.[41] On the 26th, the battle of Fleurus was fought, during which the *Entreprenant* and its observers remained in the air for some nine hours. Jourdan's adjutant general, Morlot, accompanied Coutelle for a part of this extended ascension, and their dispatches conveying information of the Austrian movements were frequently dropped from the car for delivery to the French general headquarters.[42]

The value of the information thus obtained has been a matter of conjecture. Guyton de Morveau, present at the battle, reported that dispatches sent by Morlot contributed to Jourdan's decisions in making his dispositions.[43] The deputy Lakanal, early investigator of the semaphore-telegraph system for the Republican armies,[44] was enthusiastic in his praise of the balloon observations, declaring that an army in battle without such equipment would be comparable to a man fighting a duel blindfolded.[45] Jourdan, however, seems to have been silent on the subject, and Coutelle was moderate in his estimate.[46] After

[40] Tissandier, p. 136; Testimony of Guyton de Morveau before the Convention, *Le Moniteur*, September 27, 1794; Selle de Beauchamp, pp. 39-40. Selle de Beauchamp was at this time enlisted in Coutelle's company. He later received a commission as an officer.

[41] Tissandier, pp. 136-137; Coutelle, p. 7; Report of de Morveau, *Le Moniteur*, September 27, 1794.

[42] Selle de Beauchamp, pp. 46-48; Tissandier, p. 137.

[43] Guyton de Morveau to the Committee of Public Safety, June 27, 1794, Aulard, *Recueil des actes*, XIV, 563.

[44] See Lakanal's Report to the National Convention on the initial telegraph experiments, *Le Moniteur*, July 29, 1793.

[45] Lakanal's Report, quoted in Richard, p. 626.

[46] In the account of his service, Coutelle recorded:

"I will not claim, as do those who praise or blame with exaggeration everything that is new, that the balloon won the battle of Fleurus. On this memorable day, every corps did its full duty. What I can say is, that well served by my glass, despite the oscillation and movement caused by the wind, I was able to distinguish clearly the corps of infantry and cavalry, the parks of artillery, their movements, and in general, the massed troops." Quoted in Tissandier, p. 137. A similar statement occurs in a letter to Beauchamp, dated July 3, 1794. Selle de Beauchamp, pp. 87-88.

Fleurus, the company followed Jourdan's advance into Belgium, and was in service during the remainder of the campaign, notably at Liége and Brussels.[47] A new balloon, the *Martial*, had been constructed at Meudon [48] under de Morveau's direction, and was sent to Coutelle for service with his company. This new aerostat was cylindrical rather than spherical, and proved unsatisfactory. Because of its shape it presented more resistance to the wind, and was thus too unsteady for effective observations.[49] As winter approached, the company took up quarters at Borcette, near Aix-la-Chapelle, where a balloon depot with aeronautic supplies and a gas furnace was established under Coutelle's supervision.[50]

In the meanwhile the Committee of Public Safety ordered on June 23 the enrollment of a second company of balloonists to be organized and trained by Nicolas Conté; [51] and in the following October created a permanent training school of military aeronautics at Meudon, to be known as the *École Nationale Aérostatique*.[52] The purpose of the latter was to train men for the French balloon service, construct and repair equipment, and develop and improve the technique of the new tactical arm. The school was composed of sixty balloonists, divided into three sections of twenty men each, with a lieutenant, a sergeant and two corporals attached to each section. A quartermaster and military storekeeper were appointed with assistants to serve all three sections, and were vested with responsibility for all materiel, supplies, and equipment. The entire establishment was placed under a director, assisted by a sub-director, who

[47] Report of de Morveau, *Le Moniteur*, September 27, 1794; Selle de Beauchamp, pp. 51-52.

[48] The Chateau de Meudon was the base of the early experiments and tests involving the balloon projects for the armies, and has, with intervals of interruption, remained a center of French military aeronautics up to recent times.

[49] Coutelle, pp. 7-8; Richard, p. 629.

[50] Coutelle, p. 8; Tissandier, p. 138.

[51] Decree of June 23, 1794, Aulard, *Recueil des actes*, XIV, 467.

[52] Decree of October 31, 1794, Aulard, *Recueil des actes*, XVII, 702-704. The activities and administration of the *École Aérostatique* can be traced in the laws, decrees, debates and correspondence of the Directory, published in A. Debidour, *Recueil des actes du Directoire exécutif* (Paris, 1910, *et seq.*), I, 295-296, 782; II, 147, 191, 275, 347, 407, 561, 635, 697; III, 163, 207-210, 365, 467, 480; 615; and IV, 303, 749-751.

supervised the training and administration. Conté, who had been designated to form the *2me Compagnie d'Aérostiers*, was named to head the school.[53]

The 2nd Company, created on paper by the decree of June 23, was not fully organized and trained until the following year. Coutelle was recalled from Borcette to command the new unit, which was to report for duty with its new balloons, *Hercule* and *Intrepide*,[54] with Pichegru's Army of the Rhine.[55] Ascensions were conducted throughout the siege of Mayence, and later at Mannheim, after which the company remained with Pichegru's forces until the winter of 1795, when quarters were established at Frankenthal, with a depot similar to that previously set up at Borcette.[56] The following year the 2nd Company was ordered to Molsheim, near Strasbourg, where it was attached to General Moreau's headquarters. Ascensions took place throughout the campaign of 1796, principally at Rastadt, Stuttgart, and Donauwerth.[57]

In the meanwhile the 1st Company, now under Lhomond since Coutelle's transfer to the 2nd, continued in service with Jourdan's army, and saw active service until the battle of Würzburg, in the campaign of 1796. In this battle Jourdan was defeated by the Archduke Charles, and the balloon and its entire company were captured by the Austrians.[58]

[53] Decree of October 31, 1794, Aulard, *Recueil des actes*, XVII, 704. Conté's services with the school were recognized and rewarded by his elevation to the rank of brigadier general of infantry. Decree of the Directory, June 16, 1796, Debidour, II, 635.

[54] Curiously enough, the best known of the Federal balloons during the American Civil War was named the *Intrepid*. The Chief Aeronaut, Lowe, who thus named it, possibly took the name from the early French aerostat, with the history of which he seems to have been familiar. Lowe's first balloon to be used in military experiments was also named *Enterprise*, which, though not an accurate English equivalent of Coutelle's *Entreprenant*, suggests that Lowe also took this name from the French aeronautic service. See pp. 162, ff., and 238, ff.

[55] Tissandier, p. 139; Coutelle, p. 8. Lhomond had succeeded Coutelle in command of the 1st Company.

[56] Tissandier, p. 140; Selle de Beauchamp, pp. 53-55, *passim*.

[57] Letonné, p. 411; Moedebeck, p. 231. Moreau had superseded Pichegru in command of the Army of the Rhine.

[58] Letonné, p. 408; Tissandier, p. 141; E. Cazalas, "Le ballon militaire capturé à Wurzbourg en 1796," *Carnet et sabretache* [*Revue militaire rétrospective*], Series ii, May 1909, 307-317.

The following year the 2nd Company was transferred from Moreau's Army of the Rhine to the Army of the Sambre-et-Meuse under Hoche. It remained inactive, however, throughout the period of hostilities that year, and performed no service despite repeated applications of its officers for active duty.[59] The cause of this inactivity is generally attributed to Hoche's disinclination to believe in the value of the service. His disapproval is clearly shown in his letter to the Minister of War on August 30, in which he requested that the balloon unit be removed from his army for the reason that it was "entirely useless."[60] One writer has asserted, however, that the difficulty was largely caused by the lack of initiative and energy on the part of Captain Delaunay, who replaced Coutelle, the latter having retired temporarily to Meudon after an attack of fever.[61]

In April, 1797, the Capitulation of Leoben restored the personnel of the 1st Company, captured at Würzburg the previous year, to the French service. Lhomond and his lieutenant, Plazanet, proceeded to Meudon and requested their old commander to reorganize the company.[62] By the time reorganization was complete, plans for Napoleon's expedition to Egypt were under way, and the future emperor listed the unit among the special troops to accompany his army in this campaign.[63] Coutelle, now raised to a colonelcy, commanded the aeronautic detachment, with Lhomond, also promoted to the rank of major, second in command. The balloonists, however, were not destined to perform aerial duty in the Egyptian Campaign. For some reason not disclosed by available sources, the balloon and its attending equipment were not taken ashore when the French army disembarked at Aboukir, but remained with the vessels in the roadstead. The subsequent destruction of the French fleet in the battle of the Nile likewise accounted for the aeronautic

[59] Tissandier, p. 141.

[60] Hoche to the Minister of War, August 30, 1797, *ibid.* The letter is also published in Letonné, in *Revue de génie militaire*, XXV, 412, and in Sircos and Pallier, p. 268.

[61] Letonné, pp. 410, 412.

[62] Sircos and Pallier, p. 269.

[63] "Tableau des corps des troupes rassemblés à Toulon," April 14, 1798, *Correspondence de Napoléon Ier* (Paris, 1860), IV, 82.

apparatus attached to the expedition, and the balloon detachment thus performed no regular service, though its personnel were used to advantage on other special technical details.[64] Napoleon, however, had several large montgolfiers constructed, which he sent up before large gatherings at Cairo to impress the native population with the military and scientific prowess of the French.[65]

By 1798 the French balloon service was definitely on the decline. Governmental interest in military aeronautics seems to have diminished after the open disapproval of Hoche, and though Napoleon had ordered the 1st Company to Egypt, he does not appear to have developed any particular liking or interest for military balloons.[66] Pichegru also had no faith in the apparatus, having stated to the American minister to Holland as early as 1795 that the balloonists had not proved of any value to the French armies.[67] It is difficult to determine whether the government was influenced by these opinions, and consequently discontinued the aeronautic service as a result. Whatever the causes may have been, the Directory, without stating specific reasons for its action, passed an act on January 18, 1799

[64] Detailed accounts of the balloon detachment in the Egyptian Campaign appear in Marc de Villiers de Terrage, "Les aérostiers militaires pendant la campagne d'Egypte," *L'Aéronaute*, XXXIV (1901), 85-91, and in Maurice Wolff, "Bonaparte et les aérostiers militaires en Egypte," *Revue aérienne*, V (1912), 158-161.

[65] *Ibid.*, 159-160; De Villiers de Terrage, in *L'Aéronaute*, XXXIV, 107-111; Napoleon to General Cafarelli, January 12, 1799, *Correspondence de Napoléon Ier*, V, 325.

[66] Napoleon's reliance on his cavalry for information, and the rapid mobility of his tactics may have rendered balloon observation impractical in his campaigns. One of his most devoted marshals, however, seems to have had some interest in balloon observation. A. G. Macdonell in his *Napoleon and His Marshals* (New York, 1934) tells an amusing story of how "an ingenious civilian gentleman arrived in the Pas-de-Calais with an observation balloon to sell," when Ney was in command of the 6th Corps of the Army of the Coasts of the Ocean in 1803. The balloon seller must have been persuasive, so Macdonell relates, for he succeeded in extracting 30,000 francs from Ney, and never delivered the balloon. Ney's interest in the apparatus may have been aroused by his having witnessed Coutelle's operations with Jourdan's army at Fleurus. He was the only one of the seven marshals in 1803 who had fought in this battle. Macdonell, pp. 93-94.

[67] Diary of John Quincy Adams, entry of February 16, 1795, Charles Francis Adams (ed.), *Memoirs of John Quincy Adams* (Philadelphia, 1874), I, 76-77.

disbanding the two balloon companies.[68] In conformity with
this decree the 2nd Company was ordered disbanded by the
Minister of War on March 4, 1799.[69] The 1st Company, still
in Egypt when the decree was issued, was mustered out of
service on its return to France in 1802. The *École Nationale
Aérostatique* was finally closed, and for more than a quarter of
a century the practice of military aeronautics seems to have
ceased in France.[70]

One attempt to revive the service was made by Major
Lhomond in 1808, a fantastic plan of invading England from
the air. Still enthusiastically interested in his adopted profession, Lhomond proposed that Napoleon send a fleet of a hundred immense montgolfiers, each one hundred meters in diameter with cars sufficiently large to carry a thousand men, two
cannon with caissons, twenty-five horses, and forage and food
supplies for ten days, as well as fuel for fires to generate the
heated air for ascensive power. Napoleon sarcastically inquired
if the scientist Monge would " take the trouble " to make a full
scale test of the plan.[72] Naturally nothing ever came of the
scheme.

After the discontinuance of the French aeronautic activities in
1799, the employment of military balloons in Europe assumed
a course of intermittent trials, proposals, and experiments, most
of which were productive of little or no practical results. It may
be noted that in most of these experiments, offensive tactics,
such as bombing from the air and the spreading of propaganda,
replaced the function of reconnaissance and observation. An
attempt was made by the Danes in 1807 to break the English

[68] Decree of the Directory, January 18, 1799, Moedebeck, p. 232.
[69] Milet-Muneau to the Commander of the Army of Mayence, March 4, 1799,
ibid., 231.
[70] Tissandier, p. 143; Sircos and Pallier, p. 270.
[71] Napoleon to Gaspard Monge, October 28, 1808, *Correspondence de Napoléon Ier,* XVIII, 13.
A similar and even more absurd scheme had been devised and presented in
1797 by a physician named Thilorier, who offered to construct a " portable
camp " and a montgolfier large enough to transport to England an army of conquest. *Le Moniteur,* November 27, 1797. Thilorier had, during the preceding
year, interested the Government in an invention for propelling boats against the
wind, and had been encouraged to make his invention available for use on the
rivers Oise and Seine. Decree of the Directory, August 19, 1796, Debidour, III,
396-397.

blockade by bombing from the air the British fleet off Copenhagen. A dirigible balloon operated by hand power was built for this purpose by an aeronaut named Kolding, but the propelling mechanism was found to be impractical, and the project resulted in failure.[72] The following year Denmark again made use of aerostats for war purposes, this time for the spreading of propaganda in the territory of her neighboring enemy, Sweden. For a month balloons were sent up daily from the castle of Cronberg, bearing copies of an " inflammatory address to the Swedish nation, instigating them to revolt against their then sovereign, and urgently recommending them to the extraordinary virtues of the King of Denmark." These pamphlets, landed on the Swedish coast by the free balloons, were seized and suppressed by police officials. Thus the second venture of the Danish government in military aeronautics proved unsuccessful.[73]

Russia entered the field five years later during Napoleon's campaign of 1812 with a plan to attack the French army from the air. On this occasion Alexander I employed a German engineer, Leppig, to construct a huge, fish-shaped aerostat, to be propelled by fins, and capable of carrying men and large quantities of explosives.[74] A contemporary witness has said that one of its objectives was to hover over the French army and destroy Napoleon's general headquarters and, incidentally, the Emperor also.[75] The propelling machinery, as in the case of Kolding's attempt, could not be made to work, and though the immense envelope was inflated, a process that required five days, it was never raised.[76] One other incident that may be

[72] Freiherr vom Hagen, in *Zeitschrift des Deutschen Vereins zur Förderung der Luftschiffart,* I, 354.

[73] Adams' Diary, entry of September 29, 1809, Charles Francis Adams (ed.), *Memoirs of John Quincy Adams,* II, 33.

[74] Freiherr vom Hagen, p. 354; Moedebeck, pp. 321, 322; Charles Oscar Paullin, " Early Use of Balloons in War," *United Service Magazine,* n. s., XXXIX, 533.

[75] General Philip de Segur, *History of the Expedition to Russia Undertaken by the Emperor Napoleon in the Year 1812* (Eng. trans., Philadelphia, 1825), II, 9.

[76] Freiherr vom Hagen, p. 355; Paullin, in *United Service Magazine,* XXXIX, 533; Moedebeck, p. 322.

A Russian account of this balloon, with an illustration from a drawing in the

listed as aeronautic activity during this campaign was the reported use of a fire-balloon by the Russians as a signal for the
burning of Moscow.[77] The French invaders seem to have confined their aeronautics to wishing, according to General Caulaincourt's journal, which reports that two of Napoleon's officers,
Count Daru and Marshal Duroc, expressed a keen desire for a
balloon to carry the Emperor to safety during the disastrous
retreat from Moscow. Napoleon humorously agreed that such a
piece of equipment " would not be de trop." [78]

In 1830 military aeronautics were temporarily revived in
France during the campaign in Algeria. Jean Margat, a professional balloonist of some reputation, was employed by the
Minister of War to serve with the African expedition. His
service began rather inauspiciously. While his train was en
route to Africa in the brig Vittoria, his carboys of sulphuric
acid were accidentally broken by the rolling of the ship, and
caused a fire that resulted in damage amounting to some 80,000
francs. Of his operations after his arrival in Algeria, little or
nothing is revealed in available sources.[79] It appears, however,
that he made one ascension under fire during the campaign,
and for this received a citation of praise from the general staff
of the Army of Africa.[80]

The Revolutions of 1848 gave rise to more experiments in
the use of balloons in warfare. The first instance at this time
occurred during the siege of Milan, in which the defenders of
the city sent up paper montgolfiers to carry printed pamphlets,
proclamations, and general propaganda, which burning time
fuzes detached from the free envelopes and scattered throughout the surrounding territory.[81] In the siege of Venice in 1849

Archives of Alexander I, appears in Alexander Rodnykh, Istoriia vozdukhop-
lavanea i letanea v Rossii, II (St. Petersburg, 1912), 18 ff.

[77] De Segur, II, 31.

[78] Armand Auguste Louis de Caulaincourt, With Napoleon in Russia (New
York, 1935), pp. 234, 236.

[79] Dollfus and Bouché, Histoire de l'aérostatique, p. 59.

[80] Facsimile of citation signed by Brigadier General Tholozé, General Staff,
Army of Africa, July 24, 1830, ibid.

[81] Dollfus and Bouché, p. 59. As early as 1842 the Italian patriot Mazzini
had amused Jane Carlyle by discussing the possibility of freeing Italy from
Austrian domination by means of military dirigible balloons. He had been

18

the Austrians resorted to similar tactics, but the balloons carried bombs rather than bombast, and the process was more scientifically devised. At the suggestion of an artillery officer, Uchatius, two aerial torpedo battalions were organized, each equipped with wagons, wind-screens, small pilot balloons, a hundred large montgolfiers, and a hundred bombs. These units set up their apparatus in positions windward from the city and released the pilot balloons to obtain direction. The courses of the small envelopes were plotted on bombing charts, and the wind velocity calculated. These computations completed, the bombs were then attached to the large montgolfiers with explosive boosters and releasing fuzes timed to discharge the projectiles over the city.[82] In addition to those released by the land forces, a number of these novel bombing devices were sent aloft from frigates at anchor behind the Lido.[83] In spite of the detailed preparations, the effect of this new form of attack seems to have been slight, and several writers have discredited the scheme as a complete failure.[84]

approached by Muzio Muzzi, who claimed to have invented an apparatus for propelling balloons in any direction, and was willing to sell the plans to Mazzini for the use of the patriots against Austria. Upon hearing these schemes, the wife of the great critic disavowed any desire to begin a war by " impersonating fallen angels," and added that Mazzini was as " credulous and ignorant as a two-year old child." Jane Carlyle to Jeannie Welsh, October 22, 1842, Leonard Huxley (ed.), *Jane Welsh Carlyle: Letters to Her Family, 1839-1863* (New York, 1924), p. 34. Two years later Muzzi came to America and patented his apparatus. U. S. Patent No. 3799, October 16, 1844, U. S. Patent Office, Washington. Two decades later, Solomon Andrews of Perth Amboy, New Jersey, built his *Aeron*, using Muzzi's principles, and offered the aerostat to the War Department for service in the Civil War. See Volume II.

[82] Moedebeck, p. 221; Paullin, in *United Service Magazine*, XXXIX, 534.

[83] Alexandre LeMasson, *Venise en 1848 et 1849* (Paris, 1851), p. 224; Gugliemo Pepe, *Narrative of Scenes and Events in Italy from 1847 to 1849, including the Siege of Venice* (trans. from the unpublished Italian MS, London, 1850), II, 116.

[84] George Macaulay Trevelyan, *Manin and the Venetian Revolution of 1848* (London, 1923), p. 231; LeMasson, p. 224; Dollfus and Bouché, p. 59; Pepe, II, 116. General Pepe, who commanded the Venetian forces in the defense, asserts that only one bomb fell within the city, and that the aerial bombardment served only to amuse the defenders.

Further evidence of the failure of the Austrian air bombardment can be found in an entry dated July 12, 1848, in P. Contarini, *Memoriale veneto storico-politico, 1848-1849* (Venice, 1874); also quoted in Antonio Pilot, " Palloni austriaci su Venezia nel 1849," *Rassegna storica del risorgimento* XIV (1927),

In England up to this time little progress had been made other than a number of proposals from civilians and army officers recommending the application of aeronautics to the military service.[85] In 1803 Major John Money published a tract urging that the British army adopt balloons, but extravagant claims of their value and an acrid allusion to the hide-bound conservatism of " old generals " doubtless hindered rather than furthered his cause, and the suggestion was ignored.[86] In 1806 aeronautical tactics of a sort, the first in British history, were practised by Lieutenant Lord Cochrane, who towed kites from his brig *Pallas* off the French coast for the purpose of releasing propaganda pamphlets inland, much to the annoyance of the Napoleonic government.[87] Captain Thomas Cooper in 1809 gave space to the use of balloons in one of his volumes of instructions for officers,[88] and Charles Rogier brought out a pamphlet discussing the merits of aerial bombing a decade later.[89]

735. See also contemporary anonymous letter written from Treviso, published in *Raccolta di atti e decreti e nomine . . . del Governo provisorio di Venezia* (Venice, 1849); also quoted in Pilot, pp. 723-725.

See also E. M., "Ai margini della storia. Il bombardamento aereo di Venezia nel 1849," *Revista aeronautica,* III (Rome, 1927), pp. 173-176; W. de Fonvièlle, "Le bombardment aérostatique de Venise en 1849," *L'Aéronaute,* XXXIX (1906), pp. 94-96.

[85] An excellent, detailed treatment of the development of military aeronautics in England can be found in Gamble, *The Air Weapon,* I, Chapter i.

[86] John Money, *A Short Treatise on the Use of Balloons and Field Observateurs in Militaary Operations* (London, 1803). Money claimed that the battle of Fleurus had been won because of the balloon observations, and asserted that the colonial war in America could have been won by England had balloons been known and used at the time. He also assailed the conservative military authorities, declaring:

" I would not consult old generals whether balloons or observateurs could be of use to the army, for I know that the answer would be ' that as we have done hitherto very well without them, then we may still do without them,' and so we did without light artillery, riflemen, telegraphs, &c. . . ." *ibid.,* p. 18.

[87] Thomas Cochrane, Tenth Earl of Dundonald, *The Autobiography of a Seaman* (London, 1860), I, 201.

[88] Thomas M. Cooper, *The Military Cabinet: Being a Collection of Extracts from the Best Authors . . . to Give Young Officers Correct Notions in Regard to Many Subjects Belonging to or Connected with the Military Profession* (London, 1809), I, 88-89.

[89] Charles Rogier, *A Word for My King and Country: A Treatise on the Utility of Rocket Armament, Assisted by Balloons* (London, 1819).

The first governmental notice to proposals of this sort seems to have occurred in 1846, when a board of two officers was appointed to determine the feasibility of Samuel Warner's bombing balloon. They declared the plan impractical, with the result that Warner published a defense of his invention in which he claimed that it had not been fairly tested.[90] Attempts several years later by the distinguished aeronaut Coxwell to interest the government likewise came to nothing.[91] In 1854, Colonel Lefroy [92] of the Royal Artillery recommended that a balloon be built at the School of Military Engineering for official experiments and tests. This time a board of nine officers was named to consider the plan, and again an unfavorable report was rendered.[93] At the time of Colonel Lefroy's recommendation, England was engaged in the Crimean War, a conflict that invited more suggestions involving the use of aeronautic equipment, notably a proposal to bomb Cronstadt from the air by means of a large balloon laden with explosives.[94] This scheme was likewise rejected, but the following year, Lord Panmure, Secretary of State for War, wrote privately to Field Marshal Lord Raglan, in command of the British forces in the siege of Sebastopol, advising the use of a balloon for observation during the investment. "It appears to me," wrote Panmure, "that if you are to assault, a reconnaissance by means of a balloon would be a means of ascertaining the inner defenses and obstacles which you may have to encounter. I shall have all ready, so that if you telegraph for one, it shall go to you instantly." [95] Lord Raglan died before this letter reached

[90] Samuel Alfred Warner, *Fair Play's A Jewel: A Narrative of the Circumstances Connected with My Mode of National Defense against the Whole World* (London, 1849).

[91] Henry Coxwell, *My Life and Balloon Experiences* (London, 1899), pp. 2-5.

[92] Later Major General Sir John Henry Lefroy, K. C. M. G., C. B., F. R. S.

[93] Gamble, I, 42. The youngest of this board was sixty-five years old, with some forty-four years of service in the army. There was doubtless ample conservatism in the attitude of the examining body.

[94] G. E. Grover, "On the Uses of Balloons in Military Operations," *Papers Connected with the Duties of the Corps of Royal Engineers,* n. s., XII (1863), 73.

[95] Panmure to Raglan, May [date omitted] 1855, Sir George Douglas and Sir George Dalhousie Ramsay (eds.), *The Panmure Papers* (London, 1908), I, 197.

him, and his successor apparently never acted on the suggestion.[96] The Russian army, however, is said to have made use
of captive balloons during the siege,[97] but lack of original
sources prevents an estimate of the work performed.

Three years later during the Indian Mutiny, still another
plan for aerial bombing was laid before the War Office by a
Scot named Gillespie, whose invention was calculated to assist
prominently in suppressing the rebellion. An engineer officer
appointed to investigate the scheme declared it impracticable,
although it had been warmly approved by the eminent Scottish
physicist, Edward Sang.[98] Regardless of these and other numerous recommendations, the British government did not undertake
serious experimentation with military balloons until after the
American Civil War, when additional agitation by interested
officers, particularly Captain Frederick F. E. Beaumont of the
Royal Engineers, who had witnessed the operations of the
Federal Balloon Corps in America, induced the War Office to
experiment with and eventually adopt this tactical arm.[99]

In 1859 the French balloon service was again revived in the
Italian Campaign against Austria.[100] Napoleon III commissioned one M. Prevet to organize an aeronautic detachment, and
the latter applied to the brothers Louis and Eugène Godard,
prominent French aeronauts, to undertake the arrangements.[101]

[96] Gamble, I, 43. Lord Raglan died on June 28, 1855.

[97] Grover, in *Papers of the Royal Engineers*, n. s. XII, 74.

[98] *Ibid.*, pp. 73-74. For the scientific works of Edward Sang, see J. C. Poggen-
dorff, *Biographisch-Literarisches Handwörterbuch zur Geschichte der Exacten
Wissenschaften*, IV, Part ii (Leipzig, 1904), p. 1305. A sketch of Sang's life and
accomplishments may be found in the *Proceedings of the Royal Society of Edinburgh*, XXI (1896-1897), xvii-xxxi.

[99] Major General Whitworth Porter, *History of the Corps of Royal Engineers*
(London, 1889), II, 190-192.

[100] There is some evidence that a peace-time organization of balloonists was
maintained in the French army during the Second Republic. A letter in the War
Department Archives of the United States, from a French applicant for aeronautic
service with the Union Army in 1862 states that the writer had three years'
service as " Engineer of Aeronauts " in the 4th Division of the French [Repub-
lican] Army. His references and recommendations preclude the possible suggestion that he invented the story in order to reinforce his qualifications for
employment. A. J. B. de Morat to Chief of Topographical Engineers, August
3, 1862, MS D536, Letters Received, Bureau of Topographical Engineers, War
Department Division, National Archives, Washington, D. C.

[101] For an account of the life and work of Eugène Godard, who was the better

According to the Godards, Prevet refused to allow them sufficient funds to construct a balloon designed for the rigors of war
service, but required them to take the field with their own
civilian apparatus which was unsuited for military operations.[102]
The gas balloons taken on the expedition continually lost their
charges of hydrogen, and since there were no provisions for
replenishing the supply in the field, the aeronauts were required
to use their inferior montgolfiers.[103] Ascensions with this poor
equipment before Milan convinced the French high command
of the need for a regular war balloon if the air operations were
to be successful. Accordingly, Napoleon III ordered that such
apparatus be constructed in Paris at once.[104] The montgolfiers,
in the meantime, were kept in service throughout the campaign.
Ascensions were made with them at Castelnodolo, from the
Castiglione Hills, and from MacMahon's headquarters before
and during the battle of Solferino. The new balloon, *Imperial*,
built at Paris in the meanwhile, arrived at the theater of war
after the Peace of Villafranca, and thus took no active part in
the campaign.[105]

The military value of the operations performed seems to
have been slight. One French historian of the campaign has
declared that all the observers discovered was a party of three
Austrian soldiers near the village of Pozzolengo.[106] He added
that the aeronautic equipment comprising the Godards' train
was inadequate and incomplete from the moment the detachment joined the French army.[107] The Godards themselves expressed the opinion that the results of the observations were
insignificant, though the moral effect on the French troops was

known of the two brothers, see T. Saint-Felix, *Notice biographique sur Eugène
Godard, aéronaute de son majesté l'Empereur à la campagne d'Italie* (Paris,
1901).

[102] G. E. Grover, " On Reconnoitering Balloons," *Papers of the Royal Engineers*
n. s., XII, 89.

[103] Cesar, Baron de Bazancourt, *La campagne d'Italie de 1859* (Paris, 1870),
II, 99, 100. According to Bazancourt's figures, the Godards had in their train
four hydrogen balloons, four montgolfiers, and two parachutes.

[104] Dollfus and Bouché, p. 92; Grover, " On Reconnoitering Balloons," *loc.
cit.*, p. 89.

[105] *Ibid.;* Dollfus and Bouché, p. 92.

[106] Bazancourt, II, 99.   [107] *Ibid.*, II, 124.

excellent.[108] The fact that Eugène Godard sued for claims disallowed for his services seems to indicate that the French government was not satisfied with the results obtained.[109] On the other hand, a tactical study of the campaign prepared by an officer of the British Royal Artillery in 1861 attributes some importance to the work of the balloonists;[110] but this is offset by the fact that the standard, critical narrative of the campaign from French sources completely ignores the subject.[111]

In America, the employment of military aeronautics remained practically unknown until the Civil War. The science of penetrating and navigating the upper air, however, had attracted attention very soon after the successful experiments of the Montgolfiers and Charles in France. Thomas Jefferson mentions experiments with fairly large balloons in Philadelphia as early as May, 1784.[112] The following year, an American, Dr. John Jeffries, made the first aerial voyage across the English Channel in company with the French aeronaut, Jean Pierre Blanchard;[113] and experiments with hydrogen balloons and envelopes inflated with coal gas were conducted at the College of William and Mary in 1786.[114] The first successful ascension by a man in the United States did not take place until January 9, 1793, when Jeffries' protegé, Blanchard, went aloft at Philadelphia.[115] Though these examples indicate that interest in

[108] Grover, " On Reconnoitering Balloons," *loc. cit.*, pp. 89-90. Grover states that the details concerning the Italian Campaign in his article were given him by the brothers Godard.

[109] London *Times,* January 13, 1862.

[110] F. Miller, "A Study of the Italian Campaign of 1859," *Minutes of Proceedings of the Royal Artillery Institution,* II (Woolwich, 1861), 256.

[111] Major Ferdinand Le Comte, in his *Relation historique et critique de la campagne d'Italie de 1859* (2 Vols., Paris, 1860), makes no mention of the balloon operations of the French army.

[112] Jefferson to James Monroe, May 28, 1784, Paul Leicester Ford (ed.), *The Writings of Thomas Jefferson* (New York, 1894), III, 496.

[113] Diary of John Jeffries, entry of January 17, 1785, in B. J. Jeffries, " First Aerial Voyage Across the English Channel," *Magazine of American History,* XIII (1885), 68-72. See also John Jeffries, *A Narrative of Two Aerial Voyages of Dr. Jeffries with Mons. Blanchard* (London, 1786).

[114] Rev. James Madison to Thomas Jefferson, March 27, 1786, *William and Mary College Quarterly,* 2nd Ser., V, 83.

[115] Philadelphia *General Advertiser,* January 9 and 10, 1793. Jean Pierre Blanchard, *Journal of My Forty-fifth Ascension, Being the First Performed in America* (Philadelphia, 1793).

aeronautics had quickly spread to America, the application of the science to military tactics there was slow to develop.

One of the first Americans to mention the balloon as a possible military instrument was the versatile Benjamin Franklin, who referred to the subject as early as November, 1783, in a letter to Sir Joseph Banks, President of the Royal Society.[116] Several months later he again mentioned this topic in his correspondence, pointing out the value of the balloon as a possible weapon of invasion.[117] Philip Freneau's allusions to French aerial " frigates " and to the British " fighting balloons " soon followed Franklin's comments, but these instances appear to be the only documentary evidence disclosed by careful search indicating that the balloon was considered in the light of military value by Americans in the early period.[118] No series of pamphlets and tracts appeared in America comparable to those published in Europe recommending the newly invented machine for war purposes; nor was there in this country any noticeable number of suggestions offered to the War Department involving the trial and adoption of balloons. As far as can be discovered from available source materials, only two serious attempts were made to interest the Government in the use of military balloons throughout the seventy-odd years from the Montgolfiers' invention to the opening of sectional hostilities in 1861. Both of these, coming nearly a half-century after the creation of the French balloon companies, were rejected by the military authorities.

The Seminole War in Florida produced the first official recommendation to employ balloons with American forces.[118a] Hostilities, begun in 1835 over the removal of the Seminoles to

[116] Franklin to Banks, November 21, 1783, John Bigelow (ed.), *The Complete Works of Benjamin Franklin*, VIII, 377.

[117] Franklin to John Ingenhausz, January 16, 1784, *ibid.*, pp. 432-433. Also see above, p. 4, n. 12.

[118] Franklin was living in France at the time his two letters were written, and may have been influenced by opinion that had already been expressed there. Vilette's letter to the *Journal de Paris* had appeared a month before Franklin wrote to Banks.

[118a] Cf. F. Stansbury Haydon, " First Attempts at Military Aeronautics in the United States," *Journal of the American Military History Foundation*, II (Washington, 1938), 131 ff.

the West, had dragged on for several years despite the combined efforts of a large force of regular troops and militia to compel the Indians to submission.[119] The continuation of the war without apparent prospect of conclusion led a Colonel John H. Sherburne in 1840 to suggest to Secretary of War Joel Poinsett that balloons might be assigned for service with the government forces in Florida.[120] Sherburne had personal knowledge of the Indians as well as of the terrain involved in operations against them, having served in 1837 on a delegation sent to treat with the rebellious tribes on terms of peace.[121] He could thus discuss the Seminole affair with some color of authority.

The difficulty of locating the elusive bands of hostile red men, Sherburne claimed, was largely responsible for the failure of the army to bring the war to a close.[122] For this reason he presented his plan to the War Department as a means of solving the problem, and the scheme involved was not without some merit. He recommended that a balloon be attached to each column operating against the Indians. By making night ascensions unknown to the enemy, the location of their camps could be accurately determined by observing the campfires, and with instruments, their direction and distance could be calculated. Forces could then be sent out to surround and surprise the encampments; resistance would thereby be effectually reduced, and the war concluded. Sherburne further asserted that despite the wooded country involved, the light of fires could be detected from the air. Since the ascensions were to be made under cover of darkness, the element of surprise was insured, for the Indians would not be aware that their camps were being observed in this manner.[123]

[119] For a full narrative of the war, see John T. Sprague, *The Origin, Progress, and Conclusion of the Florida War* (New York, 1848). A summary of the military operations and policy appears in Emory Upton, *The Military Policy of the United States* (Washington, 1904).

[120] Sherburne to Poinsett, September 8, 1840, MS, Misc. File No. 284 [AGO Serial] War Department Division, National Archives [cited hereinafter as File 284, WDD NA].

[121] General T. S. Jesup to Poinsett, November 10, 1837, *American State Papers, Military Affairs*, VII, 887.

[122] This assertion is given some weight by statements in the report of General Winfield Scott, April 30, 1836, *ibid.*, VII, 278, and in Upton, *Military Policy*, 173.

[123] Sherburne to Poinsett, September 8, 1840, MS, File 284, WDD NA.

Secretary Poinsett promptly acknowledged the suggestion and promised to give the plan his consideration.[124] Sherburne, however, seems to have been confident of its adoption, and proceeded to make preliminary arrangements to procure the necessary apparatus. He later wrote to Poinsett announcing that he had located, after some difficulty, a balloon-maker who had for sale a complete set of equipment for the price of $600.[125] The individual offering the equipment turned out to be the pioneer aeronaut and scientist Charles Ferson Durant, whose ascensions and experiments in the 1830's had gained for him the reputation of being the first professional American aeronaut.[126] Durant had also agreed to make an additional balloon and appendages for $900, and had committed the offer to writing at Sherburne's request.[127] The Colonel also remarked by way of persuasion that he had laid the plan confidentially before the former Secretary of War, Benjamin F. Butler,[128] for his consideration and opinion. Butler, he claimed, was highly enthusiastic over the idea, and was quoted as having said that he thought the plan was " the best that the Secretary can adopt in the present state of things in Florida." It was further stated that arrangements for rapid inflation in the field were available, a consideration of primary importance if balloon operations should be attempted. Sherburne concluded by offering to take charge of the equipment, deliver it to the army in Florida, and instruct the necessary officers in its use.[129]

To this communication Poinsett replied that although he applauded the zeal with which Sherburne had pursued his object, he could not decide so quickly upon purchasing the

[124] Poinsett to Sherburne, September 9, 1840, MS, File 284, WDD NA.

[125] Sherburne to Poinsett, November 10, 1840, MS, File 284, WDD NA.

[126] A brief sketch of Durant's life and work appears in *Dictionary of American Biography* (New York, 1928-1932), V, 540. A scrapbook of newspaper and journal clippings relating to his work is in the New York Public Library. See also E. A. Dime, "America's First Aeronaut," *Air Travel,* I (January, 1918), 214-217, 239.

[127] Memorandum of C. F. Durant, October [date omitted], 1840, MS, File 284, WDD NA.

[128] Poinsett's immediate predecessor in the cabinet. He should not be confused with the Union general of identical name.

[129] Sherburne to Poinsett, November 10, 1840, MS, File 284, WDD NA.

equipment recommended. Recent dispatches from the theater of war, he added, gave reason to hope for an early termination of the "harassing and protracted contest." In the meanwhile he promised to consult the military commander in Florida to obtain his opinion of employing balloons in the manner proposed, "for unless the commanding officers there approve the suggestions, the adoption of the measure would be inexpedient," and the Secretary concluded his letter with the pointed comment that "such means can only succeed in willing hands."[130]

Poinsett thereupon referred the scheme to General W. K. Armistead, then in command in Florida. The latter declared the terrain in which his forces were operating to be unsuited for balloon reconnaissance, and positively refused to use the proposed apparatus.[131] This rejection does not appear to have discouraged Sherburne. The plan was again submitted to the War Department several months later, and was this time referred for opinion to General Edmund P. Gaines, an officer who had commanded the forces in Florida in 1835, and was now in charge of the Western Division of the Army. Gaines approved the plan and declared that "much good may result from the use of the balloon with night glass and compass . . . not only in Florida but in the large undulating prairies of the far west, where thousands of Indians . . . are often near the position or route of our troops without being seen by them."[132] But the General qualified his approval of the balloon by admonishing the Department not to continue the error of underrating the strategy of the red men, and stated an opinion that the Indians would soon discover the aerial observatory of their enemies, and would exert every effort to destroy it. "But should they fail in this," he concluded, "they would doubtless confuse us by kindling fires on one side of our position and taking their rest in another direction, leaving their fires burning to confuse and deceive us."[133]

[130] Poinsett to Sherburne, November 18, 1840, MS, File 284, WDD NA.
[131] Poinsett to Sherburne, January 28, 1841, MS S3199, File 284, WDD NA.
[132] Gaines to Adjutant General Roger Jones, January 25, 1842, MS Letterbooks of the Western Division, U. S. Army, Vol. 140-A [Letters Sent, April, 1841-May, 1846], pp. 189-190, WDD NA.
[133] Ibid.

The qualified approval of Gaines, however, was of little use. He was not connected with the army in Florida, and wrote his opinion merely in an advisory capacity. The war was finally brought to a close some three months later, and with its conclusion ended the first attempt to apply aeronautics to warfare in the United States.

The second effort to interest the Government occurred during the Mexican War, when John Wise,[134] a widely known aeronaut of Lancaster, Pennsylvania, proposed to clear the way for the capture of Vera Cruz by reducing the fortress of San Juan de Ulúa by means of percussion torpedoes dropped from a captive balloon.[135] In July, 1846, the American high command had begun to consider a plan of campaign against Mexico City, in which the taking of Vera Cruz was an essential preliminary objective.[136] The city was guarded from naval attack by the formidable fortified castle of San Juan de Ulúa, an immense, well-armed structure of masonry built on a coral reef half a mile out from the shore. Because of the presence of reefs and dangerous banks no vessel could approach the fort within a mile and a half from the north; to the east, the reefs also extended as a barrier, and the approach from the south and west was equally dangerous if not impossible, since vessels attacking from such direction would be raked by a cross-fire from the batteries on the fortress and from the guns of the shore batteries at Vera Cruz. Only one possible avenue of attack lay open, that from the southeast, and this was covered by heavy guns from that salient of the castle.[137] The fortress has been more than once referred to as a " second Gibraltar; "[138] and Com-

---

[134] See biographical note on Wise, Chapter iii, *infra*, p. 57.

[135] John Wise, *A System of Aeronautics, Comprehending the Earliest Investigations and Modern Practice and Art* (Philadelphia, 1850), p. 257; see also idem, *Through the Air, A Narrative of Forty Years' Experience as an Aeronaut* (Philadelphia, 1873), p. 386.

[136] Justin H. Smith, *The War with Mexico* (New York, 1919), I, 349.

[137] See P. Blanchard and A. Dauzato, *San Juan de Ulúa ou relation de l'expédition française au Mexique* (Paris, 1839), pp. 294-296. See also engineer plan of the work, Smith, *War With Mexico*, II, 21, and map of the city, fortress, reefs, and surrounding territory, *ibid.*, p. 24.

[138] Blanchard and Dauzato, *San Juan de Ulúa*, p. 296; George L. Rives, *The United States and Mexico, 1821-1848* (New York, 1913), II, 378.

modore David Conner, who was consulted at the time of the projected American attack, voiced the opinion that if well garrisoned, the fort could resist successfully any naval attack.[139] Richard Pakenham, formerly British minister to Mexico, and familiar with the strategic situation, expressed the belief that a combined naval and land attack on Vera Cruz and Ulúa would be a " very hazardous undertaking." [140] Because of the strength of the objective and the obvious risks involved in taking it, the American military authorities spent several months—from July to November, 1846—in discussion before a decision and plan of campaign was reached. It was ultimately decided to send a land force to invest the city from the inland side, and to employ a fleet on the gulf side for blockading purposes rather than for an attack on the grim castle guarding the city from the sea.[141] In November, 1846, General Winfield Scott was ordered to command the land expeditionary force.[142]

While the discussions were taking place over the question of how to take the city and its protecting fortress, the balloonist Wise came to the conclusion that he could utilize his aeronautic knowledge and ability to good effect and save the Government the losses and expense entailed by a direct assault or investment, as well as materially increase his own reputation and prestige. He devised a plan to reduce the castle of Ulúa by attacking it from a direction that could not possibly be reached from the fort's batteries: that is, from the air. In order to test public opinion and bring out any objections to the feasibility of his plan that he might have overlooked before submitting it to the War Department,[143] Wise caused a full account of the scheme to be published in a Lancaster newspaper, under the title of " Easy Method of Capturing the Castle of Vera Cruz." [144]

Wise began the discussion of his plan with a reiteration of

[139] Smith, *War with Mexico*, I, 349.

[140] *Ibid.*, I, 350.

[141] *Ibid.*, I, 350-351, 354.

[142] Secretary of War William L. Marcy to General Scott, November 23, 1846, *House Document No. 60*, 30 Cong. 1 Sess., p. 372.

[143] Wise to Marcy, December 10, 1846, MS W315, Letters Received, Secretary of War, War Department Division, National Archives [cited hereinafter as LRSW].

[144] Philadelphia *Ledger*, October 26, 1846.

the difficulty of taking the desired objective. "It is acknowledged," he wrote, " to be an extraordinarily well fortified point of defense, almost impregnable to the common mode of warfare, and at best cannot be taken in that way without great sacrifice of life and ammunition." [145] He then turned to the details of his solution of the problem. The materiel consisted of a large balloon of twilled muslin, properly varnished, and capable of lifting 20,000 pounds, exclusive of its own weight, network, car, and cable. The inflation of the envelope could be accomplished either on land or on a vessel at sea, as the circumstances might require.[146] The car was to be so constructed as to carry some 18,000 pounds of percussion torpedoes and shells, with 2,000 pounds space for ballast and crew. To control the aerial battery Wise proposed to maneuver it on a cable five miles long, so as to send it up from a point well out of range of the Mexican batteries, and allow it to ascend at least 5,000 feet, thus insuring its safety from musketry and gunfire from the fortress. The position of the balloon in regard to elevation and distance from the anchoring point could be easily controlled by a judicious handling of the ballast. As the ship should be lightened by the discharge of bombs and torpedoes, a constant elevation could be maintained by releasing gas.[147] This aerial battery, suspended over the fortress, well out of reach of the defenders, was calculated to silence the Mexican batteries and secure capitulation in a very short time. In fact, Wise believed he had arranged for all contingencies, and declared that

with this aerial warship hanging a mile above the fort, supplied with a thousand percussion bombshells, the castle of Vera Cruz could be taken without the loss of a single life to our army, and at an expense that

---

[145] *Ibid.*; John Wise, *Through the Air*, p. 386; *idem, A System of Aeronautics*, p. 257.

Wise appears to have been a realist in matters of military consideration; he considered the sacrifice of ammunition to be of equal importance with the loss of human life.

[146] Philadelphia *Ledger*, October 26, 1846; Wise, *Through the Air*, p. 389; Wise, *A System of Aeronautics*, p. 257.

[147] *Ibid.*, pp. 257-258; Wise, *Through the Air*, p. 389; Philadelphia *Ledger*, October 26, 1846.

would be nothing to what it will be to take it by the common mode of attacks.[148]

Wise then concluded his article by expressing his willingness to prepare the balloon for the proposed attack, and offered " most cheerfully " to " undertake its directorship into actual service, at a moment's warning." [149]

The publication of this rather startling or at least novel plan of warfare drew considerable comment from the public. A Philadelphia newspaper observed several months later after the plan had circulated in the press, that " this new method of besieging a fortress has been discussed in every vein of serious-ness, wit, or contumely, as the idea seemed feasible, funny, or absurd to various minds." [150] In the same article a former governor of Kentucky was quoted as having declared at a public dinner that he thought the plan " an admirable one, and the inventor a man of military genius." But the former state execu-tive added that he thought it would be a " very troublesome matter to enlist the volunteers for that service." [151] The same paper in an editorial several days after the first publication of Wise's plan in the Lancaster press, referred favorably to the project, and added a jocular allusion to the political aspirations of those successful in war:

As Mr. Wise understands the operation, he is the very man to entrust with this important undertaking. Should he take this redoubtable fort-ress, he will acquire a renown that will go far to rival the claims of General Taylor to the next Presidency.[152]

---

[148] *Ibid.*; Wise, *Through the Air,* p. 389; Wise, *A System of Aeronautics,* p. 258.

[149] Philadelphia *Ledger,* October 26, 1846; Wise, *Through the Air,* p. 390; Wise, *A System of Aeronautics,* p. 258.

[150] Philadelphia *Ledger,* December 5, 1846. Comment in the *Scientific Ameri-can* was unfavorable.

" Mr. Wise, the Aeronaut," ran one article, " proposes in a Lancaster, Penn-sylvania, paper, to capture the Castle of San Juan de Ulúa . . . more easily said than done." Another issue of the same journal remarked that " however wise Mr. Wise may be on this subject, he will find our Uncle Sam too wise to employ so much gas on the San Juan castle." *Scientific American,* November 6, 1846; *ibid.,* January 30, 1847.

[151] Philadelphia *Ledger,* December 5, 1846.

[152] *Ibid.,* October 26, 1846.

Regarding such comments as an indication that the public believed his scheme to be practical, Wise then addressed it to the Secretary of War, explaining that he considered that the plan had been " tested by public opinion " and " analyzed in the popular crucible." As persuasive argument for his project, Wise asserted that " so far as any well-founded objections having appeared against its practicableness, I have some of the best minds of the country to sustain the project. . . . As to any objections that may or can be raised against its feasibility, I am ready to rebut by mathematical, philosophical, and practical argument." [153] One point raised in opposition, that of the ex-governor of Kentucky, that it would be difficult to secure volunteers for the balloon crew, Wise conclusively refuted by stating that he would need ten men at the most, and that he had that many volunteered from Lancaster already.[154]

Wise, however, failed to mention who " the best minds in the country " were; and altogether, his tone, though dignified and doubtless sincere, was perhaps a trifle too self-assured to convince the War Department officials. There is no evidence that the letter was ever answered.[155] Vera Cruz was invested and eventually taken by General Scott, and the second plan and attempt to introduce balloons into the American military system failed. At this point it may be interesting to note that during the investment of this city which Wise wished to capture by aerial tactics, served a young lieutenant of engineers, George B. McClellan [156] who, fifteen years later as commanding general of the armies of the United States, cooperated with Wise's rival, T. S. C. Lowe, in creating the first American balloon corps in 1861.

Although the few attempts to initiate military aeronautics in

---

[153] Wise to Marcy, December 10, 1846, MS W315, LRSW.

[154] *Ibid.*

[155] The " Letters Sent " books of the War Department contain no copy or reference to any reply sent to Wise. The " Letters Received " book in which Wise's letter is calendared bears no reference to disposition, this fact indicating that the letter was not sent to some subordinate bureau or department for consideration and answer.

[156] Cf. William Starr Meyers (ed.), *The Mexican War Diary of George B. McClellan* (Princeton, 1917).

the United States were unsuccessful, the development of civilian balloning progressed steadily throughout the middle decades of the 19th century. The period of the thirties witnessed increasing interest in aeronautics. Charles F. Durant, the first professional aeronaut, drew considerable attention by his activities in Baltimore and New York.[157] In Boston the aeronaut Lauriat was also active during this period, with a number of ascensions and free flights to his credit, the longest of which was calculated at forty miles.[158] Wise had also entered the field professionally, beginning a long and distinguished career as one of America's foremost aerial pioneers.[159] Likewise Samuel A. King of Philadelphia, later to become a prominent figure in the profession, joined the ever-increasing ranks of balloonists.[160]

In the decade of the fifties, ballooning had become even more widely known. Aeronauts appeared to make ascensions and give exhibition flights at numerous county fairs and urban gatherings throughout the country, and constant publicity in the press soon popularized their work far and wide. It was, proclaimed an editorial heading in one of the eastern dailies, " The Day of Balloons."

In many of the towns of the interior [the editor continued] ascensions of balloons are taking place, lives are lost, and new men, after not more than one or two ascents with others, set up on their own account. . . . Thousands are moved to the attempt because of the descriptions of the newspaper press. . . . We doubt if for mere amusement any such ascensions should be countenanced. But in nine cases out of ten this is not mere amusement with those who ascend. Each one hopes to make some discovery, or to acquire that knowledge by which he can hereafter make a longer and more important voyage.[161]

---

[157] Dime, in *Air Travel*, January 1918; Adrian [Michigan] *Watchtower*, June 15, 1855.

[158] *Ibid.*; George Henry Sargent, *Lauriat's* (Boston, privately printed, 1922), pp. 50-54.

[159] Wise's career is fully described in his *System of Aeronautics*, and *Through the Air.*

[160] Anon. [Samuel A. King?], *The Balloon. Noteworthy Aerial Voyages, from the Discovery of the Balloon to the Present Time. With a Narrative of the Aeronautic Experiences of Samuel A. King* (New York, privately printed [American Aeronautic Society of New York], 1879).

[161] Philadelphia *Ledger*, October 5, 1859.

Among the many aspirants to aeronautic recognition and success who soon attained no small degree of fame in their profession were individuals who served later with the Union and Confederate armies as military balloonists. In New England, James Allen, partner of Samuel King, became a well-known and successful balloonist, and later served with distinction in the Balloon Corps of the Army of the Potomac.[162] The air-minded dentist, William H. Helme of Providence, who later tried to introduce an improved type of montgolfier for the use of the Federal army, also made a number of experiments and flights, and published his results in the press. He also worked with J. W. Black of Boston to initiate aerial photography, and made the first photographs from the air in this country.[163] The German, John H. Steiner, who was later to serve in the Union balloon corps, was a prominent figure in the West, and was also the competitor of Eugène Godard, the French military aeronaut of Louis Napoleon's Italian campaign, in one of the early balloon races in this country.[164] Still another prominent figure who later saw service with the balloon corps of the Army of the Potomac was William Paullin, of Philadelphia. Paullin achieved a wide reputation in the United States, and became one of the first of his profession to extend his activities to South America, the West Indian Islands, and Mexico.[165] The aeronauts B. F. H. Lynn of Erie, W. D. Bannister of Michigan, and S. M. Brooks of St. Louis, were also active during the period and attracted attention in the press.[166] John La Mountain of Troy, New York, the future free-lance balloonist of the Federal army, and exploiter of the upper currents in his military free

---

[162] See Chapter ii, *infra*.

[163] Boston *Journal*, July 5, 1860; *ibid.*, October 13, 1860; Providence *Journal*, July 12, 1860; *ibid.*, August 17, 1860; *ibid.*, August 21, 1860; *ibid.*, October 6, 1860; clippings in James Allen's Scrapbook, I, 37, 39, 41, 43-44; II, 97, 99.

[164] Cincinnati *Gazette*, October 20, 1858; Toronto *Globe*, August 4, 1859; unidentified clippings in James Allen's Scrapbook, I, 33, 35, 47.

[165] James Grant Wilson and John Fiske (eds.) *Appleton's Cyclopedia of American Biography* (New York, 1888), IV, 682; see also Chapter vii, *infra*.

[166] Adrian [Michigan] *Evening Expositor*, June 15, 1855; Adrian [Michigan] *Watchtower*, June 15, 1855; Detroit *Daily Tribune*, July 25, 1859; Peoria [Illinois] *Transcript*, October 22, 1859; Erie *Dispatch*, May 17, 1859 (Lynn was an editor of the *Dispatch* as well as an aeronaut); *Leslie's Illustrated Newspaper*, October 23, 1858.

reconnaissances, was rising in the scale of aerial notoriety,[167] and Charles C. Coe of Rome, New York, constructed the largest balloon ever attempted in this country, a monster envelope of more than one million cubic-foot capacity.[168] In the South, the future Confederate military balloonists, Charles Cevor and Richard Wells, were also before the public eye,[169] and the French aeronaut, Alexander de Morat, who later applied for service with the Union army, gave public exhibitions and ascensions, and engaged in a race with Wells.[170]

Long distance voyages by air gave proof of the aeronautic progress made during this decade. Balloonists of ambition were no longer content to make captive ascensions before rural and urban crowds, but extended their activities to include free flights over long distances, to the testing of theories of air currents, and to experiments in air travel. Most active in this respect was John Wise, who carried on his work with a great interest in its scientific aspects, published his findings, and made numerous voyages.[171] The most spectacular free flight of the period was made by Wise, La Mountain, and O. A. Gager, in which a distance of some 1,100 miles was covered in less than twenty hours. The three aeronauts accompanied by a newspaper reporter ascended in the balloon *Atlantic* from St. Louis, Missouri, at 6:45 p. m. on July 2, 1859, and after being borne steadily eastward by upper currents, landed at 2:20 the following afternoon at Henderson, Jefferson County, New York, having established an enviable record in this country for speed and distance.[172] A few months later La Mountain, in company with

[167] See Chapter iv, *infra*, pp. 82-83.

[168] Rome [New York] *Sentinel,* undated clipping in T. S. C. Lowe's scrapbook in the Lowe Papers. This book contains other clippings relating to Coe and his balloon.

[169] New Orleans *Picayune,* February 16, 1858; Savannah *Evening Express,* March 9, 1860; *Harper's Weekly,* April 14, 1860; New York *Atlas,* October 20, 1860; unidentified clippings in James Allen's Scrapbook, II, 11, 31, 91, and in Lowe's Scrapbook.

[170] New Orleans *Picayune,* February 16, 1858. Unidentified clippings in James Allen's Scrapbook, II, 29.

[171] Wise, *A System of Aeronautics, passim.* See Chapter iii, *infra.*

[172] St. Louis *Democrat,* July 2, 1859; Buffalo *Courier,* July 4, 1859; Albany *Atlas,* July 4, 1859; John Wise, *et al., Full Particulars of the Greatest Aerial*

J. A. Haddock, went aloft from Watertown, New York, and was blown into the wilderness of northern Canada, a voyage which nearly cost the aeronauts their lives.[173] The numerous free flights and the celebrated Henderson-St. Louis voyage soon aroused interest in the possibilities of practical air travel. An editorial in a Detroit paper voiced this interest in a prophecy that is remarkably " modern " in aspect:

Ballooning is evidently destined to become a marked feature of the age. The restless spirit of the times demands a faster and more pleasant mode of conveyance than we have at present. Young America discarded steamboats long ago, and is becoming thoroughly disgusted with the annoyances and delays appurtenant to railroad travel. Let old fogies still drag along at the snail's pace of thirty miles an hour . . . let broad-shouldered rustics still patronize the iron horse . . . let who will . . . wearily sail on the misty deep . . . but the coming year 1860 will witness another sight . . . " Air lines " of travel will exist otherwise than upon handbills. . . . Distance will be annihilated, and extension will cease to be one of the properties of matter. The Pacific Railroad question will no longer bother politicians. Through from New York to San Francisco in twenty-four hours. . . . Hurrah for Wise and La Mountain—for Bannister and Balloons! [174]

Nor did the speculations on air travel cease with dreams of continental transportation. The ambition to cross the Atlantic to Europe by air stirred the mind of many an aeronaut of the period. As early as 1836 the English aeronaut Charles Green had asserted that a balloon crossing of the Atlantic was possible,[174a] and in 1844 Edgar Allen Poe had perpetrated his famous balloon hoax, the result of which was the publication in a reputable New York paper that the great feat had actually been accomplished.[175] Thereafter the question continued to be a subject of speculation and discussion in many quarters. Leading aeronauts, including Wise, La Mountain, and others, made

Voyage on Record, From St. Louis, Mo. to Adams, New York, in Nineteen Hours (New York, 1859).

[173] See Chapter iv, infra, pp. 82-83.

[174] Detroit Daily Tribune, July 25, 1859.

[174a] J. E. Hodgson, The History of Aeronautics in Great Britain (London, 1924), p. 758.

[175] New York Sun, April 13, 1844; Wise, A System of Aeronautics, pp. 235-246.

plans and publicly announced their intention to make the attempt.[176] On two occasions Wise tried to persuade Congress to appropriate funds for carrying out this project,[177] and in 1860 T. S. C. Lowe, future creator of the Balloon Corps of the Union army, built his huge " airship " and actually attempted to start on the long-heralded trans-Atlantic voyage.[178]

Progress in the development of equipment had also gone on apace. Balloons for captive ascensions and free flights ranged in size from 25 feet in diameter to Lowe's monster trans-oceanic ship which measured 130 feet through the center of its envelope. Gas capacities varied from 20,000 feet or less up to more than 700,000.[179] The *Atlantic*, in which the St. Louis-Henderson flight was made, measured 60 feet in diameter, and contained 65,000 feet of gas when fully inflated.[180] In lifting power, the envelopes in use ranged from 200 pounds for the small balloons up to Lowe's trans-Atlantic ship, which was designed to raise 22 tons. Aeronauts had also ventured aloft to remarkable heights, and the attainment of altitudes of more than 20,000 feet was recorded on several occasions.[181] Improvements in the construction, cordage, and other auxiliary equipment had also been developed. Wise had invented the rip-panel, a safety device that was soon adopted for use in the profession generally.[182]

By and large aeronautics had taken a definite place in the life of the nation. Balloons and aeronauts were no longer rarities. They had become part of American life, familiar to various classes of society, and drew interest and attention from all sections of the country. " Ballooning as a National Institution,"

---

[176] Wise in the Lancaster *Daily Evening Express,* undated clipping in Lowe's Scrapbook. New York *Herald,* April 24, 1860; New York *Tribune,* November 15, 1859; La Mountain in the New York *Tribune,* undated clipping in Lowe's Scrapbook. Steiner also proposed to make a trans-Atlantic flight in 1859.

[177] Wise, *A System of Aeronautics,* pp. 244-246; *idem, Through the Air,* pp. 373-375; 426-430.

[178] See Chapter v, *infra,* pp. 156-161.

[179] Clippings in Lowe's Scrapbook; see also Chapter v, *infra.*

[180] Wise, *Full Particulars of the Greatest Aerial Voyage,* p. 3.

[181] Clippings in Lowe's Scrapbook; see also Chapter v, *infra.*

[182] Wise, *Through the Air,* p. 634.

38

heading an article that summed up the progress of the science,[183] defined well the status of this professional endeavor in 1860.

That year marked what might be called the first hey-dey of American aeronautics. It also marked the rapidly approaching crisis of those irresistible forces, political, social and economic, which had slowly but surely torn the nation asunder for more than thirty years. These forces, piling up like thunderheads against a glowering sky, were soon to burst forth into the devastating tempest of civil war; and in the deluge to come, aeronauts of North and South were to be swept into the tide of martial enthusiasm that spread into all professions. Hitherto, aeronautics had served the ends of human pleasure, amusement, and to some extent, scientific investigation. War was finally to give ballooning in America its first practical utility.

[183] Unidentified clipping in Lowe's Scrapbook.

CHAPTER II

JAMES ALLEN: FIRST TRIAL AND FAILURE

Concurrent with the rapid growth of popular interest in aeronautics throughout the United States, the gathering clouds of sectional conflict and disunion banked thicker and blacker on the national horizon. The thunder of the coming tempest was first heard on December 20, 1860, when the long threatened recourse to secession became accomplished fact with the withdrawal of South Carolina from the Union. By February of the succeeding year the Gulf States had followed, and the Confederate States of America came into being, with their seat of government at Montgomery, Alabama. Attempts at compromise and conciliation marked the next few critical weeks, towards the close of which the attention of both sides became focused on the Federal garrison in Fort Sumter. Despite the extreme gravity of the situation, hope for an amicable settlement of the sectional difficulties had not been wholly abandoned, when at half-past four on the morning of April 12 " a flash as of distant lightning in the direction of Mount Pleasant, followed by the dull roar of a mortar," [1] was noted from the ramparts of Sumter, announcing that the time for negotiation was at an end: the storm of civil war had finally burst.

Three days later President Lincoln's proclamation calling for seventy-five thousand militia to execute the laws of the United States was dispatched to the governors of the twenty-four States still left in the Union. The call was received with enthusiastic response, and in the mobilization of militia and volunteers that followed, the state governments vied with one another to lead in furnishing men and equipment. [2] Rhode Island, destined to

[1] Captain James Chester [3rd U. S. Artillery], " Inside Sumter in '61," *Battles and Leaders of the Civil War* (New York, 1884), I, 66.

[2] Cf. correspondence of the governors of the various States, in *The War of the Rebellion: A Compilation of the Official Records of the Union and Confederate Armies* (Washington [Government Printing Office], 1880, *et seq.*), Series III, Vol. I, 70-85, *passim* [hereinafter cited as *O. R.*, with series number in arabic preceding the abbreviated title].

produce the first military aeronaut in United States service, strove hard to be first in placing her troops in the field.[3]

The table of quotas prepared by Secretary Cameron to accompany Lincoln's proclamation allotted one regiment to be furnished by Rhode Island.[4] On April 17 Governor William Sprague notified Cameron that this regiment was organized and ready to move.[5] On the following day the First Regiment, Rhode Island Detached State Militia, accompanied by a light battery of the Providence Marine Corps of Artillery,[6] under command of Colonel Ambrose Everett Burnside, left Providence by detachments.[7] Attached to this force was James Allen,[8] the well-known "New England Aeronaut," who became the first military balloonist to take service with American troops.

James Allen was born at Barrington, county of Bristol, Rhode Island, on September 11, 1824. He was the son of Sylvester Allen, a sea captain, who had moved with his family to Providence while James, his ninth child, was still very young. A few years after settling in Providence, the elder Allen was lost at sea, and his children were compelled to pursue any occupation in order to obtain a livelihood. James was in turn a hand in a cotton mill, a farm laborer, and finally served as a merchant seaman. After three years at sea he was persuaded by his mother to settle down to some occupation on shore, with the result that he entered the printing trade in 1841. Five years later he was employed by one of his brothers who was engaged in the jewelry business, but poor health caused him to leave his brother's establishment in Providence and retire to Wilmington, Delaware. Here he witnessed a balloon ascension for the

[3] Governor Sprague to Secretary Cameron, April 17, 1861, 3 *O. R.*, I, 82.

[4] Table of State Quotas [accompanying Lincoln's call for troops], 3 *O. R.*, I, 69.

[5] Sprague to Cameron, April 17, 1861, 3 *O. R.*, I, 82.

[6] This unit bore the official designation of 1st Light Battery, Rhode Island Volunteers. It was often referred to as the Rhode Island Marine Artillery, having been made up of troops from the Providence Marine Corps of Artillery, a militia organization originally composed of sea captains and seamen. Cf. Moses King, *Pocket Book of Providence, Rhode Island* (Providence, 1882), pp. 64-65.

[7] 1 *O. R.*, II, 315; Sprague to Cameron, April 18, 1861, 3 *O. R.*, I, 84.

[8] Augustus Woodbury [Chaplain, 1st R. I. D. M.], *Narrative of the Campaign of the First Rhode Island Regiment in the Spring and Summer of 1861* (Providence, 1862), Appendix "D," p. 175.

first time, and the spectacle so impressed him that he determined to become a professional aeronaut. He secured and read all the literature on aeronautics available to him, and finally made the acquaintance of Samuel A. King, of Philadelphia, one of the foremost aeronauts of his day. In the spring of 1857 Allen made his first ascent with King, a free flight of ten miles. He evidently displayed considerable skill and aptitude in assisting King in the management of the balloon, for the latter soon offered him a partnership. With Allen's acceptance the firm of Allen and King was established, the aerial activities of which drew wide attention for approximately four years. James Allen's first ascension alone was made from Exchange Place in Providence on July 4, 1857.[9]

During the four succeeding years the partners made continual public ascents and exhibitions in nearly every state in New England. Allen's reputation as a leading balloonist spread rapidly, and by the opening of hostilities he was known and admired over the entire New England and adjacent states.[10]

[9] This sketch of James Allen's earlier career is taken from James Grant Wilson and John Fiske (eds.), *Cyclopedia of American Biography* (New York, 1888), IX, 210. There is no published life of Allen, and unfortunately the account of him in the aforementioned work is highly colored with family pride and is unreliable for the period covering Allen's war service. The sketch was prepared by surviving members of Allen's family, who evidently sought to secure to his memory a majority of the credit for the work of the Union balloon corps. The material drawn from this source and included in this chapter contains only details of Allen's life that are not likely to be exaggerated. These details are not available elsewhere.

[10] Worcester [Mass.] *Daily Bay State*, July 23, 1857; [*] Worcester [Mass.] *Daily Transcript*, July 23, 1857; [*] Manchester [N. H.] *Mirror*, July 27, 1857; [*] Manchester [N. H.] *Daily American*, July 27, 1857; [*] Paterson [N. J.] *Daily Guardian*, August 5, 1858; [*] Providence *Daily Post*, June 14, 1858; Boston *Ledger*, June 24, 1859; Boston *Courier*, July 6, 1859; Boston *Journal*, July 5, 1858; Boston *Herald*, July 5, 1858; Boston *Transcript*, July 5 and 7, 1859; Boston *Saturday Evening Gazette*, July 16, 1858; [*] New Bedford *Mercury*, August 26 and 27, 1859; [*] New Bedford *Evening Standard*, August 24 and 25, 1859; [*] Boston *Journal*, August 27, 1859; Norwich [Conn.] *Morning Bulletin*, September 30, 1859; [*] Providence *Daily Post*, October 25, 1859; Providence *Journal*, October 25, 1859; Lowell [Mass.] *Daily Citizen and News*, July 5, 1860; [*] Boston *Journal*, July 5 and October 13, 1860; Boston *Herald*, July 6, and October 10, 1860; Lancaster [N. H.] *Democrat*, July 17, 1889. [*] Numerous accounts of the activities of Allen and also of his partner King appear in two volumes of scrapbooks, containing among other items, a large number of unidentified clippings from various newspapers, preserved and

So well-known did he become in this region that he was frequently given the title of " the New England Aeronaut." [11] He seems to have dissolved his partnership with King shortly before the outbreak of the war, and returned, at least temporarily, to his old trade as a printer, becoming a member of the firm of Allen and Hunt, in Providence.[12]

When President Lincoln's proclamation after the fall of Fort Sumter was published in Providence, Allen, who appears to have been a member of the Marine Corps of Artillery, volunteered his services and aeronautic equipment with the First Regiment, not as an artilleryman, but as a balloonist.[13] At the same time, his fellow townsman, the dentist-aeronaut Dr. William H. Helme, also volunteered for war service, enrolling as a private in Company " C " of Burnside's regiment.[14] According to a sketch of this organization published later by the Adjutant General of Rhode Island, both Allen and Helme " were authorized to act as aeronauts in connexion with the movements of the Regiment." [15] The status of Helme is perfectly clear. He was a volunteer infantryman who, because of his special knowledge and experience, was authorized to take part in whatever aeronautic operations that might be undertaken with Allen's

arranged by Allen and members of his family. These scrapbooks are now in the Division of Aeronautics, Library of Congress, and will be cited hereinafter as James Allen's Scrapbook, with volume numbers and pagination indicated. Newspaper items used as sources from these books, not independently located in original files, will be indicated with an asterisk [*].

[11] Providence *Daily Post,* April 19, 1861; *ibid.,* April 26, 1861.

[12] Washington *National Republican,* June 10, 1861; *Scientific American,* June 22, 1861; unidentified clippings in James Allen's Scrapbook, I, 45.

[13] Providence *Daily Post,* April 17 and 26, 1861; Providence *Sunday Telegram,* July 13, 1884 [*]; Lancaster [N. H.] *Democrat,* July 17, 1889 [*]; James Allen to T. S. C. Lowe, April 1, 1863, 3 *O. R.,* III, 301; Brigadier General Henry L. Abbot, " Early Experience with Balloons in War," *Professional Memoirs, Corps of Engineers, United States Army* (Washington, 1912), IV, 680.

[14] Rhode Island, Adjutant General's Office, *Official Register of Rhode Island Officers and Soldiers Who Served in the United States Army and Navy From 1861 to 1866* (Providence, 1866), p. 17; Woodbury, *Campaign of the First Rhode Island Regiment,* p. 190; rosters of the 1st RIDM, in Providence *Post,* April 20, 24, and 27, 1861, Providence *Press,* April 30, 1861, and Providence *Journal,* April 19, 1861.

[15] Rhode Island, Adjutant General's Office, " First Regiment Rhode Island Detached Militia," in *Official Register,* p. 2.

balloons. But the exact status of Allen is vague and difficult to determine. He is mentioned as a member of the Light Battery of Marine Artillery attached to Burnside's First Regiment in several press items.[16] He appears as such on the roll of this organization in Chaplain Woodbury's history of the regiment,[17] and the chaplain's information seems confirmed in the above mentioned sketch issued by the Rhode Island Adjutant General's Office.[18] But his name is not included in the official roster of the unit published by the same office,[19] nor is he listed on the rolls of the battery in the local press at the time of the unit's departure for Washington.[20]

It is evident, regardless of the vague nature of Allen's military status, that his aeronautic materiel was officially attached to Burnside's regiment. An officer of the Topographical Engineers who later conducted experimental reconnaissances with one of Allen's balloons, refers to the aerostat as " the balloon belonging to the Rhode Island Regiment." [21] Similar information appears in a letter of Colonel John N. Macomb, also of the topographical service, and later in charge of the balloon corps after it was definitely organized by Lowe, who mentions in a communication relating to his accounts the purchase of zinc for generating gas for the balloon attached to the Rhode Island Regiment. He added that this unit had in its train the first balloon used for military purposes.[22] Therefore, in the light of all evidence available, Allen appears to have been a member of

---

[16] Washington *National Republican*, June 10, 1861; Providence *Post*, May 30, 1861; unidentified clippings [2] in James Allen's Scrapbook; see also Benjamin Perley Poore, *The Life and Public Services of Ambrose E. Burnside* (Providence, 1882), pp. 106-107.

[17] Woodbury, *Campaign of the First Rhode Island Regiment*, pp. 175, 179.

[18] Rhode Island, Adjutant General's Office, " First Regiment Rhode Island Detached Militia," in *Official Register*, p. 2.

[19] *Idem., Official Register*, pp. 685-686.

[20] Providence *Journal*, April 19, 1861; Providence *Post*, April 20, 24, and 27, 1861; Providence *Press*, April 30, 1861.

[21] Lieutenant Henry L. Abbot to Captain A. W. Whipple, July 9, 1861, MS W853, enc. 1, Letters Received, Bureau of Topographical Engineers, War Department Division, National Archives [cited hereinafter as LRTE].

[22] Colonel John N. Macomb to Major J. C. Woodruff, February 23, 1863, MS [number omitted], Letters Received, Secretary of War, War Department Division, National Archives [cited hereinafter as LRSW].

the Marine Corps of Artillery before the war, who was in the group from this organization which made up the Light Battery, but was detached and carried as a supernumerary with specialist or technical status. According to a contemporary Providence newspaper, the time allowed for the arrangement of details between the State authorities and the War Department before the departure of Governor Sprague with the first units of the Rhode Island troops was so short as to render impossible the completion of any definite arrangements for the employment of Allen and the two balloons he had offered for service. Consequently Governor Sprague took upon himself the responsibility of securing Allen's aeronautic services for the Rhode Island Regiment without the official sanction of the Federal military authorities.[23] Acting upon the orders of Sprague, Allen left Providence with his two balloons on April 19, four days after Lincoln issued his call for troops, with Washington as his destination.[24]

He reached Philadelphia the following day, and finding that rail communication between that city and Washington was interrupted because of the destruction of track and bridges by Confederate sympathizers, he proceeded to Perryville, the railroad to this point being still intact. In the confusion and hurry attending the departure of the Rhode Island troops, Governor Sprague had neglected to give Allen any written orders or official papers of identification. Consequently he was stopped by the military authorities at Perryville, the embarking point for water-convoy of large bodies of Union troops en route for Washington, and was compelled to return to Philadelphia. Arriving here on April 23 he communicated with the commanding officer of the marine battery attached to Burnside's First Regiment, and received orders to join this unit at Easton.[25]

[23] Providence *Daily Post*, April 26, 1861.

[24] Providence *Daily Post*, April 19, April 22, and April 26, 1861; Providence *Press*, April 20, 1861.

[25] James Allen to Ezra S. Allen, April 24, 1861, published in Providence *Daily Post*, April 26, 1861.

Ezra S. Allen later served with his brother in the balloon corps of the Army of the Potomac. In June 1861, when the Rhode Island Regiment was in camp near Washington, the citizens of Providence contributed a large supply of ice

Allen's activities and movements for the next month are apparently unknown. He may have remained with the Rhode Island troops awaiting orders, or he may have spent the time endeavoring to secure Federal confirmation of his appointment as a military aeronaut.[26] He was with the battery of Marine Artillery in Washington towards the end of May, and wrote a long and enthusiastic description of the condition and activities of these troops to his home newspaper.[27] Several press notices announced at this time that he had just been appointed " aeronautical engineer " in the army by the Secretary of War.[28]

It appears from the sources available, that Allen did not begin active experiments or demonstrations with his balloons until after the first week of June. On the 9th he prepared his larger aerostat for inflation, and although he is reported to have had portable generating apparatus with the aeronautic equipment brought from Providence, the balloon was taken to a gas main on the corner of Massachusetts Avenue and Third Street, where it was inflated and made ready for its first experimental ascension. The novel sight of a military balloon being prepared for service attracted considerable attention, and a large crowd of interested spectators witnessed the inflation. It was reported at the time that Major Albert J. Myer, Chief Signal Officer of

and delicacies for the troops. Ezra Allen was among the contributors and also accompanied these supplies to Washington for distribution to the troops. Providence *Evening Press,* June 5, 1861; General Orders, No. 18, HDQ, 1st R. I. Detached Militia, June 2, 1861, in Woodbury, pp. 152-153.

[26] A careful combing of the Providence, Philadelphia, Baltimore and Washington newspapers fails to reveal any notices of Allen during this period. The fact that the press announced his appointment by the Federal government towards the end of May [see succeeding reference, note 28] rather suggests that he may have been endeavoring to secure an official appointment during this time.

[27] Allen to the Providence *Daily Post,* May 27, 1861, published May 30, 1861.

[28] Providence *Daily Post,* May 30, 1861; Washington *National Republican,* June 10, 1861; Baltimore *American,* June 11, 1861; *Scientific American,* June 22, 1861; unidentified clippings in James Allen's Scrapbook, I, 45.

The records of the Secretary of War, the Headquarters of the Army, the Department of Northeastern Virginia, and the Bureau of Topographical Engineers, preserved in the War Department Division of the National Archives, do not disclose evidence of any official appointment or record of Allen's status at this time. Many documents are missing from these collections, however, and since Allen was soon afterwards placed under the supervision of regular officers, it is reasonable to assume that about this time he was given some form of official recognition by the Federal military authorities.

the regular army, had been designated to supervise Allen's operations.[29]

Some details are available concerning the balloon thus made ready for experiments and tests in aerial observation. The envelope was thirty-five feet in diameter, and fifty feet in height from the valve to the car. Oiled linen was used throughout in its construction.[30] A portable windlass, through which were operated some 5,000 feet of guy ropes; cordage, car, and other auxiliary attachments formed part of Allen's equipment.[31]

As soon as the inflation of the envelope was completed, the balloon, carefully ballasted with sandbags, was towed by hand to the camp of the First Rhode Island Regiment, then in preparation for active service, at Caton's Farm about a mile north of the Capitol, near Glenwood. Here an experimental ascension to an altitude reported to have been 5,000 feet, the full length of the mooring guy, was made.[32]

The First Rhode Island had received orders to proceed from its camp near Washington and join the column under Major General Robert Patterson, which was marching from Chambersburg on Harper's Ferry.[33] Burnside's regiment cleared its camp area on June 10, the day after Allen's first ascension, and preceded by the light Marine Battery, entrained at the Baltimore

[29] Providence *Journal*, June 11, 1861; New York *Tribune*, June 10 and 15, 1861; New York *World*, June 10, 1861; New York *Evening Post*, June 10 and 11, 1861; New York *Commercial Advertiser*, June 10, 1861; New York *Morning Courier and Enquirer*, June 10, 1861; Philadelphia *Ledger*, June 12, 1861; Philadelphia *Daily News*, June 11, 1861; Baltimore *American*, June 11, 1861; Washington *National Republican*, June 10, 1861; Detroit *Free Press*, June 12, 1861; *Scientific American*, June 22, 1861.

[30] Abbot later described the fabric as silk. "Early Experience with Balloons in War," *Professional Memoirs, Corps of Engineers*, IV, 680.

[31] Cincinnati *Commercial*, June 10, 1861; Providence *Journal*, June 10, 1861; Providence *Daily Post*, May 30 and June 10, 1861; Providence *Press*, June 10, 1861; New York *Tribune*, June 15, 1861; New York *Commercial Advertiser*, June 10, 1861; New York *Evening Post*, June 11, 1861; Philadelphia *Ledger*, June 12, 1861; Baltimore *American*, June 11, 1861; Washington *National Republican*, June 10, 1861; *Scientific American*, June 22, 1861.

[32] The sources for this paragraph are the same as cited in notes 29 and 31, preceding.

[33] General Winfield Scott to Patterson, June 8, 1861, 1 *O. R.*, II, 671.

and Ohio depot, from which it was conveyed by rail to Baltimore and thence to Chambersburg and Greencastle, where it joined the First Brigade of Patterson's column.[34] It was reported that the balloon would accompany this movement,[35] and one newspaper stated that attached to the Light Battery, the aeronautic apparatus would be used to reconnoitre the position and force of the Confederate troops at Harper's Ferry.[36] Burnside makes no mention of Allen or his balloon in his report to General Scott, but states that the battery took with it all its baggage, and that the entire command was accompanied by a long wagon train.[37] It is possible that the aeronautic equipment was taken as part of these impedimenta, for Allen's inflation of his balloon and its immediate removal to the Rhode Island camp the day before the departure, coupled with the test ascension shortly before the troops moved out, all give some weight to the assumption that the balloon train was intended to be part of the maneuvering force. On the other hand, the statements in the press may well have been conjecture or hearsay picked up by over-enthusiastic war correspondents. The point, however, is worth consideration, for if the aeronautic equipment did accompany the troops on this movement, and were used for the purpose reported, then this incident should be recorded in history as the first occasion that a military balloon was used with troops on active service in this country. Unfortunately lack of documents to establish this point makes the incident a possibility rather than a fact. The First Rhode Island returned to Washington on June 19, where it occupied quarters at Camp Sprague until it marched with the Second Brigade of Hunter's Division to the ill-fated field of Manassas a month later.[38]

Regardless of whether Allen accompanied the movement of Burnside's troops in the maneuver of Patterson's column, it is

[34] Burnside to Scott, June 22, 1861, 1 *O. R.,* II, 715.

[35] New York *Tribune,* June 10, 1861; Providence *Daily Post,* June 10, 1861; Detroit *Free Press,* June 12, 1861.

[36] Unidentified clipping in James Allen's Scrapbook, I, 45.

[37] Burnside to Scott, June 22, 1861, 1 *O. R.,* II, 715-716.

[38] *Ibid.,* p. 716; "Organization of the Army Commanded by General Irvin McDowell at the Battle of Bull Run," 1 *O. R.,* II, 315.

evident that the interest of the regular military establishment was finally awakened to the possibilities of aerial reconnaissance soon after the return of the Rhode Island Regiment to Washington. By this time T. S. C. Lowe had arrived in the city and had made his celebrated demonstration with his balloon and telegraph instruments, and ranking officers of the Topographical Engineers had approached the veteran aeronaut, John Wise, with an offer of employment. On July 9 Captain A. W. Whipple of the Topographical Engineers, later a major general of the line destined to receive a mortal wound at Chancellorsville, communicated with his immediate chief, Major Hartman Bache, and related the results of further operations with Allen's balloon in the outer defenses of Washington. Even at this early stage of progress with the new equipment, an evil which impeded the operations of the balloon corps later throughout the war was recognized by Whipple and pointed out as requiring remedy: the custom of detailing squads at random from any adjacent regiment to serve as the ground crew of the balloon in operation.[39]

Whipple on July 8 had instructed one of his junior officers, Lieutenant Henry L. Abbot, to reinflate Allen's larger balloon and make a reconnaissance of a Confederate outpost near the Union lines. Abbot had at his disposal ample materials for the manufacture of gas, probably the zinc mentioned in Macomb's accounts was part of this supply, but Allen's generating apparatus was found to be crude and very defective. The process of producing the gas was slow and laborious, a task rendered more tedious by the inexperience of the detail assigned as the ground crew. All night was occupied in generating enough gas to fill about half the large linen envelope, during which time Abbot reported what he took to be the ascension of a number of fire balloons used as signals from the Confederate outposts in the vicinity opposite. At daybreak, with Allen's assistance, Abbot attempted to make an ascension with the half-filled balloon, and rose to an altitude of about 500 feet. But the unfilled lower portion of the envelope acted as a sail, and caught every cur-

---

[39] Whipple to Bache, July 9, 1861, MS W853, LRTE.

rent of wind, causing the balloon to incline towards the earth, thus reducing elevation considerably. Not only did the half-filled state of the envelope cause the balloon to lose elevation, but it also caused the car to oscillate to such an extent that Abbot was unable to use his field glasses, the motion rendering the focusing of a lens impossible. Allen objected to the ascent with the envelope only partially filled, fearing that some accident might result, and upon his remonstrances, Abbot descended, convinced that the aeronaut was right. " I am satisfied that the balloon must be *filled* to be advantageously used," he wrote to Captain Whipple. " I am now making every effort to generate the necessary gas with our defective tanks, and still more defective system of labor." [40]

In this comment, Abbot in turn raised the question of the organization of balloon details, one which was never properly settled throughout the war.[41] The detail of ground troops on this occasion was a composite platoon of men from four different regiments, all unfamiliar with and untrained in the work required of them. Abbot stated that had he had a detail of trained men with competent non-commissioned officers to direct them, he could have effected an ascension, despite the faulty generating equipment, by sunset of the afternoon that the reconnaissance was attempted. To make matters worse, the following day a new and more inexperienced detail was sent to replace the one which had served during the previous attempt, the assignment of men to this duty having been but for one day. Abbot expressed confidence in the value of the balloon as an instrument of reconnaissance; he referred favorably to the use of aerostats by Napoleon III in the Italian campaign two years before; and he pointed out to Whipple the advantages of map-making from the air. The sum of these considerations, he said,

establishes that they [balloons] may be made useful in the coming campaign . . . but without trained soldiers to direct them, I feel sure that but little ought to be expected of them. . . . It is essential that

---

[40] Abbot to Whipple, July 9, 1861, MS W853, enc., LRTE. See also Philadelphia *Ledger*, July 11, 1861.

[41] See Chapter viii, *infra*, pp. 297-302.

the officer of our corps in charge should have a few soldiers under his exclusive command. These men, if mounted, would be highly useful as an escort to the officer in his ordinary scouting and reconnoitering beyond the pickets, and there is always difficulty in getting good details from troops designed for other services.[42]

With this plea for some form of real organization in the operations of the balloon detachments, Abbot also submitted for approval a project for a more improved system of generating hydrogen in the field.[43]

General Irvin McDowell, in command of the Union army then in preparation for the advance into Virginia, had been interested to know if the balloons with Burnside's regiment would be of use in the coming campaign. For this purpose

[42] Abbot to Whipple, July 9, 1861, MS W853, enc., LRTE.

[43] *Ibid.* One contributing factor to the difficulty of securing trained troops was the total absence of enlisted men in the Corps of Topographical Engineers. Since the balloons were placed under this branch, troops from this arm would have been the logical men to assign and train for this special service. With the organization of the Corps as it then existed such an arrangement was impossible. In 1859 the Corps consisted of 40 officers and no enlisted men. [Tabular Report "A," "Organization of the Regular Army of the United States," *Report of the Secretary of War, 1861,* Sen. Doc. 2, 36 Cong., 2 Sess., pp. 592-593.] This strength remained the same even after the outbreak of the war, until August, 1861, when 12 additional officers and one company of enlisted men were added. [*Report of the Secretary of War, 1861,* Sen. Doc. 2, 36 Cong., 2 Sess., p. 128.] Under the added duty produced by wartime operations, the corps had less than enough men to perform the routine duties of its regular service; with the added burden of duties imposed by the demand for maps, surveys, and so on, of prime necessity to commanders operating in the field, the assignment of topographical engineer troops for balloon duty became impossible.

Naturally the commanders of volunteer regiments who furnished details for balloon service wanted the return of these men to duty with their own units as soon as possible, and cannot be blamed for demanding that the tours of such extra duty performed by their men be cut to a minimum. This resulted in a new and inexperienced detail reporting to the aeronauts every few days.

The root of the trouble could have been effectively removed by the organization of a regular unit of troops on permanent balloon duty, trained by professional aeronauts. Why this was not done is beyond answer. In his letter to Major Bache, Whipple urgently appealed to his chief to persuade the Secretary of War to secure an amendment to the Army Bill of 1861 which would provide for the permanent assignment or recruiting of a sufficient force of men for the Topographical Engineers. It is evident from his letter and of that from Lieutenant Abbot to Whipple, that this provision would have included men for permanent balloon duty. Whipple to Bache, July 9, 1861, MS W853, and Abbot to Whipple, July 9, 1861, MS W853, enc., LRTE.

Whipple had ordered Abbot to make the tests on July 8. In addition to his statements to Whipple, above recorded, Abbot reported that " no dependence could be made on the gas generating apparatus, and that the only available manner in which to utilize the balloons in the movement would be to have them filled with coal gas at Alexandria, and conducted by men holding guide ropes, using one balloon as a supply reservoir for the other.[44] McDowell approved the plan, and ordered Abbot to make the necessary preparations.

The two deflated balloons were taken to Alexandria, where they were filled early on the morning of July 14. Ground crews consisting of sixty men from the late Colonel Ellsworth's 11th New York Zouaves were assigned to tow them to Falls Church, where they were to be held in readiness to accompany the advance of General Daniel Tyler's Division when the grand movement should take place. One of the balloons, according to Abbot, was an old, worn aerostat constructed of cotton, and as soon as it was fully inflated, it suddenly burst and collapsed, falling in a tangle of limp and crumpled fabric to the ground. Captain Tyler, quartermaster who had made the arrangements for filling the envelopes, was reported to have said, after the accident occurred, " Abbot, I did not join the Army to be a bird." [45] The other balloon, with which the previous experiments had been made, was inflated successfully, and the ground crew manned the tow ropes and took up the march to Falls Church. In a short while a stiff breeze sprang up, and it soon freshened into a strong wind, accompanied by sudden heavy gusts. " We worried along," Abbot later wrote,

nearly to the point where our branch road diverged when suddenly a furious gust occurred. The detail, struggling and shouting, was slowly pulled toward the river in spite of their efforts until the balloon in one of its stately plunges struck a telegraph pole. There was a puff of gas, and our work was ended.[46]

A large hole was torn in the linen envelope, and the second

---

[44] Abbot, " Early Experience with Balloons in War," *Professional Memoirs of the Corps of Engineers,* IV, 680.
[45] *Ibid.*, p. 681.   [46] *Ibid.*

balloon followed the first to the ground, a tangled heap of deflated linen.[47]

Far different in time, place, and essential detail is Allen's own version of the damaging of his balloons, a version given to a newspaper reporter thirty-one years after the event. In the middle of June, 1892, the War Department was preparing to organize several balloon trains as part of the Signal Corps equipment, and sent Captain Richard E. Thompson of that branch of the army to interview Allen concerning the cost of building the balloons for these trains. Naturally Allen's Civil War service had led to this interview, which in turn produced a flow of reminiscences from the veteran aeronaut.

According to the story given out and published by a reporter covering this interview for the Providence *Journal*, the failure of Allen's balloons took place not on July 14 at Alexandria and on the road to Falls Church, but a week later, on the disastrous field of the first Manassas. Allen left Washington with the Rhode Island troops and proceeded to the scene of action with a cart containing the two balloons and a set of gas generating apparatus, which was vaguely described as being several boilers and a quantity of scrap iron. During the battle Governor Sprague is said to have gone to Allen and ordered an ascension with one of the balloons to observe the Confederate movements and position. Allen states that at the time a gale was blowing, and he knew that disaster would attend any attempt to ascend while the high wind prevailed. Therefore he declined to comply with Sprague's orders. Sprague then became incensed at this refusal to obey, and dispatched an orderly to Colonel Burnside, who was now in command of the brigade comprising all the Rhode Island troops and two other regiments, informing the brigade commander of Allen's disobedience. An order was returned from Burnside, directing Allen to inflate one of the balloons and ascend at once. Yielding to this

---

[47] See also accounts of the accident to the balloons in Boston *Transcript*, July 16, 1861; Providence *Post*, July 16, 1861; New York *Herald*, July 16, 1861; New York *Tribune*, July 15, 1861; New York *World*, July 16, 1861; Philadelphia *Ledger*, July 16, 1861; Cincinnati *Commercial*, July 18, 1861; unidentified clipping in James Allen's Scrapbook, I, 45.

summary pressure against his better judgment, Allen got a balloon ready and inflated it. According to the account given out
to the reporter, the wind tore it into shreds. A second envelope
was then filled, and just as preparations were being completed
to make the ascension, the men holding the mooring ropes were
swept off their feet, compelled to let go, and the balloon escaped, in what direction, the story fails to mention. Shortly
after this unfortunate incident, the narrative continued, the
whole Union army was in turn swept as if by a gale in the
direction of Washington, and Allen never did learn what became of the escaped balloon. Because of this failure during the
battle, he concluded, he was disgraced and discharged from the
army.[48]

Although originating from a participant in the alleged
events, this version of the affair is not at all convincing. Its
similarity in some ways to the published accounts in the press
describing the damage on July 14, is rather too marked. Two
such incidents within seven days appear to be unlikely. It is
within the realm of possibility that the wind damaged the envelopes on the 14th, and then they were immediately repaired
and put in service again to meet similar injury a week later, but
probability is against such rapid repair of so serious damage,
and also against the likelihood of two such similar events occur-
ring so close together. Further, it is most unlikely that the
inflating apparatus, that had taken all night to generate gas
enough to fill only half of one envelope, could have become
suddenly so effective as to inflate two balloons in rapid succession. Again, Abbot's account which gives the details of the
events of July 14, was also from an eye-witness and participant,
and it agrees to the letter with contemporary newspaper
accounts published before the battle.

More than likely after thirty years Allen confused the details
in his memory. It is possible that he protested against the
movement to Falls Church because of the prevailing wind.
Abbot's account of his protests against sending up a half-filled
envelope in the wind seems to give some shade of veracity to

the narrative of his objections to ascending during the gale described during the battle of Bull Run. There are other possibilities of a confusion of events after so many years. In some respects, Allen's story resembles the accident that overtook John Wise's balloon on the road to Manassas, on the same day that Allen's reported difficulties took place.[49] And two years later, in 1863, one of Allen's balloons was damaged by wind after it had been ordered aloft against the protest of the aeronaut, who was then in charge of the Union balloon corps.[50] After thirty-one years, the chances for a confusion of these several similar occurrences are great, and doubtless Allen in 1892 muddled the events in his memory.[51]

Regardless of whether the damage occurred on the 14th on the road to Falls Church, as most of the evidence indicates, or on the field at Manassas on the 21st, the fact remains that Allen's efforts to operate his balloons in the military service were considered failures by the authorities. Abbot later declared that he had become convinced that the balloon experiments with the equipment Allen provided were destined to certain failure, and quoted General McDowell in similar terms.[52] Allen also stated in a subsequent letter to Lowe that his " operations were pronounced unsatisfactory " by the Topographical Engineer officers. Consequently he left the service and returned to Providence where he remained, following with interest the progress that Lowe was now making in military aeronautics, until he was called in the spring of 1862 to join Lowe's corps, then preparing for the Peninsular Campaign.[53]

Although considered a failure in his initial attempts, the subsequent career of James Allen with the Army of the Potomac more than vindicates his ability as an aeronaut and a military balloonist. With his brother, Ezra S. Allen, he served with Lowe's corps on the Peninsula, in the latter part of the Mary-

---

[49] See Chapter iii, *infra*, pp. 70-71.   [50] See below, Volume II.

[51] Allen was at this time sixty-eight years old, and it is not unusual at this age, after hundreds of ascensions and countless incidents during the war and later, that he should have confused the events in his mind.

[52] Abbot, " Early Experience with Balloons in War," *Professional Memoirs, Corps of Engineers*, IV, 681.

[53] Allen to Lowe, April 1, 1863, 3 *O. R.*, III, 301.

land Campaign of 1862, at Fredericksburg, and in the campaign of Chancellorsville. When Lowe was taken ill shortly before the battle of Malvern Hill,[54] Allen assumed charge of the balloon corps and remained its acting head until the army was withdrawn from Harrison's Landing. Again, he became chief of the corps after the Chancellorsville disaster, and kept this rank until the balloon train was disbanded during the march to Gettysburg.[55]

As an individual he seems to have possessed excellent qualities that fitted him for the dangerous and arduous work of an aerial observer. An officer of the 6th New York Cavalry, who knew both the Allens, testified that they served faithfully amid all the exposures and dangers of the various campaigns " with distinguished skill and ability." [56] Colonel Cyrus B. Comstock, whose difficulties with Lowe produced the latter's resignation in 1863, and an officer whose attitude was never friendly to the balloon establishment, stated that if it should be necessary to use balloons again, he would re-employ the Allens for service.[57] This statement was fully indorsed by General Gouverneur K. Warren, Chief Engineer under Hooker, and the last officer to superintend the balloon corps before its suppression.[58] General Burnside, Allen's old commander, under whom he again served at Fredericksburg, later declared that the aeronaut's service had been faithful and efficient, and that he was " glad to indorse the opinions of General Warren and Colonel Comstock." [59] Other officers with whom he came in contact also spoke well of Allen in their official correspondence.[60]

Similar testimony of Allen's later ability and qualities may be found in the contemporary press. One of the journals of his

---

[54] This battle made such an impression on James Allen, who was acting chief of the balloon corps at the time, that he named a son in honor of the action, Malvern Hill Allen. See clippings in James Allen's Scrapbook, II, *passim.*

[55] See Chapter xi, *infra,* and Volume II.

[56] G. D. Crocker [Chaplain, 6th N. Y. Cavalry] to Allen, March 29, 1864, James Allen's Scrapbook, I, 53.

[57] Statement of Colonel Comstock, April 9, 1864, *ibid.,* I, 53.

[58] Warren's indorsement, *ibid.*   [59] Burnside's indorsement, *ibid.*

[60] Lieutenant Colonel W. A. Roebling to Major General A. A. Humphreys, February 24, 1863, MS, Letters Received, HDQ 2nd Army Corps, Army of the Potomac, War Department Division, National Archives.

home city described him as a man " of good temper, courage, quickness of perception, and that order of mind which enables him to act in an emergency." [61] A war correspondent who met him on the Peninsula wrote to his paper that he found Allen to be a " gentlemanly, intelligent, and valuable officer," and that " his presence in the Balloon Corps is still another illustration of the honorable way in which Rhode Island is represented in every department of the government service." [62]

The reasons for his early failure can only be speculated upon. In the first place, he took the field with balloons designed for civilian ascensions, which were intended for use only when weather conditions were good, and were not suited for the rigors and emergencies of war service. This equipment, too, had been long in use, and was probably not in the best of condition. One of the balloons was described by Abbot as " an old cotton affair "; the other, which was ruined by the telegraph pole, was in good condition.[63] But in the case of the latter, Allen had been compelled to use untrained men for his ground crew, a circumstance that accentuated the unsuitability of the apparatus, and placed the aeronaut at a disadvantage when atmospheric conditions were unfavorable. And finally his generating equipment was definitely inadequate for the purposes of field operation. When placed in charge of the balloons and generators Lowe designed especially for war service, Allen's performance drew praise from his associates and superiors. It may be said in conclusion that Allen himself did not fail. His apparatus, not designed for the use imposed upon it, coupled with the lack of trained assistants to handle it, failed.

[61] Providence *Post*, April 19, 1861.

[62] War Correspondent to [Providence?] *Press*, May 24, 1862, clipping in James Allen's Scrapbook, I, 49.

[63] Abbot, " Early Experience with Balloons in War," *Professional Memoirs, Corps of Engineers*, IV, 680.

CHAPTER III

JOHN WISE: SECOND TRIAL AND FAILURE

While James Allen was expressing his desire to further the cause of the Union by offering his services as a military balloonist with Governor Sprague's Rhode Island volunteers, his celebrated contemporary, John Wise, the veteran Pennsylvania aeronaut,[1] was likewise inspired with an ardent enthusiasm for the same cause, differing, however, in that he volunteered his services, not as a balloonist, but as the captain of a picked company from his home town of Lancaster. In view of his wide reputation and long experience as an aeronaut, and his intense interest in aerostatics from the aspect of a scientific profession,[2]

[1] By 1861 Wise had twenty-six years' practical experience as an aeronaut. He was born in Lancaster, Pennsylvania, on February 24, 1808. He received elementary and high school education at local schools and also came under the tutelage of the Reverend Augustus Muhlenberg, a Lancaster divine, whose library afforded opportunity for more advanced reading. Young Wise originally intended to study for the ministry, but this inclination soon gave way to a lasting interest in aerostatics. While still a boy he experimented with kites, parachutes, and small paper montgolfiers. He served an apprenticeship to a cabinetmaker for several years, and then set up in trade as a pianoforte-maker. From this experience he became proficient in manual craftsmanship that he later applied to the technical problems of balloon making. In 1835 he began his career as a professional aeronaut, and made his first ascent at Philadelphia on May 2 of that year. From then on he steadily gained a wide reputation, and conducted ascensions and free flights from more than forty towns in some ten states. In 1843 he petitioned Congress for a grant of $15,000 with which to attempt a trans-Atlantic flight, and again renewed the appeal in 1853. He also conducted numerous experiments of a scientific nature, invented improvements in aeronautic equipment, and corresponded with the eminent English balloonist Charles Green regarding parachutes and safety devices. He was associated with La Mountain, and with Dr. Helme, of Providence. At the outbreak of the war Wise was widely known and recognized as the outstanding balloonist of his day.

A full account of his life and aeronautic work is available in his own book, *Through the Air*, which is largely an autobiographical narrative of his experiences, published six years before his death. This work is prefaced by a brief outline of his early life before he became a balloonist. The sketches of Wise in *Dictonary of American Biography* [XX, 428-429] and in Appleton's *Cyclopedia of American Biography* [VI, 581] are brief, but adequate.

[2] Wise, more than any other of his contemporary aeronauts, with the exception of Lowe, constantly approached his work and profession from the scientific point

it is surprising that he did not follow the example of Allen and Lowe in attempting to further his chosen profession by offering to serve as a military balloonist. Stranger still does this appear, against the background of his earnest suggestion to the War Department fifteen years before, that his services be employed in the Mexican War. Whatever may have been his reasons, his initial attempt at service in the Civil War was that of a volunteer officer of the line. He also shares with his former associate, John La Mountain, the distinction of having been sought out by the military authorities and offered an appointment as balloonist with the Federal army. In contrast to the indifference with which Lowe's early efforts to secure recognition and government employment were received by the ranking officers of the army, despite the support of President Lincoln and Professor Joseph Henry of the Smithsonian Institution, the action of the Chief of Topographical Engineers in sending for Wise and offering him an appointment constituted a distinction indeed.

At the time that Allen was making his way to Washington with his two ill-fated balloons, Wise was busy in Lancaster raising and drilling a company of selected men for the duration of the war. Early in May, 1861, he addressed a letter to Secretary Cameron, tendering the services of this unit. " I am forming a company here of picked men," he wrote, " now nearly full and well drilled in the rifle manual. These men would serve for three or five years. . . . Would such a company be

of view. " Professional ballooning," he wrote in 1873, ". . . I have followed to this day, more from the scientific attractions than from mere love of adventure and money making." His career justifies this statement. He continually carried on meteorological and atmospheric experiments, made tests to demonstrate the value of various types of parachutes, invented the rip-panel, and other improvements in equipment. He also made experiments with falling bodies and with the effect of altitude on sound waves. He tested theories of air currents, recorded data on atmospheric conditions at various elevations, on electrical phenomena, and made other investigations. He corresponded with Professor Joseph Henry of the Smithsonian Institution, was long a member of the Franklin Institute of Philadelphia, and served as librarian of the latter institution in 1870. Wise also published two books, his *System of Aeronautics,* and *Through the Air,* both of which contain much information of scientific interest. He also contributed articles to the *Journal of the Franklin Institute* and to the *Scientific American,* and wrote numerous articles and letters of a scientific nature for the contemporary press. His column, " Systematic Ballooning," frequently appeared in the New York *Tribune.*

acceptable? Will you authorize me to fill it and render it to your service—the service of our Country?" [3] The earnestness of this appeal of the fifty-three year old aeronaut who had risked his life scores of times in the pursuit of his profession and the advancement of his science, now offering to take his chances as a line officer in the coming campaigns, speaks well of Wise's sincere desire to do his part in the cause that he had chosen. Safe in years beyond the limits of conscription, [4] and in the full enjoyment of an enviable national recognition as foremost aeronaut in America with the possible exception of Lowe, Wise could have remained comfortably at home and enjoyed the benefits of success that had crowned his twenty-six years of aeronautic experience and investigation. Or he could have sought a spectacular post with the army as a balloonist, claiming recognition because of his long experience and technical knowledge. Despite these possibilities, his conviction led him to seek a different path, that of line service with the other thousands who were flocking to the colors in the early period of martial enthusiasm.

Wise had not, in his desire to lead a Lancaster company in the rapidly gathering Union legions, fully informed himself of the administrative system that had been adopted in the marshalling of this army. The raising of militia and volunteer troops had been placed in the hands of the governors of the several States, who in turn supervised the organization and equipping of the officers and men, and tendered organized regiments and batteries to the Secretary of War for service. [5] Consequently when Wise's offer of his company was received at

---

[3] Wise to Secretary Cameron, May 3, 1862, MS W72, Letters Received, Secretary of War, War Department Division, National Archives [cited hereinafter as LRSW].

[4] The maximum age at which men could be drafted by Congressional acts of conscription for militia service or for service with the "national force" was 45 years. Acts of July 17, 1862, March 3, 1863, and February 24, 1864, *United States Statutes at Large* (Sanger ed.), XII, 597, 731, and XIII, 7.

[5] General Orders, No. 15, AGO, May 4, 1861, *Official Records of the Union and Confederate Armies* (Washington [Government Printing Office], 1880 *et seq.*), Series iii, Vol. I, 151-157 [cited hereinafter as *O. R.*, with series number in arabic preceding abbreviated title]. "Act to Authorize Employment of Volunteers," July 22, 1861, *United States Statutes at Large*, XII, 268-271.

the office of the Secretary of War, its author was referred by Cameron to Governor Andrew G. Curtin of Pennsylvania, " who," the Secretary advised, " under call for soldiers during the war, may qualify you." [6] Accordingly Wise submitted his offer to Governor Curtin, and was waiting his turn in the mass of administrative details attending the organizing and equipping of Pennsylvania's quota when he was called for service as a balloonist by the Bureau of Topographical Engineers.

On June 10, the same day that James Allen was inflating his balloon from the gas main on Massachusetts Avenue in Washington, preparatory to making his initial experimental military ascension, several blocks away in the office of the Topographical Bureau Major Hartman Bache,[7] acting chief of the corps, was dictating a telegram to John Wise, asking for the estimated cost of a balloon of five hundred pounds lifting power and the amount of pay the aeronaut to operate it would demand for his services. The telegram was addressed to Philadelphia, Bache for some reason being under the impression that Wise was in this city.[8] Receiving no reply, Bache again wired the following day to G. Castor Smith, of the Lighthouse Office in Philadelphia,[9] asking if Wise were in the city, and explaining that he had received no answer to his message the day before.[10] Smith sent word that Wise was believed to be in Lancaster, but that he had located another aeronaut, William Paullin,[11] in Philadelphia.[12] Bache then telegraphed Wise in Lancaster, re-

[6] Cameron to Wise, May 4, 1861, MS W87, enc. 1, LRSW [Irregular]; MS Letterbooks, Secretary of War, XLIV, 24.

[7] It is appropriate that Major Bache [1797-1872] should have been prominently connected with the first experiments with military balloons in this country. He was the grandson of Benjamin Franklin, the first known American to have suggested the possibility of balloons in warfare.

[8] Bache to Wise, June 10, 1861, MS Letterbooks, Bureau of Topographical Engineers [War Department Division, National Archives], XXII, 440 [hereinafter cited as LBTE].

[9] Bache was at this time, in addition to his other duties, Engineer of the 4th Lighthouse District.

[10] Bache to Smith, June 11, 1861, MS LBTE, XXII, 442.

[11] A well-known aeronaut of the period. He later served with the Balloon Corps of the Army of the Potomac under Lowe.

[12] Smith to Bache, June 12, 1861, MS S458, Letters Received, Bureau to Topographical Engineers, War Department Division, National Archives [cited hereinafter as LRTE].

peating his offer,[13] and immediately received answer that the cost of a balloon fitting the description would be $300, and that his services in its operation in the field would be " gratis " in the cause of the Union.[14] It appears at this juncture that Captain A. W. Whipple, of the Topographical Bureau, who had been detailed to investigate the question of balloons for war service, was first consulted by his chief, Bache, before Wise's offer was accepted and orders given the aeronaut to construct the apparatus and report for duty.[15] Evidently Whipple considered a balloon of the specifications described in Bache's original communication to Wise inadequate for military purposes. By this time Lowe had appeared on the scene and had demonstrated his aerial telegraph from his balloon *Enterprise*, which, although not primarily designed for war service, was of excellent materials and craftsmanship. Whipple had seen this aerostat during Lowe's demonstrations in Washington,[16] and was, perhaps, impressed by its dimensions and workmanship. He had also seen something of Allen's balloon, having detailed one of his subordinates, Lieutenant Abbot, to conduct experiments and reconnaissances with it. At any event, Whipple and Bache must have deliberated for some days after receiving Wise's answer, for it was not until June 26 that Bache reopened negotiations with the balloonist. He telegraphed asking for an estimate of the cost of a balloon " of best India raw silk, capacity twenty thousand cubic feet, cordage of linen, everything complete." Bache also asked how long its construction would require.[17] Wise promptly replied that such a balloon with all appendages complete would cost $850, and would require two weeks to build from the time of order.[18]

The same day that he telegraphed his estimate to Bache, Wise followed up his message with a letter giving details that explained the figures quoted in his telegram. The estimate was calculated on the basis of cost only, he assured Bache. He had

[13] Bache to Wise, June 12, 1861, MS LBTE, XXII, 443.
[14] Wise to Bache, June 12, 1861, MS W505, LRTE.
[15] Bache to Whipple, June 13, 1861, MS LBTE, XII, 445.
[16] T. S. C. Lowe to Secretary Stanton, June 4, 1863, 3 *O. R.*, III, 256.
[17] Bache to Wise, June 26, 1861, MS LBTE, XXII, 468.
[18] Wise to Bache, June 26, 1861, MS W546, LRTE.

" no desire to make profit on a war balloon," but earnestly wished " to introduce a means of Topographical and Hydrographical Engineering into governmental use that is not appreciated, even if known, to the engineering world." [19] These statements appear to fit in with Wise's character as manifested in his long scientific investigations, as well as in his offer to serve in the war as an officer of the line. He further urged the consideration of a portable inflating apparatus, and added that an observation balloon with a source of gas in the field would be of " inestimable value in time of peace, to say nothing of its service as a means of reconnaissance in time of war." [20] In the event that the Government should order the balloon, Wise continued, he promised to make it as successful as his long experience of twenty-six years in the profession would permit, and as a further indication of his technical knowledge in this field, he asked permission to send Bache a copy of his book, *A System of Aeronautics*.[21] Bache must have accepted a copy of the treatise thus offered, for a quotation from it is discussed in correspondence between Bache and Whipple some weeks later.[22] In conclusion Wise referred Bache to Professor Henry of the Smithsonian Institution for testimony of his professional standing, since Henry's opinion on such matters was then considered the best recommendation the authorities could wish.[23]

The figures quoted in Wise's telegram and the statements in his letter of June 26 must have convinced Bache. Five days later it was publicly announced in the press that Wise had been appointed in the government service as a military balloonist.[24] At this time Lowe was pressing his application for service with the army, and the Cincinnati *Commercial*, whose editor, Murat

[19] Wise to Bache, June 26, 1861, MS W839, LRTE.
[20] *Ibid.*
[21] *Ibid.* The full title of the work is *A System of Aeronautics, Comprising the Earliest Investigations and Modern Practice and Art* (Philadelphia, 1850).
[22] Whipple to Bache, August 6, 1861, MS W920, LRTE.
[23] Wise to Bache, June 26, 1861, MS W839, LRTE.
[24] Boston *Transcript*, July 1, 1861; New York *Commercial Advertiser*, July 1, 1861; New York *Evening Post*, July 1, 1861; Philadelphia *Daily News*, July 2, 1861; Philadelphia *Ledger*, July 2, 1861; New York *Times*, July 3, 1861; Lancaster *Daily Evening Express*, July 1, 1861; Baltimore *American*, July 2, 1861; New York *Herald*, July 2, 1861.

Halstead, had backed Lowe's South Carolina flight, and was even now interceding in his behalf with Secretary Chase,[25] charged the Secretary of War with political partiality in this minor appointment, declaring that "Professor Lowe is displaced by Wise, a Pennsylvanian of no scientific acquirements— another of Cameron's official malfeasances."[26] This charge was completely unfounded. Cameron's political axe-grinding as chief administrator of the Federal armies needs no exposition or apology here; but in justice to both Wise and the Secretary, it must be said that Cameron had no part in the former's appointment. In fact, Wise states that he had "written to the Secretary of War in the beginning of our internecine troubles" concerning his service as a line officer, "but received no encouragement from this quarter."[27] The initiative in the matter came from the Topographical Engineers, largely from Major Bache.

In accordance with his promise as to time of construction, Wise completed the balloon on July 16.[28] The following day he received a telegram from Captain Whipple asking how soon the equipment and its operator could be in Washington.[29] McDowell was advancing on Centreville and, from all indications, a major action in the next few days was imminent.[30] Wise, accompanied by his son Charles as assistant, left Lancaster with his balloon the same day that Whipple's message was received,[31] and arrived in Washington next morning, where he reported to Major Bache and remained awaiting orders from Whipple, who was then with McDowell's column advancing into Virginia.[32] The next day Whipple telegraphed orders from Centreville for Wise to inflate his balloon and report at once to McDowell's headquarters.[33]

[25] See Chapter v, *infra*, pp. 168-170.
[26] Cincinnati *Commercial*, July 4, 1861.
[27] Wise to Bache, June 26, 1861, MS W839, LRTE.
[28] Lancaster *Daily Evening Express*, July 17, 1861.
[29] Lancaster *Daily Evening Express*, July 18, 1861.
[30] General Orders, No. 17, Department of Northeastern Virginia, July 16, 1861, 1 *O. R.*, II, 303-305. [McDowell's movement and preliminary combat order.]
[31] Lancaster *Daily Evening Express*, July 17 and 18, 1861.
[32] Bache to Whipple, July 18, 1861, MS LBTE, XXII, 504.
[33] Whipple to Bache, July 19, 1861, MS W850, LRTE.

In the meanwhile, before Wise arrived in Washington with
his balloon, Whipple had doubted whether the Lancaster aeronaut would finish construction in time for the general movement of the army into Virginia, and had, when Wise's balloon
had not arrived by July 17, notified Lowe to join McDowell's
column.[34] Whipple at this time being with the advancing army,
Lowe had difficulty in securing authority for detailing a detachment of troops to assist him in preparing the balloon for service
and in towing it to McDowell's headquarters. He was in the
midst of preparations single-handed when Wise reached
Washington with his new balloon.[35]

The plan for using a balloon for reconnaissance in the
Manassas Campaign had come to the notice of Major Albert J.
Myer, Chief Signal Officer of the regular army, on July 15, and
this officer requested Whipple to arrange for him to take charge
of Lowe's balloon operations in the field. "I learn from Captain Woodruff [36] that you will probably make balloon ascensions on the other side [of the Potomac]," Myer wrote to
Whipple on the same day.

Get Captain Woodruff your chief to apply to the Secretary that I may
be ordered to arrange the signal communications from the balloon to
the earth. Whether this is done by electricity or by aerial signals, it is
equally my duty to arrange it. We can keep up communication easily,
and to points some miles distant from that of ascension. You will do
me a favor as well as doing the country, I hope, some service if your
will cause the application to be made and inform me when it is made.[37]

Whipple thereupon asked Bache to recommend that Myer take
charge of the balloon party in the coming campaign.[38] Bache
accordingly gave full support to the suggestion and addressed
a direct request to the Adjutant General of the Army, Lorenzo
Thomas, in which Whipple's communication was enclosed, in-

[34] Lowe to Stanton, June 4, 1863, 3 O. R., III, 256.
[35] Lowe to Bache, July 29, 1861, MS, Papers of Thaddeus S. C. Lowe [cited
hereinafter as Lowe Papers]; also printed in 3 O. R., III, 256.
[36] Israel C. Woodruff, at this time assistant chief of the Bureau of Topographical Engineers.
[37] Myer to Whipple, July 15, 1861, MS LBTE, XXII, 501.
[38] Whipple to Bache, July 16, 1861, MS LBTE, XII, 501; Whipple to Bache,
July 16, 1861, MS W840, LRTE.

dorsed and approved.[39] True to the traditional red tape of military administrative correspondence, the matter was then carried up the line to the General-in-Chief, Winfield Scott, before Myer could be detailed as he and the engineer officers desired.[40] " Old Fuss and Feathers " evidently concurred in the matter, and after the numerous communications and indorsements relating to it had been weighed, sorted, and filed, Myer was given the necessary orders that would permit him to take charge of Lowe's balloon party.[41]

But by this time Wise had reached Washington with the new balloon specifically ordered for the campaign, and Lowe was compelled to step aside, even at the moment of inflating his balloon for the field. He removed his envelope from the gas main, and that of his Pennsylvania rival was connected to the valve in its stead.[42]

Some details of the construction of this new balloon, the first to be designed and built in the United States for the Government solely for military use,[43] are appropriate in the present narrative. Its gas capacity was 20,000 cubic feet. The envelope was made of raw india silk, in double thickness. Four hundred and fifty square yards of this material had gone into its construction. The net and cordage were of fine linen, fashioned from Italian flax thread, and rigged so as to be peculiarly adapted to military and topographical ascensions. The hoops, or concentration rings,[44] were of tough hickory, wrapped and re-wrapped with linen cord and fish glue to render them strong

[39] Bache to Thomas, July 16, 1861, MS LBTE, XII, 500.

[40] Bache's indorsement [n. d.], Whipple to Bache, July 16, 1861, MS LBTE, XXII, 501.

[41] Special Orders, No. 117, HQA, July 17, 1861, 1 O. R., LI, i, 421.

[42] Lowe to Bache, July 29, 1861, MS, Lowe Papers; also printed in 3 O. R., III, 256.

[43] Although experiments with Allen's balloons had taken place before Wise arrived for duty, the former's equipment consisted of ordinary balloons for civilian use, and had neither been designed for war purposes, nor had been ordered by the Government. The same is true of Lowe's first experimental military balloon, *Enterprise*.

[44] The concentration rings were the large hoops to which the netting and cordage were attached, and to which the car was in turn fastened. They served to bring the netting together at the bottom of the envelope and provided for the suspension of the basket.

and durable, yet sufficiently elastic. The car, or "basket," was constructed of strong willow and cane, with a bottom of thick sheet-iron to protect the aeronaut and observers from rifle fire below, in the event that they might come under fire at musket range.[45] This detail of the balloon's construction is of particular interest. By this time armor plate had been adapted to ships of war and floating batteries. The armored railroad battery had also made its appearance. And even the medieval body armor of five centuries before had been revived in the form of steel vests and breastplates for the use of the Union troops.[46] But this instance seems to have been the first and only time thus far that "armor plated aircraft" has been placed in active service with American forces.[47]

Wise also planned a further precaution against capture of the balloon and its occupants during action. As part of his ballast he was to carry a number of percussion grenades and bombs. In the event that the infantry lines behind which he would ascend should be driven back before the balloon could be towed to safety, thus enabling enemy troops to reach the position of the ground station and seize the mooring ropes, Wise was prepared to rain down his grenades upon such an unwelcome ground crew and force them to release the ropes, thus allowing him to rise out of reach and trust to a favorable current to take him and the balloon to a position of security. In case this scheme should fail, the simple method of cutting the guys where they were attached to the concentration ring could be resorted to; but the bombing arrangement was designed to permit the aeronaut to save his mooring cables.[48]

[45] Wise to Bache, June 26, 1861, MS W839, LRTE; Wise in the Lancaster *Daily Evening Express,* July 17, 1861.

[46] 1 *O. R.,* I, 17, 18, 63; Captain James Chester, "Inside Sumter in '61," *Battles and Leaders of the Civil War* (New York, 1884), I, 67, 68; Philadelphia *Ledger,* May 6, 1861; Baltimore *American,* May 6, 1861; New York *Tribune,* June 28, 1862. Cf. also F. Ştansbury Haydon, "Confederate Railway Battery at Jacksonville, Florida, 1863, not the First Use of Railway Ordnance in the United States," *Journal of the American Military History Foundation,* II (1938).

[47] Although the British "Salamander" combat planes during the World War were equipped with light armor plate on their fusilages, this protective equipment was not applied to the American planes.

[48] Wise in the Lancaster *Daily Evening Express,* July 17, 1861.

Wise also considered the theory of free ascensions under favorable wind conditions to obtain vertical and close observations of enemy positions, a system that his contemporary, La Mountain, exploited with much publicity several months later.[49]

Although Whipple had requested Bache on July 19 to have Wise proceed at once to Centreville with his aeronautic equipment, it appears that official orders for this movement were not issued by Bache's chief, Woodruff, until the next day. Wise accordingly inflated his balloon at the Columbian Armory and made preparations for the departure.[50] He was joined by a detail of twenty picked men from Colonel William F. Small's 26th Pennsylvania Infantry,[51] who had been assigned to balloon duty by express order of General McDowell on the 17th. Bache in the meanwhile kept an eye on all details that would render the movement a success, and on the 20th requested Captain Theodore Talbot, one of the assistant adjutants general of the Department of Washington to direct that a wagon and team be placed at the disposal of the balloon party to facilitate transportation to the advanced lines of McDowell. Whipple had again telegraphed the day before that the balloon was needed with the forward eschelon of the advancing army, and asked that Wise's party be sent forward as quickly as possible. Fearing that the request for a wagon and team might be interpreted as a revocation of the assignment of Small's troops, Bache assured Talbot that both were needed, and that it was not the intention of the officers concerned to substitute the wagon and horse for the ground troops detailed to the party.[52]

There seems to have been delay in completing all the details necessary to the preparations prior to departure. Wise did not

---

[49] *Ibid.*

[50] Lancaster *Daily Evening Express,* July 24, 1861.

[51] Colonel Small and his men seem to have been directed by coincidence to assist in the early balloon operations. When Wise's movement to Manassas failed and Lowe attempted to reach the battle with his balloon on his own responsibility, Colonel Small graciously assigned another detail to assist him, although he was not proceeding under official orders. Small later made ascensions in one of the balloons of Lowe's corps, and also furnished troops for ground detachments. See *infra,* pp. 192, 355-356.

[52] Bache to Talbot, July 20, 1861, MS LBTE, XXII, 508.

finish his arrangements and get under way until two o'clock [53] on the morning of the 21st, the day on which McDowell's army was routed at Manassas.[54] According to a local press notice, the plan called for Wise to march on Friday night, July 19, and reach Centreville before morning. He was expected to be ready for ascension by daylight on Saturday, the 20th, and take observations at a point two miles from the Confederate main line of resistance, carrying up with him an officer who was to signal observed movements and other details of importance to McDowell's headquarters.[55] Thus when he did not start for the front until nearly thirty hours after the time originally appointed, the plan of operations was miscarried from the very beginning.

At two o'clock Sunday morning Major Myer joined the balloon party and took command. The train, with its twenty men from the 26th Pennsylvania under the direction of Sergeant Edward Bird, accompanied by an escort wagon drawn by four horses, took up their line of march from the Columbian Armory and proceeded out Pennsylvania Avenue towards the river. The inflated balloon was towed by the detail, who held the mooring ropes in their hands and maneuvered the huge, swaying envelope past telegraph poles and wires that lined the way. Fortunately a bright moon, unobscured by clouds, gave light enough to help them in their task. The dodging of poles and wires, however, delayed the party despite the clearness of the night, and gray dawn was breaking when the unit passed through Georgetown. Progress after leaving the city was even more difficult. The long Georgetown aqueduct bridge had to be crossed, after which the line of march followed the Chesapeake and Ohio Canal, a route that worked particular hardship on the towing detail. More than once the men had to wade and swim the canal in order to tow the balloon around project-

[53] Wise, in his account of the operations of this date, gives the hour as one o'clock. This narrative was published twelve years later, and the war correspondent's account at the time, giving the hour as two o'clock, is more likely to be accurate.
[54] War Correspondence of the Lancaster *Daily Evening Express*, July 24, 1861 [cited as War Correspondence]. John Wise, *Through the Air*, p. 554.
[55] Lancaster *Daily Evening Express*, July 22, 1861.

ing trees and thick clumps of tall bushes that lined the towpath. After five miles of heavy going, the party cut across country to the Fairfax road, and there joined the wagon which had crossed the Potomac on the ferry and had followed the road to Fairfax Court House.[56] The route from this point was marked intermittently with pickets and videttes, who gave assistance whenever the occasion required. Although the footing was better than it had been along the canal, the road was lined with trees and wound through patches of woodland. Often the balloon had to be raised high above the tops of tall trees, and the mooring ropes carefully maneuvered to avoid entanglement in the branches. The troops, who were by this time becoming more practiced in the novel procedure, worked manfully and with energy to make as good progress as possible. They were ably directed by Wise and his son Charles, who supervised the handling of the cables. Occasional creeks had to be forded, rickety bridges crossed, and mudholes in the roadbed became more frequent as the route led deeper into Virginia. By noon of the fateful Sunday, the party had not gone more than halfway.[57]

Even at this distance from the scene of action, the sound of battle was plainly audible. The thunder of artillery was distinctly heard through the woods that bordered the road, and the continuous roll of sustained musketry in massed volleys filtered across the intervening miles from Bull Run's raging battlefield. Above the forest-broken horizon the smoke of battle hung like a thick pall. The sound of conflict in front aroused the slowly moving party to a high pitch of excitement. Every effort was made to increase the snail's pace at which the train was advancing and press on rapidly to the distant battlefront. A large beech tree overhanging the road beyond a small

[56] The reader may wonder why the balloon and the towing detail did not also cross on the ferry and follow the main road. The only reasonable explanation seems to be that this route was the terminal of McDowell's main line of communications with Washington, and was naturally burdened with much traffic. The slowly moving balloon party would have possibly caused a serious blocking of the road close to Washington. The one wagon would offer no such difficulty. The route by way of Georgetown and the canal was also shorter in actual distance, though the difficulties in traversing it vitiated its value as a quicker way.

[57] War Correspondence, July 24, 1861; Wise, *Through the Air*, p. 554.

bridge was hastily felled by the excited pioneers who accompanied the party. But the rate of march, though quickened somewhat, was too slow to satisfy Major Myer, who now gave the order that led to the complete failure of the movement.[58]

He ordered the balloon tied to the wagon in order to hasten the march, a plan that Wise emphatically disapproved. Giving way to the prevailing excitement that animated the whole party, and to the inescapable authority of Myer's rank, Wise yielded against his better judgment. The balloon was made fast to the wagon, the horses whipped up to a smart trot, and the vehicle proceeded at a rapid pace towards Fairfax Court House, with the balloon swaying and lurching in the air above. The big envelope brushed violently against trees and overhanging branches, fortunately escaping, however, injury to the silken fabric. This evidently gave Myer the impression that the envelope was strong enough to withstand even rougher treatment, an impression that soon led to disaster. A short distance from Fairfax the party ran into a wooded patch that proved too thick for the tactics adopted. When the teamsters attempted to drive through the narrow passage between the trees that lined the road, the balloon became wedged in the thick upper foliage, and remained stuck in the branches as though in a vise. Wise's efforts to extricate it carefully by hand proved futile, and in a last impatient effort, the horses were driven forward with a dash, in an attempt to free the envelope by force.[58a] The results

---

[58] *Ibid.*, pp. 554-555; War Correspondence, July 24, 1861.

[58a] The celebrated war correspondent of the London *Times,* Dr. [later Sir] William Howard Russell, rode out to the battle, taking the same road that had been followed by Myer's balloon party. He apparently overtook the detachment just after the balloon became tangled in the trees, and he recorded his impressions of the sight in one of his dispatches to the *Times*:

". . . The road was devious and difficult. There were a few persons on their way, for most of the Senators and Congressmen were on before us. Some few commissariat wagons were overtaken at intervals. Wherever there was a house by the roadside the negroes were listening to the firing. All at once a terrific object appeared in the wood above the trees—the dome of a church or public building, apparently suffering from the shocks of an earthquake and heaving to and fro in the most violent manner. In much doubt we approached, as well as the horses' minds would let us, and discovered that the strange thing was an inflated balloon attached to a car and wagon, which was on its way to enable General McDowell to reconnoiter the position he was then attacking—

could have been readily foreseen, except for the desire that impelled Myer and his party to press on to the battle. Large rents were torn in the envelope, the gas escaped, and the balloon was now useless until repairs could be made. In his haste to get the balloon to the field as soon as possible, Myer produced the opposite result, thus rendering aerial observations in the battle of Bull Run impossible.[59]

With the balloon now damaged beyond all possibility of field repairs, and with no gas available even had Wise been able to close the rents in the envelope, Myer placed the train under Wise's command, with instructions to return to Washington, make repairs, and bring the balloon back to the front as soon as possible. It is evident that the Major thought the battle would last several days. Myer then proceeded alone to the front, and reported as signal officer to McDowell's headquarters.[60] He served with distinction and gallantry throughout the remainder of the action, and was cited in McDowell's dispatches.[61]

Wise and his party retraced their way to Washington, and reached the Columbian Armory about four o'clock in the afternoon. His men, having performed twenty-four consecutive hours of most laborious and exacting duty, were completely exhausted from fatigue and lack of sleep. Wise, however, despite his fifty-three years, allowed himself but little rest. He set about repairing the damage at once, and by the 23rd the balloon was again ready for service.[62] He did not, however, comply with Myer's order to march again for Manassas. On the evening following his exhausted arrival back in Washington, the shattered remnants of McDowell's routed army were streaming in wildest disorder and panic back into the Capital

just a day too late. The operators and attendants swore as horribly as Anna's warriors in Flanders, but they could not curse down the trees, and so the balloon seems likely to fall into the hands of the Confederates." Russell to the London *Times,* July 22, 1861, published August 6, 1861, p. 7, col. 5. Also reprinted in the New York *Herald,* August 20, 1861.

   [59] War Correspondence, July 24, 1861.
   [60] *Ibid.*; Wise, *Through the Air,* p. 555.
   [61] McDowell's Report, Battle of Bull Run, August 4, 1861, 1 *O. R.,* II, 322-323.
   [62] War Correspondence, July 24, 1861.

in such a fashion as to justify a later soldier-historian's description of the first Manassas as "a special-constable affair on a
gigantic scale." [63]

While the ranking officers and other military authorities
were rallying and reorganizing the scattered regiments of Mc-
Dowell's army, Wise remained in Washington at the Bureau
of Topographical Engineers awaiting instructions.[64] Whipple
had accompanied the rout back to the capital, and again took
up his duties at Arlington Heights. On the 24th he telegraphed Bache asking where were Wise and his balloon.[65]
Bache informed him of Wise's presence and readiness at the
Bureau,[66] and Whipple asked that Wise be instructed to inflate
his balloon and bring it over to Arlington at once, under
charge of Lieutenant Abbot. He further suggested that Major
Myer also accompany the party in order to ascend with Wise
and signal to headquarters what could be observed of the
enemy's movements from the vicinity. An unidentified balloon,
he added, had been hovering over Fort Corcoran all day.[67] In
reply to Whipple's second telegram, Captain Woodruff replied
that Wise was inflating his balloon and would proceed at once.
The strange envelope seen over Fort Corcoran was, according
to Major Myer, Professor Lowe's balloon.[68]

Wise and his party arrived at Arlington the same day, the
Wednesday following the battle, and ascensions were made
during the afternoon. Confederate advanced detachments were
observed reconnoitering beyond the Washington defenses, and
Wise asserted that he saw several pieces of artillery bearing on

[63] Colonel William Wood [8th Royal Mounted Rifles, Canadian Army],
Captains of the Civil War (New Haven ["Chronicles of America Series," Yale
University Press, Vol. XXXI], 1921), p. 54.

[64] Bache to Whipple, July 24, 1861, MS LBTE, XXII, 510.

[65] Whipple to Bache, July 24, 1861, MS W838, LRTE.

[66] Bache to Whipple, July 24, 1861, MS LBTE, XII, 510.

[67] Whipple to Bache, July 24, 1861, MS W848, LRTE.

[68] Woodruff to Whipple, July 24, 1861, MS LBTE, XXII, 510. The title
"professor" was commonly accorded to professional aeronauts by the newspaper writers of the day. Lowe was habitually referred to as Professor Lowe,
not only in the press, but also officially by the military officers with whom he
served. Allen, Wise, La Mountain, and Steiner all frequently appear in the
press as "professors."

the Capital in position about five miles from the city.[69] Whipple appears to have been occupied with other duties at the moment, for he was not present when these ascensions were made.[70] Having completed his observations, Wise returned with his party to Georgetown, and that evening was instructed to move the balloon forward early next morning for observations at Ball's Cross Roads.[71]

At sunrise, Friday, July 26, Wise and his detail set out across the Georgetown aqueduct bridge. The men were encumbered with their knapsacks and rifles, so that free movement of their arms in managing the mooring ropes was seriously impaired. Perceiving that the crew would be more efficient if these impediments were removed, Wise ordered the men actually engaged in handling the cables to strip off their knapsacks and place them in the empty car of the balloon, along with their heavy muskets. With the men thus unencumbered, Wise believed the movement could be accelerated and made safe from accidents. But misfortune seems ever to have dogged Wise's maneuvers. A stiff breeze was blowing downstream on the Potomac, and its pressure caused the big, swaying envelope to get slightly out of control despite the precautions taken. Before the course could be changed to avoid collision with the telegraph lines that bordered the route, the mooring ropes were caught against the wires, which were of rough, twisted strands. The pressure of the wind and the straining of the buoyant gas caused the wires to saw through the manilla ropes that held the envelope, and presently before the men could relieve the pressure, the cables parted and the balloon, loaded with haversacks and with glittering bayonets protruding from the car, sailed off in the direction of the Confederate lines.[72]

Wise viewed the accident with consternation, but did not lose his presence of mind. He quickly ordered the few men

[69] Wise, *Through the Air,* p. 555.

[70] Whipple to Bache, July 26, 1861, MS W860, LRTE.

[71] Lancaster *Daily Evening Express,* July 26, 1861; also see map, *Atlas to Accompany Official Records of the Union and Confederate Armies, 1861-1865* (Washington [Government Printing Office], 1891, *et seq.*), I, Plate vi.

[72] Special Correspondence, Lancaster *Daily Evening Express,* July 29, 1861.

who still had their muskets with them to fire into the envelope and bring it down in order to save it, even though damaged, from capture. Several minié balls were sent through the large gas bag,[73] but the holes thus made were too small to permit sufficient gas to escape to lower the envelope. A number of troops in the vicinity of Arlington, who evidently heard the firing, and then saw the runaway balloon overhead, loosed a series of volleys into the envelope and accomplished Wise's purpose. The balloon came to earth in the vicinity of General Lee's old home, where a party of soldiers from an adjacent New York regiment took it in hand, incidentally making a hearty breakfast on the rations found in the towing detail's haversacks still in the car. In return for this windfall of food, the men folded up the collapsed envelope and tangled cordage and placed it in a wagon detailed by Assistant Quartermaster George Porter to bring back the balloon to the Columbian Armory.[74] A reporter from the Philadelphia *Press*, who witnessed the affair from a distance, saw the shining bayonets in the car, and believed that the balloon was carrying passengers. He published a rumor that the occupants of the car had been killed,[75] a story that spread into several contemporary newspapers.[76] A dispatch in a Lancaster paper soon corrected the error, and announced that the car had been unoccupied.[77]

In the meanwhile Whipple was waiting for the arrival of the balloon. When disaster overtook the aeronautic party again, word was sent to him giving bare details of the accident. He immediately telegraphed Bache, notifying the Bureau of the new failure,[78] and then proceeded to give Wise a piece of his mind in regard to balloons in general. Wise's son Charles was present at the interview, a young man of tender years, and

[73] New York *Tribune*, July 26, 1861; Lancaster *Daily Evening Express*, July 27, 1861; Detroit *Free Press*, July 28, 1861.

[74] Special Correspondence, Lancaster *Daily Evening Express*, July 29, 1861.

[75] Philadelphia *Press*, July 26, 1861.

[76] Boston *Journal*, July 27, 1861; New York *Evening Post*, July 27, 1861; Philadelphia *Press*, July 26, 1861; Lancaster *Daily Evening Express*, July 26, 1861; Baltimore *American*, July 29, 1861; Washington *Star*, July 30, 1861.

[77] Lancaster *Daily Evening Express*, July 27, 1861.

[78] Whipple to Bache, July 26, 1861, MS W860, LRTE.

according to his father, the choice of language employed by the irate captain was not of the most polite and delicate character.[79] Possibly there was some excuse for Whipple's outburst. He had waited for the balloon at Manassas, only to have it damaged on the way and fail to arrive. Again he waited for it on the road to Ball's Cross Road, only to have the same thing happen. With all that he had gone through in the late battle and the rout afterwards, his temper could not have been any too even. But whatever may have been the justification for Whipple's testiness, Wise was not really responsible for either of the two failures. The first was clearly attributable to the orders of Major Myer, and judging from the course of events at Bull Run, it was good fortune rather than misfortune that the apparatus was damaged and sent back to Washington. Had it reached the field in time it would have undoubtedly fallen into Confederate hands. Since the fleeing army could not save twenty-five of its field guns, of which twelve were the best rifled Parrotts of the regular batteries,[80] it is not likely that a balloon could have been brought off the field in the rush that followed the beginning of the rout. The second failure also, was not caused by any fault of Wise himself. He had taken sensible precautions to prevent accident, but the wind and inexperienced men, plus some element of sheer bad luck, proved more than he could cope with.

It is no wonder, then, that Wise resented the eloquent flow of profanity that accompanied Whipple's remarks. Since he was a devout, God-fearing man, and once had seriously considered the ministry as a profession,[81] he objected to such comments, particularly in the presence of his son. He rose, however, to the occasion, and returned the compliments with the remark that " the balloon part of the disastrous affair " at Bull Run " was just about as good as the fighting part." [82]

This stormy interview with Whipple, though producing

---

[79] Wise to Woodruff, June 5, 1863, MS W542, LRTE.
[80] " Statement of Artillery Lost in the Battle of Bull Run . . .," Report of General McDowell, 1 O. R., II, 328. See also *ibid.*, 347.
[81] Wise, "Autobiographical Sketch," *Through the Air*, p. 27.
[82] Wise to Woodruff, June 5, 1863, MS W542, LRTE.

somewhat strained relations between the two men, did not sever Wise's connection with the army, nor did it dampen the aeronaut's efforts to prove the value of the balloon as an auxiliary form of military equipment. In the beginning of his negotiations with the Bureau, he had urged Bache to consider the construction of a portable generator for the manufacture of hydrogen in the field.[83] He submitted plans for such a contrivance when he reported with his newly-built balloon before the battle of Bull Run,[84] and these were shortly thereafter approved by Bache,[85] to be referred to specialists in machinery and foundry work in Philadelphia.[86]

It is of interest that the generating system proposed by Wise was designed to use the water decomposition process, similar in principle to that employed by the first balloon corps commander in history, Captain Coutelle, sixty-seven years before.[87] Wise added the element of mobility to the chemical principles followed by his predecessor, and thus hoped to produce an apparatus that would be capable of accompanying an army in the field and of producing gas for inflation wherever needed.

His plan called for a contrivance after the form of a common "wagon boiler," consisting of two eighteen-inch cylindrical retorts, constructed of boiler plate, arranged to contain the oxidizing materials, iron turnings and charcoal. With these retorts was to be mounted a steam generating compartment, with a fire-box attached to furnish heat for both. The steam boiler was connected with the retorts by a stout metal tube. As soon as the oxidizing material should be heated to a white-heat, a stop-cock between the steam box and the retorts would be opened, permitting the steam to pass in and through the heated and decomposing turnings and charcoal. The action of the steam on these materials in white-hot condition would produce a supply of hydrogen, which when cooled would serve for inflation of the balloon. With the whole apparatus

[83] Wise to Bache, June 26, 1861, MS W839, LRTE.
[84] Lancaster *Daily Evening Express,* July 22, 1861.
[85] War Correspondence, July 24, 1861.
[86] Wise, *Through the Air,* p. 555; Morris, Tasker and Co. to Bache, July 24, 1861, MS M1435, LRTE.
[87] *Supra,* p. 7.

mounted on a wagon frame to be drawn by a team of horses, gas could be thus manufactured wherever there should be a supply of fuel for the retorts and steam boiler. Wise planned to have a mobile tender for carrying supplies of water and iron turnings accompany the generator. Since an abundant supply of firewood could be obtained almost anywhere the army might go, Wise believed the problem of replenishing gas in the field would be solved by his plan.[88]

The plans were submitted to the firm of Morris, Tasker and Company, expert machinists of Philadelphia, as well as to Dr. John C. Cresson, President of the Franklin Institute, who examined the practical merits of the specifications presented.[89] Wise had calculated that the field generator would produce gas in sufficient volume to inflate his balloon in four hours, or a total of 5,000 cubic feet per hour. But when the experts worked out the figures and details, they found that the materials needed for a single inflation with such an apparatus would amount to at least 7,500 pounds of iron turnings, and 22 cubic feet of water, the latter weighing almost as much as the iron. A generator capable of using this quantity of material would of itself weigh more than five tons, in addition to the tender carrying the 7,500 pounds of iron. The machinists also consulted Professor Leslie Rogers and other scientific men, and all agreed that while it was perfectly possible to produce the required amount of hydrogen with the equipment designed, the plan would be "hardly advisable, as the machinery and extra quantities of iron required will be both troublesome and expensive." [90] The figure estimated by Morris, Tasker and Company amounted to $7,000, almost ten times the cost of the balloon. The Topographical Bureau declined to consider the project, which was abandoned as "too expensive." [91]

Wise, however, still hoped to make his aeronautic operations with the army a success. Despite the impracticability of his

[88] Wise in the Lancaster *Daily Evening Express,* July 17, 1861.

[89] Morris, Tasker and Co. to Bache, July 24, 1861, MS M1435, LRTE; Wise, *Through the Air,* p. 555.

[90] Morris, Tasker and Co. to Bache, July 24, 1861, MS M1435, LRTE.

[91] Wise, *Through the Air,* p. 555.

proposed generator, and despite the discouraging accidents that attended his two major tours of active duty, he appealed to Bache on August 3, declaring that " there *is* a system of Aeronautics," and that " this system if properly instituted and practiced will produce results accordingly." In order to make aerial reconnaissance a success, he continued,

it is required to have at least four assistants who are acquainted with the nature of the balloon. With such aid the balloon may be carried to any distance within a day's march, in calm weather, through woods, over telegraph wires, over rivers by the aid of boat, and through the streets of cities. This method requires the aeronaut to be in the car of the balloon, and a practised attendant to guide the tow rope below. And this method, most effectively to be carried out, also requires that a survey of the route should be made to get the balloon to her place of use in the shortest time and the safest condition.

Under such an arrangement the balloon may be taken from the Armory grounds to Centreville in advance of the army on its line of march.

Wise also advised that a balloon of smaller dimensions be employed, preferably with an envelope of 8,000-foot capacity. Such a balloon, he pointed out, would not be so liable to damage as the large aerostat then in use, as it would be less difficult to manage in unfavorable weather.[92]

There is much to be said for Wise's arguments. Had his plan for towing the balloon been carried out, and had the route been surveyed in advance, the accident that occurred on the way to Manassas might have been avoided. Again, had he been furnished with experienced assistants, and had he operated a smaller balloon, his aerostat might not have escaped on the way to Ball's Cross Road. Both of these principles were carried out in the successful work of Lowe later on in the war.

Bache referred Wise's suggestions to Whipple,[93] who, unfortunately for Wise, was now somewhat prejudiced against all balloons. Three days before Lowe, who had been ordered to take his balloon to Arlington for ascensions, had been compelled to discharge gas because of a sudden storm.[94] For a

[92] Wise to Bache, August 3, 1861, MS W919, LRTE.
[93] Bache's indorsement, August 3, 1861, *ibid.*
[94] Lowe to Stanton, June 4, 1863, 3 *O. R.*, III, 258; Joseph Henry to Whipple, August 2, 1861, *ibid.*, p. 259.

third time Whipple had waited for the arrival of a balloon only
to be disappointed. After this, he declared he would have
nothing more to do with balloons.[95] Accordingly he recommended that " Mr. Wise's project be deferred until assured
of some possibility of success." [96] Three days later he defined
his attitude clearly in a letter to Bache, explaining his indorsement concerning Wise's suggestions:

The failure of Mr. Wise's " War Balloon " has caused great disappointment. American aeronauts have thus proved far less successful than
the French were at the close of the last century. Until we can equal
others in the beaten paths, I cannot recommend new experiments.[97]

Whipple's reference to the accomplishments of the French military balloon service in the 1790's failed to mention that the
aeronautic companies then operating had been composed of
men especially selected because of special technical knowledge,
men trained in aeronautics before they were sent into the field
with balloons; that a training school of military aeronautics
had been established for this purpose; and that the balloon
units had been highly organized both tactically and for administration. Wise, on the other hand, had been compelled to function with a detail of raw volunteer troops, completely ignorant
of the technique required of them. Had he had the advantage
of the superior French organization and foresight, his efforts
would doubtless not have been " far less successful than the
French at the close of the last century."

In the face of such criticisms, Wise must have felt keenly
what he considered lack of appreciation of his efforts. Whipple
evidently blamed him for the Bull Run affair, which was un-
deniably the fault of Major Myer; he also inferred that Wise
was responsible for the mishap on the way to Ball's Cross Road,
when in reality Wise could not be held strictly to blame. Now
when Wise recommended that measures be taken to prevent
similar failures, Whipple baldly declared that because of these
two failures which Wise was now trying to prevent in the
future, he could " not recommend new experiments." Wise

[95] Whipple to Woodruff, August 11, 1861, MS W912, LRTE.
[96] Whipple's indorsement, Wise to Bache, August 3, 1861, MS W919, LRTE.
[97] Whipple to Bache, August 6, 1861, MS W920, LRTE.

80

had received no pay, not even rations or quarters for himself or
his son. He had furnished the balloon for the army's use
without profit to himself, and without wages for the labor of
construction. And he even bore the expense of repairing the
balloon after it had been twice damaged.[98] Consequently when
he received an order to construct a balloon for the use of a civil
engineer in the West,[99] he notified the acting chief of the Bureau, Woodruff, that he was leaving Washington for Lancaster
to attend to this commission. He offered, however, to send a
trained aeronaut to resume operations, should the Bureau desire, and assured Woodruff that he would return to the Capital
within twenty-four hours, should his personal services be again
required.[100] There is no indication in the records of the
Bureau's correspondence that this letter was ever acknowledged.[101] He left Washington on August 13, a disappointed
man, but still ardently desirous of serving the Union.[102]

Thus ended the aeronautic services of John Wise, the second
balloonist to take service with armed forces in the United
States. But he continued to direct his efforts for the Federal
cause. Back in Lancaster, he returned to his former plan of
serving as an officer of the line. He again set about raising a
unit, and by October he had under arms a cavalry troop known
as the Lochiel Grays,[103] which was mustered into the United
States service as Company " F," 9th Cavalry, Pennsylvania

[98] Wise to Bache, June 26, 1861, MS W839, LRTE; Wise to Woodruff,
November 28, 1862, MS W948, LRTE; Wise to Woodruff, June 5, 1863, MS
W542, LRTE.

[99] J. H. Crever to Wise, August 10, 1861, quoted in Wise to Woodruff,
August 13, 1861, MS W970, LRTE.

[100] Wise to Woodruff, August 13, 1861, MS W970, LRTE. Evidently resenting what he felt was shabby treatment, Wise later presented a claim for his
personal expenses and the cost of repairing the balloon, adding to this sum a
charge for his son's services. Obtaining no satisfaction, he then placed the matter in the hands of a Lancaster attorney for collection. There is no evidence
that the claim was ever paid.

Wise to Woodruff, November 28, 1861, MS W948, LRTE; J. B. Livingston
to Bache, May 21, 1862, MS L435, LRTE; Wise to Woodruff, June 5, 1863,
MS W542, LRTE.

[101] The letter is filed in the Bureau's correspondence, without disposition
entry, and the outgoing letterbooks contain no copy of an answer.

[102] Wise to Woodruff, August 13, 1861, MS W970, LRTE.

[103] Lancaster *Daily Evening Express,* October 26, 1861.

Volunteers.[104]   His fellow citizens of Lancaster appreciated his services and sincerity far more than did the engineer officers in Washington.   When his company was about to leave for active service, Wise was presented with a handsome sword, a relic of the War of 1812, as an expression of general esteem and appreciation of his efforts in the Federal cause.[105]   His cavalry service was of short duration.   Wise's health became impaired from the hardships and exposure, and after several months he was compelled to retire from active duty and again settle down in Lancaster, where he began preparations for a new edition of his *System of Aeronautics.*[106]

[104] Wise to the editor of the Lancaster *Daily Evening Express,* October 14, 1861, published same date.

[105] Dr. Charles Herbst to Wise, November 8, 1861; Wise to Herbst, November 8, 1861; both letters published in the Lancaster *Daily Evening Express,* November 11, 1861.

[106] Wise to Woodruff, November 28, 1861, MS W948, LRTE. Wise continued in civil aeronautics after the war. In 1872 he revived his project for a trans-Atlantic voyage by air, but was unsuccessful in raising the necessary funds. He then moved to Louisiana, Missouri and gave up aeronautics, intending to spend the remainder of his life in farming. But his old interest in ballooning returned, and on September 28, 1879, he ascended with several companions from St. Louis, Missouri. The balloon was blown over Lake Michigan, and was not heard from again. Several weeks later the body of one of Wise's companions was washed ashore. Wise's body was not recovered.

St. Louis *Globe-Democrat,* September 29—October 7, 1879; October 15, 19, 25, and 26, 1879. See also sketches of Wise in *Dictionary of National Biography,* XX, 429, and Appleton's *Cyclopedia of American Biography,* VII, 581.

CHAPTER IV

JOHN LA MOUNTAIN: FREE LANCE BALLOONIST OF THE ARMY OF THE POTOMAC

During the period of attempt and failure that marked the earlier military activities of James Allen and the operations of John Wise, a third contestant for recognition in the field of military aeronautics entered the arena. This individual was John La Mountain, of Troy, New York. Of his early life, very little is known. Like his professional contemporary, Allen, he seems to have been a seaman turned aeronaut, but available records show very little of his career before his part in the spectacular aerial voyage with Wise and Gager in 1859. He had made a number of ascensions before this record flight, chiefly in the New England states, but beyond these few activities, his early life is almost a blank to the investigator. Even the exact date of his birth is uncertain, although the year 1830 is generally accepted.[1]

Two months after the famous flight with Wise from St. Louis to Henderson, New York, La Mountain took part in another spectacular aerial voyage which nearly cost him his life. In company with John A. Haddock, editor of the Watertown *Reformer*, he took off from Watertown, New York on September 22, 1859, intending to make only a short experimental flight. Strong northerly currents carried the balloon far over the Canadian wilderness, and landed the two aeronauts in dense forestland some 300 miles from their point of departure, approximately 150 miles north of Ottawa. After wandering about the wilderness for nearly a week, during which time they were entirely without food for four days, the stranded aeronauts were rescued in the last extremities of exhaustion and

---

[1] The sketch of La Mountain in *Dictionary of National Biography* is practically valueless for a picture of La Mountain's career. No biography or even adequate sketch of the aeronaut has ever appeared, and he left no writings of his own. The only source of information available seems to be the various newspapers of his time.

exposure by a party of lumbermen.[2] From this sensational experience La Mountain obtained much publicity and became for a time a national figure, though he was not recognized as the equal of Wise, either as an aeronaut or a scientist.[3] From 1859 to the outbreak of the war, he continued to practice his profession as a balloonist, giving numerous public exhibitions and demonstrations in various parts of the North. Like Lowe and Wise, he also aspired to cross the Atlantic by air, and on several occasions publicly announced his intention to make the attempt.[4] He seems to have been a man of unusual courage and daring, endowed with energy and vigorous physical strength.[5]

When the long threatened sectional hostilities finally became reality, La Mountain was scarcely two weeks behind Allen and Wise in offering his services to the Federal government.[5a] Probably he was moved by the same spirit of patriotism that had animated his fellow balloonists. Or he may simply have seen opportunity to enhance his reputation and secure lucrative government employment that would suit his professional tastes. Unlike his friend Wise, he applied for military service as an aeronaut, but did not rely solely on his known reputation as indicative of his ability and qualifications. He first drew up a

[2] New York *Herald,* September 29, 1859; *Ibid.,* October 4 and 6, 1859; Watertown [N. Y.] *Reformer* [extra], October 5, 1859; New York *Evening Express,* October 7, 1859; Albany *Morning Times,* October 4, 1859; Leslie's *Illustrated Newspaper,* October 22, 1859.

The account in the Watertown *Reformer* is Haddock's own personal narrative. Those in the *Express* and Leslie's are narratives by La Mountain, the latter being well illustrated by a series of La Mountain's drawings depicting the adventures of the two aeronauts. The Albany *Times* contains a telegram from La Mountain to his family in Troy, announcing his safety and giving some vivid details of the voyage.

[3] In the numerous accounts of La Mountain's work in the various newspapers there is but little evidence that he practiced his profession with a view to advancing it as a science. He does not seem ever to have been connected with any scientific institutions or prominent scientists; and he left no written evidence of his work and aeronautic experience.

[4] Clippings in Lowe's Scrapbook.

[5] Haddock described La Mountain as "a brave man, who probably does not know what personal fear is." Haddock's narrative in Watertown [N. Y.] *Reformer* [extra], October 5, 1859.

[5a] Baltimore *Exchange,* May 25, 1861.

petition recommending himself for military service, and canvassed a number of influential citizens of Troy for their signatures.[6] By May 1 he had secured a formidable list of local dignitaries including the postmaster, the editors of the two newspapers of the town, a lieutenant of the regular army, and twenty-seven other ostensibly prominent citizens who declared that

having full confidence in the skill of Mr. John La Mountain, as one of the most scientific balloonists of the country, and being familiar with his capacity, [we] have no hesitation in recommending him . . . and assuring . . . that he is a gentleman of thorough practical information, upright, honest, and indefatigable, and one who may be relied on to perform all he promises, when performance is possible.[7]

This laudatory document was then sent to the Secretary of War, and a copy also forwarded to the Military Commissioners of the State of New York. Appended to it was a disquisition in La Mountain's own hand, describing the history and practice of military ballooning from Fleurus to Solferino, which, despite several errors of fact obvious to a modern historian, showed considerable familiarity with the history of the subject. On the basis of this historical discourse, La Mountain urged that the Government adopt aeronautics as a part of the army tactics, and further, on the strength of the accompanying signed petition, he urged that he be selected to perform the service that such an adoption would require.[8] As in the case of many other letters of a similar nature, the officials in the Secretary of War's office evidently shelved the communication without taking the trouble to answer it.[9] La Mountain, however, was

[6] Petition recommending La Mountain for army service, May 1, 1861, MS L129, Letters Received, Secretary of War, 1861, War Department Division, National Archives [hereinafter cited as LRSW].

[7] *Ibid.* Among the signatures were those of James R. Fonda, Postmaster of Troy; G. L. Willard, 1st Lieutenant, U. S. Army; J. M. Francis, W. E. Kisselbaugh, and George W. Demers, editors of the Troy *Times*; J. B. Hubbel and James S. Thom, editors of the Troy *Whig*; and twenty-seven others whose positions or occupations were not indicated.

[8] La Mountain to the Secretary of War and the Military Commissioners of the State of New York, May 1, 1861. MS L129, enc., LRSW.

[9] The MS cited in Note 8 above shows neither indorsement nor disposition entry. The same is true of the letterbook in which receipt of the communication was recorded.

apparently a man of determination as well as daring, and when he received no reply at the end of a week, he again addressed the Secretary of War, repeating in substance the arguments that had been advanced in his previous application.   Having outlined the advantages of aerial reconnaissance to an army in the field, he summarized his own career as a balloonist, calling the Secretary's attention to his part in the sensational flights with Wise and Haddock.[10]   Additional weight was attached to his persuasions on the ground that he was prepared to provide special equipment for servicing his balloons in the field, and in this connection he declared,

I have a portable apparatus of my own invention and used solely by myself with which I can manufacture gas for inflation from the decomposition of water, at a barely nominal rate of expense. This with balloons and all appurtenances necessary could be carried in a single wagon in the train of the army and the balloon filled in a few hours' notice wherever there is coke or charcoal and water.

Should the Government consider his appointment in the army favorably, he concluded, he would be ready to take the field with all necessary equipment within twenty days' notice.[11] Again the Secretary's office did not grant him even a perfunctory reply and, from all appearances, La Mountain was doomed to disappointment in his efforts to interest the Government in his offers of service.[12]   He remained in Troy and does not seem to have made any further application.   Possibly he felt that the two snubs he had received made any additional efforts on his part useless.

A month later, however, a welcome yet unofficial offer of employment came from an unexpected quarter.   On June 5, Major General Benjamin F. Butler, the politician-general who later was to earn for himself the unique title of " Beast Butler " as military governor of New Orleans, wrote from Fortress

[10] La Mountain to Secretary Cameron, May 7, 1861, MS [number omitted], LRSW; Scientific American, June 15, 1861.
[11] La Mountain to Cameron, May 7, 1861, MS [number omitted], LRSW.
[12] The MS cited in Note 11 above shows neither indorsement nor disposition entry. The same is true of the letterbook in which the receipt of the communication was recorded.

Monroe, the headquarters from which he commanded the Department of Virginia, offering La Mountain employment as an aerial observer with his command.[13] According to the Troy *Times*, which doubtless got the details from La Mountain himself, Butler had expressed an opinion that the proposed balloon service would be " of the greatest importance," and " cannot fail to prove of much benefit." The General was also quoted as having promised to use his influence with the War Department to secure La Mountain an official position in the army if the balloon experiments should succeed.[14] La Mountain responded eagerly, and advised Butler that he would have his balloons and apparatus in readiness to start for Virginia the following week. His field equipment would consist of two balloons and a field generator of his own design with all necessary appendages.[15] A Philadelphia editor, having learned of La Mountain's intended army service, predicted that a new military title would be needed to fit the occasion. In a jocular tone he suggested that " Balloon Major " might be adopted, and added that regardless of title, the aeronaut would certainly enjoy the highest rank in the service.[16]

In the meanwhile, as if by peculiar coincidence, La Mountain's old associate, Gager, with whom he had made the St. Louis-Henderson voyage, happened to visit Fortress Monroe, and learned from Butler that La Mountain had been invited to serve with the General's command. Gager immediately wrote to La Mountain and urged him to hasten preparations and report for duty as soon as possible. Having returned to New York to find his friend in financial difficulties because of the necessary travelling expenses, Gager advanced funds to La

[13] La Mountain to Butler, June 10, 1861, MS, Benjamin F. Butler Papers, Division of MSS, Library of Congress [hereinafter cited as Butler Papers]. See also *Private and Official Correspondence of General Benjamin F. Butler during the Period of the Civil War* (Norwood, Mass. [Privately printed], 1917), I, 132 [hereinafter cited as *Private and Official Correspondence*]. O. A. Gager to Butler, June 22, 1861, MS, Butler Papers.

[14] Troy *Times,* quoted in New York *Commercial Advertiser,* June 11, 1861; and Philadelphia *Ledger,* June 13, 1861.

[15] La Mountain to Butler, June 10, 1861, MS, Butler Papers; *Private and Official Correspondence,* I, 132.

[16] Philadelphia *Daily News,* June 18, 1861.

Mountain and then communicated with Butler at La Mountain's request, stating that the latter was making every effort to report without delay.[17]

Delays, however, seemed inevitable. In his eagerness to perform government service, La Mountain had far overstated the readiness of his equipment, and also lacked funds with which to procure materials necessary for the manufacture of hydrogen. He proceeded to New York and attempted to secure an order from the militay authorities there for its supply, but was informed that such orders and requisitions came only from Washington.[18] Having borrowed money from his friend Gager to transport his balloons to New York City, and for passage to Washington, La Mountain then hurried to the Capital and sought an interview with Secretary Cameron, in which he asked for inflating materials amounting to about five hundred dollars. There were, he pointed out to the Secretary, no gas works at Fortress Monroe, and the operations desired by General Butler could not be effected without the supplies La Mountain requested. Cameron, though his subordinates in the War Office had ignored La Mountain's previous letters with stony indifference, received the aeronaut graciously and lent a sympathetic ear to his troubles. His answer was quite favorable, as quoted by La Mountain to Butler:

If General Butler, General McDowell, or any of the Generals should want a balloon attached to their camps, all that they would have to do would be to order whatever they might want. It is for them to say what they think they need.

La Mountain also quoted the Secretary as favoring military balloon reconnaissance, and as considering such experiments as La Mountain was about to make, "well worth a trial."

The glowing terms with which La Mountain had described his portable generating apparatus, however, proved to be utterly without foundation. In explaining the delay in reporting for duty at Fortress Monroe, and the necessity of his negotiations with Secretary Cameron, he frankly admitted to Butler that his

[17] Gager to Butler, June 22, 1861, MS, Butler Papers.
[18] *Ibid.*; La Mountain to Butler, June 25, 1861, MS, Butler Papers.

generator was not as efficient as he had represented it to be, and informed the General that

> I *dare* not rely entirely upon it, as I have had of late some trouble with my experiments, and have concluded to adopt the old method which I have used for years.

He then assured Butler that he had laid aside all other business and was determined to demonstrate the value of balloons to the army. But the difficulty was, he continued, the cost of materials to make this demonstration possible. The gist of the matter, to which he was leading up in all his explanations, was simply a request that Butler sign an order for a quantity of sulphuric acid and iron turnings, with necessary tanks and copper pipes for the manufacture of hydrogen. Since the Secretary of War had given his authorization for anything that the generals might "think they needed," the only remaining obstacle was Butler's written order. In return for this assistance, La Mountain agreed to serve without pay until he should satisfy Butler of the practicability of his operations.[19]

In this communication appealing for supplies, appeared the first official expression of the professional jealousy that marked La Mountain's attitude towards a contemorary aeronaut, T. S. C. Lowe, who was then pressing applications for military service. This attitude was fully shared by the latter, who at times repaid it with interest. Later developments in the work of the two men will demonstrate this unfortunate element far more clearly, but its first manifestation in connection with their war service was displayed in La Mountain's request for Butler's assistance, in which he expressed complete contempt for the use of telegraph with balloons. On June 18 Lowe had completed his successful demonstration of aerial telegraph in Washington in the presence of President Lincoln, Secretary Cameron, Professor Henry, and several ranking army officers. The experiment was acclaimed all over the country as a major step in the advancement of military science.[20] On June 25 La Moun-

---

[19] *Ibid.* The materials La Mountain asked Butler to furnish consisted of 60 carboys of oil of vitriol [sulphuric acid], 3½ tons of iron filings, 2 tanks, and a quantity of copper pipes, the whole amounting to a cost of $490.

[20] See Chapter v, *infra,* pp. 174-181.

tain stated to Butler that he could see neither value nor advantage in employing telegraph for communication between observers and headquarters. "I have a simple method by which I can convey intelligence from above *without any expense* whatever," was his challenge to Lowe's recent accomplishment.[21] La Mountain's opinion notwithstanding, the efficiency, value and advantages of aerial telegraph were established and proven in three major campaigns, and were regarded as a necessity by ranking generals of the army long after their critic had been dismissed from the service.[22] In view of this, La Mountain's attack on Lowe's newly acclaimed system seems to have been prompted by an attitude of jealousy toward his rival aeronaut, whose civilian work before the war he had more than once disparaged in the public press. The unfortunate relations between the two men were to assume far greater proportions as time went on.

La Mountain, after dismissing the system of telegraph communication as inferior, rested his case, and urging Butler to favor his new plan, concluded by saying that "as the matter is entirely in your hands, I will await orders for these materials."[23] There was really no reason why Butler should not approve La Mountain's suggestion, despite the aeronaut's obvious failure to make good the promises he made first to the War Department, and second in his letter accepting Butler's first proposal. After all, the latter had taken the initiative originally, and wanted a balloon for reconnaissance in the vicinity of his command. Also, the spending of the small sum of $490 of the Government's money could never have been a matter of importance to Butler's sense of economy, as his later administration of New Orleans luridly demonstrated. Accord-

[21] La Mountain to Butler, June 25, 1861, MS, Butler Papers.
[22] The campaigns of the Peninsula, Fredericksburg, and Chancellorsville. So vital to the operations of the army was the telegraph from Lowe's balloon considered during the preparations for Hooker's flank movement at Chancellorsville, that General Butterfield, Chief of Staff, gave orders that any person, regardless of rank or excuse, found tampering with these lines should be shot summarily. [See below, Volume II.]
[23] La Mountain to Butler, June 25, 1861, MS, Butler Papers.

ingly he prepared the necessary orders and forwarded them to La Mountain, with instructions to report as soon as possible.[24]

On July 12 La Mountain placed his two balloons with the tanks and other inflating apparatus on board the schooner *Arctic*, which was scheduled to clear New York for Fortress Monroe on the 15th.[25] He had employed two assistants, Albert Kendrick and George A. Waters, to take charge of the equipment and supervise its handling, while he proceeded by rail to Baltimore, from which point he took a steamer for Fortress Monroe.[26] He arrived at Old Point Comfort about the 23rd, and set about preparations for ascensions, the first of which was scheduled for the 25th.[27]

Butler's desire to secure the services of an aerial observer are easy to understand. Fortress Monroe was strategically situated for Federal operations, and provided a strong point of departure for an attack upon Norfolk and Richmond, as well as for a base for blockading the Confederate capital from the coast.[28] But at this time it was by no means a favorable position for the small garrison that held the citadel. "Fortress Monroe in Danger," ran a headline in one of the Baltimore papers towards the end of June, and the correspondent foreboded that "there can be no doubt that Fortress Monroe will soon be the point of attack." [29]

Butler stated that his command was "surrounded by enemies," of whose strength and disposition he had no certain knowledge.[30] In any land operations his flanks were cut off by

[24] Butler to La Mountain, June [day omitted] 1861, MS, Butler Papers; La Mountain to Butler, July 2, 1861, MS, Butler Papers.

[25] The *Arctic* cleared New York two days later, on July 17. Cf. Schedule of Ship Clearings, Port of New York, July 17, 1861, New York *World*, July 18, 1861.

[26] La Mountain to Butler, July 15, 1861, MS, Butler Papers.

[27] Boston *Journal*, July 24, 1861; Boston *Transcript*, July 24, 1861; Providence *Post*, July 25, 1861; New York *Times*, July 25, 1861; New York *Evening Post*, July 24, 1861; New York *Commercial Advertiser*, July 24, 1861; Lancaster *Daily Evening Express*, July 24, 1861; Philadelphia *Press*, July 25, 1861; Philadelphia *Ledger*, July 25, 1861; Philadelphia *Daily News*, July 25, 1861.

[28] Butler to Lieutenant General Winfield Scott, June 6, 1861, 1 *O. R.*, II, 667.

[29] Baltimore *Clipper*, June 29, 1861.

[30] Butler to Assistant Secretary of War Thomas A. Scott, August 13, 1861, MS, B236 [Irregular], LRSW.

the York and James Rivers, whose waters at the time could have afforded approach to Confederate gunboats.[31]   On the land side his position was open to the threat of possible investment by forces from Richmond, and the waters surrounding the fort were becoming menaced by Confederate works built on points of tactical importance, with heavy ordnance to command the channels.   As early as May, Confederate engineers had erected a strong battery on Sewall's Point, at the mouth of the Elizabeth River, opposite Fortress Monroe, and troops were reported in force behind this position amounting to 3,000 or 4,000 men.   Another heavy battery had been established on Pig Point, and reports also indicated that the Confederates were fortifying Newport News.[32]   Rumors that a force under General John B. Magruder was preparing the move on Butler's outposts near this place were brought in by scouts towards the end of June;[33] and a serious reduction in the strength of Butler's garrison occasioned by the expiration of enlistments in several regiments of three-months' volunteers,[34] and the sudden withdrawal of four regiments and a battalion from Fortress Monroe for the defense of Washington and Baltimore after the Union defeat at the first Bull Run,[35] reduced Butler's force to a point where it was " impossible to take up or hold any advanced position."   There were reported further, from the best available sources of information, Butler wrote to the General-in-Chief, some 8,000 enemy troops at Yorktown, and 10,000 at Norfolk, whereas his own command had shrunk to about 2,000 effectives.[36]   The main basis of security for Fortress Monroe

[31] See map, " Fortress Monroe and Vicinity," *Atlas to Accompany Official Records of the Union and Confederate Armies, 1861-1865* (Washington [Government Printing Office] 1891, *et seq.*), I, Plate xviii, Sheets 1 and 2.   The Confederate Government at this time, however, had no gunboats of any consequence. Butler, nevertheless, could not be sure of this fact.

[32] Butler to Winfield Scott, May 24, 1861, 1 *O. R.*, II, 651.

[33] Butler to Winfield Scott, July 3, 1861, 1 *O. R.*, II, 735.

[34] Butler to Winfield Scott, July 15, 1861, 1 *O. R.*, II, 741.

[35] Winfield Scott to Major John A. Dix, July 24, 1861, 1 *O. R.*, II, 761; Dix to Butler, July 24, 1861, *ibid.*; also in *Private and Official Correspondence*, I, 178; Winfield Scott to Dix, July 25, 1861, 1 *O. R.*, II, 762; Butler to Cameron, July 30, 1861, *Private and Official Correspondence*, I, 185.

[36] Butler to Winfield Scott, July 26, 1861, 1 *O. R.*, II, 763.

was the navy and its maintenance of uninterrupted communications with Washington by the Potomac, but with the exception of several obsolete wooden frigates, the present Union sea force was almost negligible. Such a position, surrounded by hostile territory and enemy forces of unknown size and disposition, with rumors and counter-rumors of enemy activity constantly reaching the garrison, made a means of reconnaissance from the air most welcome to the general in command. Also, the flatness of the surrounding terrain, and the presence of the rivers and bay made the region admirably suited for aerial observation.[37] These possibilities were recognized as early as June by the correspondent of the New York *World*, who wrote accordingly to his editor: " A balloon reconnaissance is now sadly wanted at this point, and I am informed that General Butler has sent North for this valuable aid to our military operation." [38]

On July 24 La Mountain completed arrangements for his initial ascension. One of his balloons and the generating apparatus were taken to the Federal outposts near Hampton, and inflation was begun in the afternoon. An engineer officer was assigned to accompany the aeronaut in his ascensions.[39] The Northern press announced that the first observations would be taken on the following evening.[40] True to the published reports, La Mountain went aloft from the outskirts of Hampton on the evening of July 25, thus effecting his first ascension in the service of the Government. Unfortunately a stiff wind was blowing at the time and he was unable to attain sufficient altitude to secure effective observation.[41] A Massachusetts volun-

---

[37] See map, *Atlas to Accompany Official Records,* I, Plate xviii.

[38] New York *World,* June 24, 1861.

[39] New York *World,* July 29, 1861.

[40] Boston *Journal,* July 27, 1861; New York *Commercial Advertiser,* July 27, 1861; New York *Evening Post,* July 27, 1861; New York *Times,* July 27, 1861; New York *World,* July 27, 1861; Philadelphia *Daily News,* July 27, 1861; Philadelphia *Ledger,* July 27, 1861; Philadelphia *Press,* July 27, 1861; Lancaster *Daily Evening Express,* July 26, 1861; Baltimore *American,* July 27, 1861; Providence *Press,* July 24, 1861; New York *Herald,* July 26 and 27, 1861. Even the Richmond papers copied the notice from the Northern correspondents, and announced the event. Richmond *Enquirer,* July 31, 1861.

[41] Boston *Transcript,* July 29, 1861; Providence *Post,* July 29, 1861; New York *Tribune,* July 29, 1861; Philadelphia *Inquirer,* July 29, 1861; Baltimore *American,* July 29, 1861; New York *World,* July 29, 1861; Philadelphia *Ledger,* July 29, 1861.

teer who was then serving as orderly and clerk at Butler's headquarters witnessed the operations and recorded that La Mountain's radius of observation was about ten miles.[42] The elevation, even with this radius, was too low to permit the discovery
of anything important,[43] except the presence of a sloop anchored
in Back River.[44] Unfavorable weather conditions interfered
with observations for the next five days, but on the 31st La
Mountain again renewed his operations. He made two ascensions from the vicinity of the Fortress, the first of which was
unsuccessful. In this attempt he reached an altitude of scarcely
400 feet, and descending, made adjustments and again went
aloft. The second venture was productive of definite results.
Ascending to a height of 1,400 feet, he was afforded an excellent opportunity for observation of Confederate activity in the
vicinity within a radius of some thirty miles.[45]

The first item of importance was the discovery of a concealed
camp in rear of the Confederate water battery on Sewall's Point.
Fifty-eight tents were seen in a large clearing behind the battery, denoting the presence of several hundred men in addition

[42] Warren Howe to [name of addressee omitted], July 25, 1861, *Passages from
the Life of Henry Warren Howe, Consisting of Diary and Letters Written during
the Civil War, 1861-1865* (Lowell, Mass. [Privately printed], 1899), p. 96.
    At the time this letter was written, Howe was a private in the 7th Massachu-
setts Battery, Heavy Artillery, detailed as orderly and clerk to Butler's headquarters. He later became a commissioned officer in the 13th Massachusetts
Volunteer Infantry.
[43] Boston *Journal*, July 29, 1861; New York *Times*, July 29, 1861; Philadelphia *Press*, July 29, 1861.
    La Mountain's own reports of these operations are missing from the files of
the War Department, and from Butler's papers. Hence the war correspondence
of the current newspapers is the only available source in addition to Howe's
brief mention in his letter. The presence of the sloop in Back River was again
mentioned in La Mountain's report of his next observations, in which he mentioned the vessel as the one reported in his dispatch of the previous week.
[44] La Mountain to Butler, July 31, 1861, MS, Butler Papers; *Private and
Official Correspondence*, I, 189.
[45] *Ibid.*; La Mountain to Butler, July 31, 1861, MS, Butler Papers; Boston
Journal, August 2, 1861; Providence *Post*, August 2, 1861; New York *Commercial Advertiser*, August 1 and 2, 1861; New York *Evening Post*, August 2, 1861;
New York *Times*, August 2, 1861; New York *Tribune*, August 2 and 3, 1861;
New York *World*, August 2, 1861; Philadelphia *Inquirer*, August 2, 1861;
Philadelphia *Ledger*, August 2, 1861; Philadelphia *Press*, August 2, 1861; Balti-
more *American*, August 2, 1861; New York *Herald*, August 2, 1861.

to the troops manning the battery, but the rumor of a force of several thousand in this area was thus effectively put at rest. Hitherto the camp had been shielded from view by a fringe of trees and brush. From the air it stood out clearly to the observer's eye. The method of estimating the strength of a force by counting the tents and huts from the air became standard practice both in the operations of La Mountain and in those of Lowe's balloon corps later. No traces of unusual Confederate activity could be detected on James River for nearly twenty miles inland, and across the bay in the direction of Norfolk, there appeared to be no occupation of the banks of Tanner's Creek. Another camp, showing about forty tents, was discovered half a mile in rear of the Pig Point battery opposite Newport News. Like that detected behind Sewall's Point, the presence of this force had been concealed behind clusters of trees. All appeared to be quiet in the direction of Yorktown, and if the report that Magruder had assembled 8,000 men there was true, there was at the time of observation no evidence that this force was preparing for any active operations towards Fortress Monroe. Only a small cluster of tents, apparently an outpost, was noted between Bethel and Yorktown. No visible encampments appeared in rear of Newport News.[46]

La Mountain's successful ascension on July 31 may be said to mark the first effective use of observation balloons in the American army. Lowe's experiments and demonstrations in late June, though an accurate aerial map had been made during one ascent, were still in the formative stage, and he was as yet only an applicant for regular service. Wise had made one or two ascensions from the defenses of Washington several days before La Mountain's operations, but there is scant evidence that his observations were of any particular significance. His subsequent accident and retirement from aeronautic service tended to lessen the importance of his work. With La Mountain's ascent, however, began a series of operations that eventually led to his appointment in the Army of the Potomac, with

[46] La Mountain to Butler, July 31, 1861, MS, Butler Papers; *Private and Official Correspondence,* I, 189. See also special correspondence of New York *World,* August 5, 1861, and New York *Herald,* August 3, 1861.

which he served with comparative success until February, 1862. Thus after the unsuccessful attempts of Allen and Wise, La Mountain achieved the distinction of rendering the first effective aeronautic service with armed forces in the United States.

In general, the information obtained in the July 31 ascension was such as to give Butler a greater sense of security, as well as to unmask to his knowledge the presence and accurate size of the camps La Mountain had observed. It was doubtless a source of satisfaction to the general commanding to know that the force behind Sewall's Point originally reported in thousands was in reality but a few hundred. Absence of available Confederate returns prevents the verifying of all the data submitted in La Mountain's report. From the proximity of the areas inspected, and from the general nature of the information stated, it seems reasonable to assume that the majority of the statements were correct. In his observations of the Tanner's Creek area, however, La Mountain was mistaken in his conclusion that no troops occupied this vicinity. A Confederate artillery camp of instruction had been established on one bank of this creek, and three field batteries were in position for training there on the same day that the aeronaut reported the area unoccupied.[47]

La Mountain's aerial activities, while producing no excitement among the Confederates in the region, were nevertheless noticed. Colonel Robert Johnston, commanding the cavalry forces scouting from headquarters at Cockletown, reported the Union balloon ascensions to Magruder's headquarters in Yorktown, but only as a postscript to the body of his report.[48]

On August 1 La Mountain continued his operations, ascending to an altitude of 1,000 feet, from which he confirmed his observations of the day before. He reported the town of Hampton entirely free from Confederate occupation, and could discover neither new nor unusual activity in any direction.[49]

[47] Brigadier General Benjamin Huger to Major General R. E. Lee, July 24, 1861, 1 *O. R.*, II, 997.
[48] Colonel Robert Johnston to Major G. B. Cosby, July 27, 1861,1 *O. R.*, II, 1005. At the end of his report, Johnston added: "I forgot to report, in mentioning yesterday's proceedings, that the enemy made two attempts to inspect us in balloons."
[49] Hartford *Daily Courant*, August 2, 1861; Detroit *Free Press*, August 2, 1861.

Two days later he effected his first aerial observation from the water, and in some respects may be said to have operated from the first aircraft carrier in American history.[50] One of his balloons was inflated and placed on board the armed transport *Fanny*,[51] and was secured to the stern by mooring ropes and a windlass. The vessel then steamed out into the channel opposite Sewall's Point, where La Mountain ascended to a 2,000 foot altitude and made a careful inspection of the Confederate position. The troops attached to the camp discovered on the 31st were seen to be actively at work erecting additional fortifications with gun pits and embrasures for pieces evidently intended to bear upon the Rip Raps, Fortress Monroe, and shipping in the Roads. These operations had hitherto been concealed from ground and ship observation by the same screen of trees that had hidden the camp detected from the air the previous week.[52] Additional troops were engaged in mounting two guns of heavy caliber in the forward or main battery, to bear upon the shipping and the Fortress.[53] Several days later

[50] Possibly the first recorded form of "aircraft carrier" might be credited to Lieutenant Lord Cochrane, of the Royal Navy, whose brig *Pallas* towed kites along the French coast during the Napoleonic wars for the purpose of spreading propaganda pamphlets. [See Chapter i, *supra*, p. 19.] Kites, however, can hardly be called "aircraft" in the modern military sense. Also in the technical meaning of the word, La Mountain's transport *Fanny* could not be termed a carrier according to modern usage. The conversion of the U. S. S. *George Washington Parke Custis*, however, into a specially constructed balloon barge for ascensions from the water, fits almost all the requirements of the definition, and can be classed as the first American aircraft carrier to serve with armed forces. [See Chapter vii, *infra*, pp. 256-258.]

[51] The *Fanny* was captured two months later by Confederate gunboats off Loggerhead Inlet, North Carolina, having taken part in Butler's attack on Hatteras Inlet in late August. Butler to Major General John E. Wool, August 30, 1861, 1 *O. R.*, IV, 586; Reports of Brigadier General Joseph K. F. Mansfield, October 5, 1861, Brigadier General Benjamin Huger, October 5, 1861, and Colonel A. R. Wright [3rd Georgia Infantry], October 2, 1861, 1 *O. R.*, IV, 595-597.

[52] *Scientific American*, August 17, 1861; Boston *Journal*, August 9, 1861; Philadelphia *Inquirer*, August 6, 1861; Washington *Evening Star*, August 11, 1861; New York *World*, August 7, 1861.

[53] Warren Howe to [his father?], August 6, 1861, *Passages from the Life of Henry Warren Howe*, p. 98; New York *Times*, August 6, 1861; New York Tribune, August 6, 1861; New York *Commercial Advertiser*, August 5, 1861; New York *World*, August 6, 1861; New York *Evening Post*, August 6, 1861;

these guns were reported in position in the main battery, protected from flank attack by strong earthworks.[54]   After these
observations were completed, the *Fanny* weighed anchor and
proceeded towards Craney Island and Pig Point, where La
Mountain again went aloft and scrutinized the Confederate
positions in this vicinity.   Here nothing new was ascertained.
The Confederates, according to the Union aeronaut, regarded
the balloon with consternation from all the positions inspected,
and ran from view, hiding in various shelters and behind trees,
from which he noted them peeping surreptitiously at his aerial
observatory.[55]   A rather different account of the Confederate
attitude towards the inquisitive Yankee observer was furnished
by a runaway negro who had escaped from Sewall's Point, after
having witnessed La Mountain's ascension from this position.
Rather than hiding from view, the garrison and engineer troops
looked up at the balloon with a profuse use of profanity; and,
said a war correspondent who interviewed the contraband, " if
Sambo is to be believed, their rage amounted to a perfect
frenzy. . . .   Nothing else has occurred which has so much
enraged them." [56]

Although La Mountain's report of observations on July 31
was in error in regard to the Tanner's Creek area, his statements
concerning Sewall's Point and his report of the two heavy guns
was entirely correct.   On July 26 General Lee had written to
Brigadier General Huger, then in command at Norfolk, stating
that he was " much gratified to learn the progress made in the

Philadelphia *Press*, August 6, 1861; Philadelphia *Daily News*, August 6, 1861;
Philadelphia *Ledger*, August 6, 1861; Baltimore *American*, August 6, 1861;
Washington *National Republican*, August 6, 1861; New York *Herald*, August
6, 9, and 11, 1861.

[54] New York *Times*, August 7, 1861.

[55] *Scientific American*, August 17, 1861; Boston *Journal*, August 9, 1861;
Philadelphia *Inquirer*, August 6, 1861; Washington *Evening Star*, August 11,
1861; New York *World*, August 7, 1861.

[56] Detroit *Free Press*, August 27, 1861. The *World* correspondent also interviewed the contraband and reported a similar story: " Their rage at ' de insult,'
as they called it, only found vent in volleys of oaths. For a time they thought
of firing on him with long range rifles, but finally concluded that ' de damn
Yankee flew too high for 'em ' (I quote the negro's words), and looked up at
him with stifled indignation." New York *World*, August 17, 1861.

defenses of Norfolk, and arrangements to prevent landing on our shores. . . . I will next week," Lee continued, " send you two 8-inch columbiads rifled, the guns being of the weight and size of the 9-inch, and they will be followed in a few days by two of the 9-inch caliber." Lee stated that the guns were for " planting a battery opposite Fort Monroe." [57] Likewise correct was La Mountain's conclusion that the guns being mounted and those which would later fill the embrasures observed, were intended to shell the Union shipping at Old Point Comfort and the Roads. In requesting the columbiads from Lee, Huger had specifically stated this ultimate purpose: " Their fire will be dangerous to shipping and will make landing of steamboats at the wharves there [Old Point] a troublesome operation." Huger's report of progress and his outline of further plans for the defenses of Norfolk and vicinity also confirm other observations reported by La Mountain on August 3.[58]

However accurate was the information obtained by La Mountain during his reconnaissance of August 3, there is no evidence of his observing the Confederate movements which preceded Magruder's attack on and burning of Hampton on the 7th.[59] On July 22, the day following the Confederate victory at the first Manassas, Magruder had ordered Colonel Robert Johnston to make a reconnaissance of the Union positions near the Fortress, Hampton, and Newport News. Magruder had learned from several sources of information that the Federal forces had been reduced in this area after McDowell's signal defeat, and instructed Johnston to verify these reports by careful cavalry reconnaissance. Johnston was also ordered to clear the country in the vicinity of the Union positions of all negroes, and to make a demonstration in force and destroy all Union outposts and parties " which may venture far from their posts and works." [60] On the 26th and 27th Johnston effected this mission with a cavalry force of some 2,000 men.[61] In the

---

[57] Lee to Huger, July 26, 1861, 1 *O. R.,* II, 1001.

[58] Huger to Lee, July 24, 1861, 1 *O. R.,* II, 997.

[59] Details of the burning of Hampton appear in the official Union and Confederate reports of this incident, in 1 *O. R.,* IV, 567-573.

[60] Magruder to Johnston, July 22, 1861, 1 *O. R.,* II, 990-991.

[61] Magruder's Report, August [2?], 1861, 1 *O. R.,* IV, 569.

vicinity of Hampton the Confederates actually came into contact with Butler's troops, and Johnston reported on the 27th that his videttes under Colonel John B. Hood had been fired on by Union troops moving in force out of this town.[62] At the approach of the Confederate cavalry, Johnston declared, "a large balloon was sent up," and the Federal forces hastily evacuated Hampton.[63] Subsequently a Richmond paper circulated a story that Butler's forces evacuated Hampton in a panic because of reports from the balloon, and claimed that the Federal aeronaut had informed Butler that a force of 10,000 men had been seen marching on Hampton. This intelligence, it was asserted, led to the hurried withdrawal from the town.[64] No evidence to support this story, however, appears in the Federal reports of the period.

Following these operations, the Yorktown garrison and part of the Confederate troops stationed at Williamsburg, in all a force of some 4,000 infantry and 400 cavalry, with two batteries of artillery and a howitzer battalion attached, effected a junction early in August under the command of General Magruder at Young's Mill and marched to Bethel Church. From this point Magruder moved on August 6 to Newmarket Bridge, repaired the span over Southwest Branch of Back River previously destroyed by Union troops, and threw his force between Butler's garrison at Fort Monroe and his outpost garrisons at Newport News.[65] The following day, according to Magruder's report, this Confederate force was "displayed . . . within a mile and a half of Newport News . . . with the hope of drawing the enemy out." [66] After the Federals declined to take up the challenge and fight, several units were detailed to burn the town of Hampton.[67]

[62] Colonel Robert Johnston to Major G. B. Cosby, July 27, 1861, 1 O. R., II, 1004.
[63] Magruder to Colonel George Deas [AG, CSA], August 9, 1861, 1 O. R., IV, 570.
[64] Richmond *Examiner*, August 12, 1861; New York *Tribune*, August 18, 1861 [Reprint from the *Examiner*]; Philadelphia *Ledger*, August 20, 1861 [Reprint from the *Examiner*].
[65] Magruder to Deas, August 9, 1861, 1 O. R., IV, 570.
[66] *Ibid.*, p. 571.
[67] Butler to Winfield Scott, August 8, 1861, 1 O. R., IV, 567; Magruder to Deas, August 9, 1861, *ibid.*, 571.

Certainly movements such as these should have been discovered from the air. In the instance of Johnston's cavalry demonstration before Hampton on July 26 and 27, we have evidence that the balloon was aloft and in position to observe. Johnston testifies that he saw it go up. That a body of 2,000 mounted troops could come so near and actually engage in skirmishes without being seen from such a point of vantage seems incredible. The absence of the balloon observer's reports of this movement may be explained in the fact that many documents are missing from the War Department files.

More puzzling is La Mountain's silence concerning Magruder's movements during the first week of August. That a force of nearly 4,000 men, with several batteries of artillery attached, could maneuver for two days within observing distance, and then openly " display " themselves within less than two miles from the Federal fortifications, without being discovered by the balloonist, seems even more than incredible. In the absence of aerial reports of these movements, several possibilities can be considered. It is not known, as in the former case, that La Mountain was aloft during the period of Magruder's maneuvers. No record of any ascension at this time appears in the available Union documents, and the Confederate reports do not mention the presence of the Yankee balloon. The weather may have been too unfavorable to have permitted ascensions during the week of August 6—9, and this fact might account for the aeronaut's apparent failure to discover and report the large scale Confederate movements. Bad weather had interfered with La Mountain's operations during the last week of July, and had rendered his first ascension ineffective. There is also the possibility that he may have made the ascensions, reported his observations, and the documents, like many others, may have been lost or misplaced in the thousands of papers preserved in the archives. Or, having made the ascensions, he may have failed for reasons unknown, to detect the troop movements. That his reports were verbal, and thus no record of them would survive, seems most unlikely, since he never used telegraph from his balloon,[68] and to have descended

[68] At times, Lowe's dispatches from the balloon by telegraph were reported

to report each observation in person would hardly have been a logical procedure. The positive, correct answer to the question seems to defy careful research.

La Mountain's next ascensions of which there are available records took place three days after Magruder burned Hampton, on August 10. These, five in number,[69] were made from the ground in the vicinity of the Fortress.[70] The maximum elevation reached was 3,500 feet, higher than the aeronaut had previously attained.[71] The weather had been unsettled during the day, with showers throughout the afternoon,[72] rendering the state of atmosphere somewhat misty.[73] Despite the slight haze that prevailed, La Mountain was able to make several observations of value.

A large Confederate camp was discovered approximately five or six miles northwest of Hampton, containing a force estimated to be between 4,000 and 5,000 men.[74] These troops were Magruder's Yorktown and Williamsburg units, which had erected works at Bethel and then moved back to Young's Mill, where they were engaged in constructing field works at the time La Mountain observed them.[75] No vessels appeared on the waters of York and Back Rivers, and no new encampments could be discerned on the banks of these streams, or at New Market Bridge. A large camp, showing from 150 to 200 tents,

verbally to the general officers. This procedure occurred prominently during the battle of Fredericksburg. See below, Volume II.

[69] New York *Tribune*, August 14, 1861.

[70] Philadelphia *Press*, August 15, 1861; Washington *Daily National Intelligencer*, August 14, 1861; "N. Y." to the editors of the New York *Commercial Advertiser*, August 11, 1861, published August 13.

[71] La Mountain to Butler, August 10, 1861, MS, Butler Papers; also MS copy forwarded to War Department [number omitted], Misc. File, "Balloon Corps," War Department Division, National Archives [hereinafter cited as File, "Balloon Corps," WDD, NA] also printed in 1 *O. R.*, IV, 600.

[72] Diary of Warren Howe, entry of August 10, 1861, *Passages from the Life of Henry Warren Howe*, p. 27.

[73] La Mountain to Butler, August 10, 1861, MS, Butler Papers; also MS copy forwarded to War Department [number omitted], File, "Balloon Corps," WDD, NA; also printed in 1 *O. R.*, IV, 600.

[74] La Mountain to Butler, August 10, 1861, MS, Butler Papers; also MS copy forwarded to War Department [number omitted], File, "Balloon Corps," WDD, NA; also printed in 1 *O. R.*, IV, 600-601.

[75] Magruder to Deas, August 9, 1861, 1 *O. R.*, IV, 572.

was located on the left bank of the James, about eight miles upstream from Newport News.[76] This camp was doubtless that of the Confederate force which fortified and occupied positions at the mouth of Watt's Creek, located on General Wool's map at six and a half miles plotted distance from Newport News.[77] It also appeared that the troops manning the Pig Point battery had been strengthened considerably, for a new camp of from 40 to 50 tents had been established in rear of the battery since La Mountain's last report on this position. Two large men-of-war were observed lying at anchor in the channel at Norfolk and, from the appearance of their canvas, were preparing to sail. La Mountain still reported no activity in the vicinity of Tanner's Creek.[78] The ground occupied by the artillery training camp must have been especially well defiladed from sight, even from the aerial observer's high elevation.

Sewall's Point was again inspected, and on this occasion a sketch was made showing the two large guns in battery on the extreme tip of the point. Lesser batteries had been established along the shore to the south, with a camp showing tents situated to their left flank, inland. The rear of the secondary batteries and the camp was protected by earthworks, as was the right flank towards the main battery.[79] These dispositions exhibited nothing particularly new, except that La Mountain concluded that the two heavy guns in the forward battery were only very large field pieces, and not seacoast ordnance.[80] If

[76] La Mountain to Butler, August 10, 1861, MS, Butler Papers; also MS copy forwarded to War Department [number omitted], File, " Balloon Corps," WDD, NA; also printed in 1 O. R., IV, 601.

[77] Map, " Military Reconnaissance, Department of Virginia," [under command of Major General John E. Wool], *Atlas to Accompany Official Records of the Union and Confederate Armies, 1861-1865,* I, Plate xviii, Sheet 1.

[78] La Mountain to Butler, August 10, 1861, MS, Butler Papers; also MS copy forwarded to War Department [number omitted], File, " Balloon Corps," WDD, NA; also printed in 1 O. R., IV, 601.

[79] Sketch of Confederate position at Sewall's Point, elevation 3,500 feet, drawn by John La Mountain, enclosed in La Mountain to Butler, MS copy [number omitted], File, " Balloon Corps," WDD, NA.

[80] La Mountain to Butler, August 10, 1861, MS, Butler Papers; also MS copy forwarded to War Department [number omitted], File, " Balloon Corps," WDD, NA; also printed in 1 O. R., IV, 601; see also dispatches in Boston *Transcript,*

the guns sent to Huger for this battery were the 8-inch and 9-inch columbiads Lee promised, La Mountain's conclusion on this point was in error. It is possible, however, that Lee might have been unable to secure the columbiads intended, and had substituted the heavy field ordnance. Troops at Sewall's Point were now estimated at approximately 1,000.[81]

Having completed his daylight observations of the region, La Mountain continued operations on the night of August 10. By sundown the unsettled weather became clear, with no winds to break the prevailing stillness.[82] Ascensions from the water were again decided upon, and this time the tug *Adriatic* was used as a base of operations. General Butler accompanied the aeronaut on board, and from the few reports available, the observations must have been regarded as successful.[83] A war correspondent of one of the New York papers commented favorably on the results, and gave La Mountain considerable credit for the usefulness of his services:

A novel feature with regard to the recent ascensions of Mr. La Mountain at Fortress Monroe has been his discovery of the enemy by means of the lights in their tents. He has in this way counted the exact number of tents in each encampment or fort, and calculating how

---

August 12, 1861; New York *Tribune*, August 13, 1861; Hartford *Daily Courant*, August 13, 1861; Lancaster *Daily Evening Express*, August 12, 1861; Baltimore *American*, August 13, 1861; Washington *National Republican*, August 13, 1861; New York *Herald*, August 13, 1861.

[81] New York *Commercial Advertiser*, August 13, 1861; New York *Times*, August 13, 1861; New York *World*, August 13, 1861; Philadelphia *Inquirer*, August 13, 1861; Philadelphia *Ledger*, August 13, 1861; Philadelphia *Press*, August 15, 1861; Detroit *Free Press*, August 13, 1861; Washington *Daily National Intelligencer*, August 13, 1861. Also reprinted dispatches in Richmond *Enquirer*, August 14, 1861, and Richmond *Examiner*, August 15, 1861.

[82] Diary of Warren Howe, entry of August 10, 1861. *Passages from the Life of Henry Warren Howe*, p. 27.

[83] Appleton's *American Annual Cyclopaedia, 1862* (New York, 1863), p. 185; New York *Tribune*, August 14, 1861; Diary of Warren Howe, entry of August 10, 1861, *loc. cit.*, p. 27; New York *World*, August 14, 1861; New York *Herald*, August 13, 1861.

The correspondents of the *World* and the *Herald* identified the vessel as the *Fanny*. Howe's diary and the other sources specifically mention the *Adriatic*.

The details of the observations, according to the *World's* correspondence, were withheld from publication, exceptional procedure in view of the extensive publication of the other data reported by La Mountain.

many men usually occupy a secession tent, the whole force may be easily ascertained. The lines of fire dotted over the surface of the earth may be traced with precision from a balloon with a strong glass.[84]

Though the accuracy of the calculations made in this way was perhaps somewhat exaggerated in the correspondent's dispatch, by and large the method of estimating the strength of enemy forces by the number of fires was generally sound. Lowe and Allen later used this system with satisfactory results.

The ascensions of August 10 were the last La Mountain made in the region of Fortress Monroe. His preparations for operations of this date had expended all his available materials for making hydrogen, and by the evening of the 12th the lifting power of his balloon was insufficient for further ascensions. Therefore he discharged the remaining gas from the envelope as a safety precaution, and appealed to Butler for facilities to extend his operations. He asked permission to return to Troy and bring back his larger balloon *Saratoga*, along with his previously discredited water-decomposition generator. The manufacture of hydrogen " chemically," as he called the vitriol-iron filing process, with the tanks and materials at his present disposal, he declared had been unsatisfactory, and with them gas could not be made available whenever the balloons might be needed, thus placing " a serious impediment in the way of successful ballooning for army reconnaissance." [85] His plans for extending his aeronautic activities also went beyond the scope of observation, and included offensive tactics. He brought forth for Butler's consideration a scheme highly modern in aspect, resembling many of the numerous plans that were submitted continually to the War Department throughout the duration of the war:

I am well convinced in my mind that at an expense not to exceed eight or nine thousand dollars, I can build a balloon in a month's time, and with it shell, burn or destroy Norfolk or any city near our camps. Ballooning can be made a very useful implement in warfare. All depends on the encouragement it receives.[86]

[84] " C. W. D." to the Editors of the New York *Commercial Advertiser,* August 17, 1861, published August 19; reprinted in Washington *National Republican,* August 21, 1861.
[85] La Mountain to Butler, August 13, 1861, MS, Butler Papers.
[86] La Mountain to Butler, August 13, 1861, MS, Butler Papers.

Reminding Butler that he could do nothing further until he had a fresh supply of materials, La Mountain again urged that he be allowed to go north for his larger balloon and generator, and requested an interview to discuss his plans with the General at once. With this communication he enclosed a bill for his expenses and the materials he had purchased in accordance with Butler's original order for supplies. Notwithstanding the figure of $490 he had submitted when he requested initial authorization, La Mountain's charges amounted to more than $1,200.[87]

Throughout the whole period Butler had been entirely satisfied with La Mountain's operations and their results. The observations of August 10 he considered of particular importance, and had had the aeronaut's report and aerial sketch of that day copied and forwarded to General Winfield Scott on August 11.[88] The statement of expenses and money due, despite the fact that it was almost three times the amount originally authorized, Butler approved without hesitation, adding his certification that the charges were just and correct and that all monies and items enumerated were "necessary for the public service . . . and that the disbursements were honestly and faithfully made."[89] He also received La Mountain's suggestion for a larger balloon and his scheme for aerial bombing with interest and favor. He forwarded a copy of La Mountain's letter to assistant Secretary of War Thomas A. Scott,[90] and advised that the plan be given a trial. Butler also described La Mountain's work at Fortress Monroe as "very successful," and further stated that the information obtained

[87] Statement, "United States Government to John La Mountain, Dr.," August 13, 1861, MS, Butler Papers; see also *Private and Official Correspondence*, I, 211.

[88] Butler to Winfield Scott, August 11, 1861, enc., MS 1128B, File, "Balloon Corps," WDD, NA; also MS [copy], Butler Papers; *Private and Official Correspondence*, 1, 206; 1 *O. R.*, IV, 600; *Military Correspondence Relating to the War of the Rebellion* (Washington [Adjutant General's Printing Office], 1875), I, 471; Benjamin F. Butler, *Butler's Book: Autobiography and Personal Reminiscences of Major General Benjamin F. Butler* (Boston, 1892), p. 278.

[89] Butler's indorsement, Statement, "United States Government to John La Mountain, Dr.," August 13, 1861, MS, Butler Papers; *Private and Official Correspondence*, I, 211.

[90] La Mountain to Butler, August 13, 1861, MS copy enclosed in Butler to Thomas A. Scott, August 13, 1861, MS 236 [irregular], LRSW.

from his ascensions had been "of great advantage." [91] The General had also taken active interest in the ascensions themselves, having accompanied personally the balloon party on the tug *Adriatic* on the evening of August 10.[92] As to the aeronaut himself, Butler wrote that "La Mountain is a daring and apparently reliable man, and thus far I have taken the liberty to encourage his efforts." And concerning the plan to destroy cities from the air, Butler stated his opinion to Secretary Scott in La Mountain's favor:

I take leave to call your attention to that portion of his communication in which he proposes to use the Balloon for warlike purposes. Is it not worth the experiment?

The proposition is a novel one and daring, but with the explanation he has given me, I think it may have an element of feasibility in it; at any rate, the cost is not large, in comparison with the results thus far. Please advise me whether I shall go farther with him, either in reconnaissances or otherwise.[93]

Record of Scott's disposition of the matter appears to be missing from the War Department files, and also from Butler's papers, but there is evidence that Scott approved the continuance of aerial reconnaissance and La Mountain's request to bring his large balloon *Saratoga* into the field. Three days later La Mountain left Fortress Monroe for Troy, and the northern press announced that "General Butler is so well pleased with the results of La Mountain's reconnaissances that the latter has gone north for a balloon of much larger size." [94]

[91] Butler to Scott, August 13, 1861, MS 236 [irregular], LRSW; MS [copy], Butler Papers; also printed in *Private and Official Correspondence*, I, 210.

[92] Diary of Warren Howe, entry of August 10, 1861, *loc. cit.*, p. 27.

[93] Butler to Scott, August 13, 1861, MS 236 [irregular], LRSW; MS [copy], Butler Papers; also printed in *Private and Official Correspondence*, I, 210.

[94] Boston *Journal*, August 17, 1861; Boston *Transcript*, August 16, 1861; Hartford *Daily Courant*, August 17, 1861; New York *Times*, August 17, 1861; New York *Tribune*, August 17, 1861; Philadelphia *Inquirer*, August 17, 1861; Philadelphia *Press*, August 17, 1861; Philadelphia *Ledger*, August 17, 1861; Philadelphia *Daily News*, August 17, 1861; New York *World*, August 17, 1861; New York *Commercial Advertiser*, August 16, 1861; New York *Evening Post*, August 16, 1861; Lancaster *Daily Evening Express*, August 16, 1861; Baltimore *American*, August 17, 1861; Washington *National Republican*, August 17,

The Richmond papers, as though equally cognizant of the Yankee aeronaut's achievements, promptly copied the northern press dispatch and proclaimed it to the Confederate capital.[95] The following week Butler issued orders to the president of the Bay Line Steamboat Company to " ship to Fortress Monroe any material and men Mr. La Mountain desires to have brought." [96] La Mountain's success and the approval of his commanding general had been definitely established, and his future as a military aeronaut seemed assured.

One incident, however, while apparently having no effect on his work or reputation, may be cited to show that he was not entirely free from petty annoyance. He had employed, in addition to Kendrick and Waters, an assistant named Francis Raveneth, who aided him in his operations at Fortress Monroe. On the day of La Mountain's first successful ascension, a committee of zealous citizens of Lansingburgh, New York, the aeronaut's birthplace, addressed a letter to General Butler, condemning Raveneth as a Confederate spy and traitor, and warning the General against his subversive activities. Raveneth was quoted as having frequently expressed his desire for the thorough defeat of the Union, and was declared to be " notoriously in sympathy with the South," and was " deemed here by all Union men to be a dangerous man at this juncture of our national troubles." The self-appointed committee of public safety then referred Butler to all the commissioned officers of Company " A," 30th Regiment New York State Militia then in Butler's department, for full corroboration of their statements. Nothing in the charges they made, the committee assured Butler, was intended to reflect on the character of " Professor " La Mountain, " as we know him to be a strong Union man and a patriot." [97]   There is no evidence that Butler paid any

1861; Providence *Press,* August 16 and 17, 1861; New York *Herald,* August 15 and 16, 1861.

[95] Richmond *Enquirer,* August 19, 1861; Richmond *Enquirer* [Semi-Weekly], August 20, 1861.

[96] Butler to the president of the Bay Line Steamboat Company, August 23, 1861, MS, Butler Papers.

[97] A. B. Lockstader [Secretary of Citizens' Committee of Lansingburgh, N. Y.] to Butler, July 31, 1861, MS, Butler Papers.

attention to this missive, but doubtless he did make some sort of inquiry. If the latter assumption is true, La Mountain must have satisfied him as to the integrity of his assistant, for Raveneth was still the aeronaut's chief lieutenant in September, and was entrusted with the collection of La Mountain's accounts due from the Government.[98]

Another incident during La Mountain's period of service may be cited to illustrate the widespread notice and interest that his operations attracted. As early as June, before La Mountain was actively engaged in making military ascensions, Captain Albert Tracy, an officer of the 10th Infantry, regular army, then stationed at Jefferson Barracks, Missouri, wrote to Butler suggesting the use of photography from balloons. Articles in the press announcing that La Mountain was to be engaged by Butler for aeronautic operations had induced Tracy to make his suggestion to Butler, which strikingly resembles a description of the use of modern aerial mosaics, now indispensable to the present day fighting force:

Why not then, obtain, at a moment in the air, a picture of the country beneath, with the camps, batteries, and perhaps well-defined impressions of any body of troops . . . of an enemy? . . . Fine stereoscopic glasses would reveal to the eye with the strongest individuality each and every object upon the plates. Of course pictures might be multiplied *ad finitum* by the usual process, and as many copies furnished in as many directions as desirable. Thus we might come to possess information of the most reliable character of the position and operations of the enemy.[99]

Butler does not seem to have followed up the idea, as there is no evidence that La Mountain attempted photography during his ascensions.

La Mountain's appointment had not only induced Captain Tracy to recommend the use of aerial photography to Butler,

[98] La Mountain to Major General John E. Wool, September 6, 1861, MS L10, Letters Received, Department of Virginia, War Department Division, National Archives [hereinafter cited as LRDV].

[99] Captain Albert Tracy [10th U. S. Infantry] to Butler, June 16, 1861, MS, Butler Papers.

The question of aerial photography in the Civil War is discussed fully in Chapter ix, *infra*, pp. 329-335.

but it also led Edward La Mountain, his brother, to apply for army service. While John was preparing his equipment for shipment to Fortress Monroe, Edward La Mountain addressed an application to Secretary Cameron asking for employment similar to that enjoyed by his brother. Claiming several years' practical experience as an aeronaut, as well as the possession of a new balloon and appendages in readiness for field service, the younger La Mountain referred the Secretary to Representative Schuyler Colfax of Indiana, a personal friend, who would vouch for his ability and integrity. His brother John also, he added, now in government service, would confirm his statements as to qualifications and professional skill.[100] This letter, like those of his brother, passed into the silence of the War Department files.[101] Although Cameron had received John La Mountain cordially, and had authorized Butler to order whatever he might need for balloon experiments, it is evident that this action was caused by Butler's original initiative in the matter. It will be remembered that John's first communications were ignored by the Secretary's staff before Butler entered the scene. Edward La Mountain likewise received the same stony indifference to his application, notwithstanding his intimate reference to Representative Colfax.[102]

The recognition and success of John La Mountain in his Fortress Monroe operations must therefore be attributed to Benjamin F. Butler's liberal views with respect to the new tactical equipment, and his hearty encouragement of the aeronaut's efforts. It is greatly to his credit that he encouraged a branch of military science then in its infancy in this country, when his administrative superiors failed or refused to recognize its possibilities.

[100] Edward La Mountain to Secretary Cameron, June 15, 1861, MS L51 [Irregular], LRSW.

[101] The MS cited in the preceding note [100] shows neither indorsement nor disposition entry. The letterbooks contain no copy or reference to a reply.

[102] Edward La Mountain made no further attempt to obtain an appointment with the army as an aeronaut. He continued to practice his profession as a civilian until his tragic death in a balloon accident at Ionia, Michigan, on July 4, 1873. Chicago *Tribune*, July 5 and 7, 1873; St. Louis *Missouri Republican*, July 8, 1873; St. Louis *Globe*, July 10, 1873; St. Paul *Dispatch*, July 8, 1873; New York *Times*, July 8, 1873.

Some description of the technical details of La Mountain's balloons and apparatus is appropriate here, but the scarcity of surviving documents prevents the inclusion of much pertinent data. The balloons in service at Fortress Monroe were of the ordinary type for civilian use, if such a category may be applied. One of them, the *Atlantic*, was constructed from parts of the large aerostat of the same name used in the celebrated flight from St. Louis in 1859, and in which La Mountain later made his perilous flight over the Canadian wilderness.[103] The rebuilt *Atlantic* was a small aerostat as compared to La Mountain's *Saratoga*, or Lowe's balloons of the *Intrepid* class, having a gas capacity of 1,200 cubic feet.[104] It was small in carrying capacity also, being designed to carry only one observer.[105] Of the inflating apparatus which first proved unreliable, but was later commissioned into service with the Army of the Potomac, little is known. From the meagre references to it in scattered letters, it may have resembled somewhat that designed by Wise.[106] Its chemical principles were similar to those of Wise's generator, but further description of its construction and functioning cannot be detailed. The cordage employed by La Mountain, like that of Wise and Allen, was of linen.[107] Other details of the equipment in La Mountain's train are lacking in the records. The contemporary press, though it frequently published notices of La Mountain's activities, was silent as to technical detail,

[103] Although the original *Atlantic* was at first abandoned in the Canadian north woods, it was later recovered and rebuilt into the aerostat that La Mountain used in the War.

[104] N. F. White [assistant aeronaut to La Mountain] to Brigadier General William B. Franklin, December 21, 1861, MS L222, Letters Received, Army of the Potomac, War Department Division, National Archives [hereinafter cited as LRAP].

[105] Brigadier General Fitz John Porter to Colonel R. B. Marcy [Chief of Staff, Army of the Potomac], September 21, 1861, MS P8 [supplementary], LRAP.

[106] La Mountain described his generator simply as "a portable apparatus for the manufacture of gas from the decomposition of water," and later mentioned with it a steam engine and boiler. Since Wise's generator was substantially the same in principle, it is possible that the two were of similar construction. [See La Mountain to Wool, September 2, 1861, MS L7, LRDV.] Wise's generator is fully described in Chapter iii, *supra*, pp. 76-77.

[107] Item 9, Statement, "United States Government to John La Mountain, Dr.," August 13, 1861, MS, Butler Papers; also in *Private and Official Correspondence*, I, 211.

except that one correspondent gave the colors of the balloons as being solid red and blue, respectively.[108]

* * *

The exact nature of La Mountain's status after he returned north for his larger balloon *Saratoga* is not clear. Obviously he had instructions of some sort to return and resume his operations in the Department of Virginia, and it is equally apparent that Butler wanted him to continue his work in this theater of Federal activity, having given orders to the Bay Line Steamboat Company to transport the aeronaut's assistants and materials.[109] But on August 11, before La Mountain left Fortress Monroe, Butler received notification that he was to be relieved as commander of the Department of Virginia by Major General John E. Wool.[110] Since La Mountain's status with the army had been purely one of personal contract with Butler, his continued connection with the military forces depended upon Wool's disposition to retain him in service. Wool arrived at Fortress Monroe and assumed command on August 17, the day following La Mountain's departure.[111] Butler, despite the change of command, was to retain charge of all volunteer troops in the Department except those stationed at the Fortress itself,[112] until August 25, when he was ordered to organize and command a force to be sent to attack and seize several Confederate batteries at Hatteras Inlet, North Carolina.[113] Having accomplished this mission, Butler was then ordered to Washington.[114] It does not appear that Butler made any arrangement with Wool, either before or after his departure from the Depart-

[108] New York *Tribune*, August 2, 1861.

[109] Butler to the President of the Bay Line Steamboat Company, August 23, 1861, MS, Butler Papers.

[110] Butler to General Winfield Scott, August 11, 1861, MS B1128, File, "Balloon Corps," WDD, NA; MS [copy], Butler Papers; *Private and Official Correspondence*, I, 206; 1 *O. R.*, IV, 600.

[111] General Orders, No. 1, Department of Southeastern Virginia, August 17, 1861, 1 *O. R.*, IV, 601.

[112] Special Orders, No. 9, Department of Virginia, August 21, 1861, 1 *O. R.*, IV, 602.

[113] Special Orders, No. 13, Department of Virginia, August 25, 1861, 1 *O. R.*, IV, 580.

[114] Wool to Winfield Scott, September 2, 1861, 1 *O. R.*, IV, 580.

ment, concerning La Mountain's continued service. In fact Wool seems to have had no knowledge of La Mountain or of his relationship with Butler's command. He later stated that he had been given "no instructions, and find none here [Fortress Monroe] on the subject." [115]

La Mountain, however, evidently believed all arrangements had been made with the new department commander, and proceeded to prepare his generator and large balloon for service. On September 2 he wrote to Wool explaining that he had been unavoidably delayed in his preparations. He had found that a larger engine and boiler were necessary for the successful functioning of his field generator. The purchase and connection of equipment required more time. Also, he had suffered severe burns about the face from the accidental igniting of hydrogen during his experiments with gas manufacture.[116] Infection had set in, and his physician advised that he delay his departure for active duty until the wounds should be healed.[117] On the 6th he again addressed assurances to Wool that he would soon be physically able to report for duty, and in the meanwhile sent his partner, Francis Raveneth, whom the zealous citizens of Lansingburgh had previously denounced as a Confederate spy, to collect the amount of his statement of expenses rendered to Butler on the 13th of August. The money was badly needed, he told Wool, for the purchase of new materials and for the payment of old bills incurred in the previous operations at Fortress Monroe.[118]

Some ten days later La Mountain reached Wool's headquarters with the *Saratoga* and his generator, only to find that the general commanding knew nothing of any arrangements for the continued aeronautic service. Wool followed the usual pro-

[115] Wool to Cameron, September 18, 1861, MS, Papers of Simon Cameron, Division of MSS, Library of Congress [hereinafter cited as Cameron Papers]; also MS [copy] W241, LRSW.

[116] La Mountain to Wool, September 2, 1861, MS L7, LRDV.
This was the first and last accident involving the use of the highly inflammable hydrogen gas that occurred throughout the war in connection with the balloon operations of the army. No accidents of this nature ever occurred in Lowe's corps.

[117] La Mountain to Wool, September 2, 1861, MS L7, LRDV.

[118] La Mountain to Wool, September 6, 1861, MS L10, LRDV.

cedure in such circumstances. He wrote to Secretary Cameron announcing La Mountain's arrival with his equipment and, declaring that he had no instructions whatever concerning the aeronaut, asked the Secretary for guidance in the matter. La Mountain was sent to deliver this letter personally to Cameron.[119]

In the meanwhile La Mountain's rival, Lowe, had continued his demonstrations and experiments before officers of the Topographical Engineers, and had finally received a civilian appointment as aeronaut in the army. He was assigned to duty with the newly organized Army of the Potomac, to operate a new balloon especially constructed for military service. His initial operations were carried out under the direction of Brigadier General Fitz John Porter,[120] who had recently succeeded Brigadier General William T. Sherman in command of the brigade which was now enlarged to become Porter's Division.[121] Lowe's activities had come to the attention of Brigadier General William F. ["Baldy"] Smith, who had witnessed some of the ascensions in the vicinity of Fort Corcoran, and was now in command of another adjacent division of McClellan's new army.[122] Smith had been quick to recognize the value of aerial reconnaissance, and the day after La Mountain arrived in Washington to present Wool's letter to Cameron, he wrote to the General-in-Chief through the assistant adjutant general, Major Seth Williams, requesting the services of a balloon and aeronaut with his division:

I deem it of much importance that a balloon be permanently attached to this division, not only for the present, but to move with it, and owing to the distance from the gas works in Washington, I request

---

[119] Wool to Cameron, September 18, 1861, MS, Cameron Papers; also MS [copy] W241, LRSW.

[120] See Chapter vi, *infra,* pp. 220, ff.

[121] William Tecumseh Sherman, *Memoirs of General William T. Sherman* (New York, 1875), I, p. 193; see also "Organization of the Division of the Potomac, August 4, 1861," 1 *O. R.,* V, 15 and "Organization of the Army of the Potomac, October 15, 1861," *ibid.,* p. 16.

[122] See Chapter vi, *infra,* pp. 221, ff. Also see "Organization of the Army of the Potomac, October 15, 1861," 1 *O. R.,* V, 17.

that an inflating apparatus be purchased to accompany the balloon. I have to ask that the Major General [Commanding] will order one to be sent.[123]

Concurrent with this request, La Mountain's arrival with a letter to the Secretary of War asking for instructions and for assignment to duty was most opportune. The exact disposition of the two communications is not known, but it is evident that both led to La Mountain's assignment to duty with the Army of the Potomac. But Lowe had now been given official status with this command, and had won the confidence and respect of a number of the general officers, particularly McClellan and Fitz John Porter.[124] The presence of his rival and former antagonist in the field of aeronautics was likely to prove troublesome and embarrassing not only to the aeronauts, but to the general officers under whom they were to operate. Since none of the army officers knew anything of the technique of ballooning, friction or difficulties between the balloonists could interfere seriously with the efficiency of their work. It was a situation where ordinary army regulations and procedure might not be applied to the letter. The hostility existing between Lowe and La Mountain must have been sensed by General McClellan at once, for he instructed his friend and subordinate, Fitz John Porter, to call the two aeronauts to his quarters and interview them together, in order to " ascertain the feelings of each party in relation to the contemplated employment of both in the United States Service." [125]

The joint interview must have had its humorous aspect. In some ways the whole affair somewhat resembles the action of an older guardian or teacher calling in two obstreperous youngsters and admonishing them to behave themselves and patch up their youthful differences. Youngsters, in fact, they were. La Mountain, at thirty-one, was Lowe's senior by two years.[126] But the serious side is apparent. Since aerial recon-

---

[123] Smith to Williams, September 20, 1861, MS S217, LRAP.

[124] See Chapter vi, *infra,* pp. 220, ff.

[125] Porter to Colonel R. B. Marcy, September 21, 1861, MS P8 [supplementary], LRAP.

[126] Porter himself was comparatively young for his rank, being just thirty-nine at the time of the interview.

naissance was entirely new to the army, full cooperation and harmony among its personnel had to be assured. And as none of the officers, from the commanding general down, had any knowledge or experience with the subject, the usual procedure of enforcing harmony by military discipline could not be solely relied on.

Fitz John Porter was a man equal to the occasion. The interview was apparently a complete success. Porter reported to McClellan that each aeronaut was aware that in all probability there would be ample field for the activity of both, and that the present operations of the army were not likely to require united action by the balloonists, or even observation from the same immediate locality. Both Lowe and La Mountain assured Porter that should the occasion arise requiring them to work together in the same area, they would cooperate heartily to further the best interests of the service. La Mountain recognized that Lowe had been engaged in the Potomac region first, and told Porter that he understood and appreciated his rival's feeling that this vicinity was his " special locality " unless more than one aeronaut should be required there. He volunteered to retire from the scene whenever there should be no necessity for the presence of both balloonists. " I think," General Porter concluded in his report to his chief,

the Commanding General can rely upon the cordial cooperation of both to forward his [McClellan's] views, and in working for the service.[127]

The result of the conference, strengthened by the favorable opinion of Porter, induced McClellan to employ La Mountain. After all, the applicant was a balloonist of established reputation, and had performed successful and valuable military service at Fortress Monroe. In addition to these considerations, one of his senior division commanders had requested the services of an aeronaut with his command.

La Mountain's equipment, however, was still at Fortress Monroe.[128] A week elapsed before it could be brought to the new theater of activity. On September 27 La Mountain was

[127] Porter to Marcy, September 21, 1861, MS P8 [supplementary], LRAP.
[128] Ibid.

officially assigned as aeronaut with the army of the Potomac and, Baldy Smith's request notwithstanding, was attached to the division commanded by Brigadier General William B. Franklin.[129] His status was that of a civilian employee, and he was carried on the roll of the Topographical Engineers at a rate of approximately $10 per day.[130] The second phase of La Mountain's war service was about to begin, and was announced to the public by a solitary notice in one of the Boston papers.[131]

During the interview with Lowe in Porter's tent at Fort Corcoran, La Mountain had again brought forward his spectacular scheme of free aerial reconnaissance that he had previously recommended to Butler. The bombing of enemy camps and positions, however, does not seem to have been considered. The idea of free observation was not original, having been suggested by Wise as early as the previous June.[132] The plan consisted of ascending at a time when the wind was blowing towards the enemy lines, and when at a satisfactory elevation, casting off from the mooring cables in order to be carried directly over the enemy positions. Here, out of rifle range, the aeronaut could take vertical observations of the enemy dispositions and activity, thus eliminating the disadvantage of oblique distortion and concealment in defiladed and tree-screened positions, factors which often interfered with captive observation. When all possible information should be obtained, the aeronaut would then jettison his ballast and rise into a higher atmospheric stratum, in which La Mountain claimed there was always a constant easterly current. Entering this current, the balloon would be carried back to its original point of departure, where gas would be released, and a safe descent to earth effected.[133]

[129] Lieutenant Colonel Edward McK. Hudson [ADC, McClellan's Staff] to William B. Franklin, September 27, 1861, cited in Lieutenant Colonel John N. Macomb to Brigadier General Seth Williams, February 24, 1862, MS M266, LRAP.

[130] *Ibid.* The payrolls showing the rate La Mountain was paid are missing in the archives. Macomb informed Williams that " Lowe's pay is $10 per day, and Mr. La Mountain's is nearly as much."

[131] Boston *Transcript,* September 30, 1861.

[133] See Chapter iii, *supra,* p. 67.

[133] Porter to Marcy, September 21, 1861, MS P8 [supplementary], LRAP;

The advantages of close vertical observation afforded by such a plan are obvious. But the difficulties of making the scheme work satisfactorily were many. With all other factors completely favorable, the relative positions of the opposing armies became a condition precedent to any possibility of the system being feasible. Only when the Union lines were east of the Confederate positions could La Mountain's scheme of using the easterly current even be considered. Such an arrangement though present in the situation around Washington in the summer and fall of 1861, could not be expected to remain permanent throughout the war; nor was it likely that the tactical situations governing future operations would continue to conform to the geographical requirements of La Mountain's plan. Then too, there were technical difficulties which seriously detracted from the efficiency of the free reconnoitering service.

In the first place, the observer had no communication with the ground troops, and any discoveries of importance that might require immediate action on the part of the observer's superiors would become useless whenever time should be the essence of successful action. Lack of adequate control in direction was also a serious disadvantage. Winds are seldom so constant and invariable as to permit direction control of a free balloon to the extent needed for such operations as La Mountain intended to make. Again, the point of landing after returning to friendly lines was always an uncertainty. Even though the counter current could be relied on to bring the balloon back, it frequently prevailed at such a high altitude that identification of the original place of departure became difficult if not impossible,[134] even when the observer was so fortunate as to be brought back over the same spot. More often he would be blown far away from his point of departure, and, after landing, might easily find himself in a place devoid of communication with the headquarters from which he operated, or where communication of his reports might require loss

Boston *Transcript,* October 10, 1861; "Carleton" to the Editor of the Boston *Journal,* December 3, 1861, published December 6.

[134] La Mountain states that the elevation on the return flights to his own lines averaged from 18,000 to 25,000 feet.

of vital time, and even nullify the value of the information obtained. A further disadvantage was to be found in the necessity of discharging gas in order to descend. This procedure inevitably left the aeronaut with a deflated envelope, often at considerable distance from his headquarters and source of gas. The time lost as a result might well destroy the value of the service performed. The system also entailed a continual waste of gas, thus adding constantly to the expense of operations, a factor in argument often raised against the use of balloons with the army.[135]

Still another weakness was the fact that the aeronaut alone was to be the observer of the enemy positions. The obvious risks in the scheme evidently prevented commanding officers from sending their trained topographical subordinates on these daring expeditions, and hence the accuracy of the information obtained depended entirely on the interpretation of enemy activity and dispositions made by an untrained civilian. In the use of captive balloons, engineer officers and the generals themselves often accompanied the aeronaut, or went aloft alone. But there is only one instance in the records that an officer took part in one of these free excursions. In this case the individual was Colonel Clinton G. Colgate, 15th Regiment of Engineers, New York Volunteers.[136] Colgate, however, was an exception, and there is no evidence that any other officers, either volunteers or regulars, ever followed his example. Possibly the losses of trained officers and engineers in the course of ordinary casualties were sufficiently great to prohibit risking further loss that might result from accident or miscarriage of La Mountain's daring plans. The question may well be raised that if vital information could be secured in this manner, was this advantage not worth the risk? The answer cannot be conclusive. The reorganization of the Federal army after the

[135] For example, Colonel Macomb's report soon after taking charge of the balloon operations, in which he declared that " Ballooning for the army . . . is a very expensive business." Macomb to Bache, October 23, 1861, MS M1310, LRTE.

[136] La Mountain to Franklin, December 21, 1861, MS L222, LRAP; see also Francis B. Heitman, *Historical Register and Dictionary of the United States Army* (Washington [Government Printing Office], 1903), II, 90.

first Manassas, with the attending system of examination for officers to determine their fitness and efficiency,[137] revealed a serious shortage of qualified men in nearly all ranks and branches.[138] The regular officer personnel was far inadequate to serve the swelling volunteer army, and the trained engineers and topographical officers were relatively few.[139] With these considerations, plus the fact that La Mountain's system required daring operations of untried practicability, may lie the answer why ranking commanders did not send their subordinates to participate in the free aerial excursions. Another logical reason may be found in the fact that the added weight of an additional observer would have been a disadvantage, since La Mountain needed all the buoyancy possible to reach the high altitudes required for his return flight in the upper current. Regardless of speculation, the bald fact remains that La Mountain always went alone, with the one exception mentioned, save for one occasion when a war correspondent accompanied him. Hence solely upon his judgment and interpretation were based the reports that he rendered.

Broadly considered, the disadvantages of La Mountain's plan overbalanced its advantages. The success that was achieved can be largely attributed to the situations occupied by the opposing armies in the late summer and fall of 1861, when both forces held their ground from more or less fixed lines and watched each other without risking any major actions. It is perfectly obvious that La Mountain's scheme would have been utterly useless for battle reconnaissance during a moving situation. Likewise any change of position by the two armies which would remove the east-west factor would similarly render the system ineffective.

General Porter, evidently recognizing the possible utility of La Mountain's system during the period of watching and wait-

[137] "An Act to Authorize the Employment of Volunteers," Section 10, *United States Statutes at Large* (G. P. Sanger, ed.), XII, 270.
[138] Cf. Fred Albert Shannon, *The Organization and Administration of the Union Army* (Cleveland, 1928), I, 186-190.
[139] In December, 1861, Secretary Cameron reported an estimate of only 107 engineer officers of all branches in an army of more than 660,000 officers and men. *Report of the Secretary of War, 1861* [Sen. Doc. 1, 37 Cong., 2 Sess.], p. 4.

ing along the Potomac lines, believed the scheme worth trying and told McClellan that he " would recommend the experiment, if balloons can be procured, and the benefit to be derived will compensate for its [a balloon's] falling into the hands of the enemy." [140] Little did the General anticipate, in recommending the scheme, that in the space of six months, he would be the involuntary and unwilling participant in one of these excursions over the Confederate works at Yorktown.[141]

La Mountain was accordingly allowed to demonstrate his novel method of reconnaissance. Although he had previously informed General Wool that he was perfecting his generator and had added a new engine and boiler for that purpose, his inflating equipment seems still to have been inadequate or out of commission. On October 1 he inflated the *Saratoga* at the works of the Alexandria Gas Company [142] and towed the balloon to the vicinity of Cloud's Mill. Here, on October 4 the first experimental free ascension was made, apparently to demonstrate the existence of the prevailing easterly current which he claimed to be constant at high altitudes. He ascended to an elevation of 18,000 feet, entered the current which carried him in an easterly direction over Washington, and landed in the vicinity of Beltsville, Maryland.[143] Though the test does not appear to have included an excursion over the enemy lines, several war correspondents announced that observations of the enemy had proved of value and importance.[144]

The sight of the balloon over the capital, coming from the direction of Confederate territory, caused a wave of excitement and speculation. Many reports were circulated to the effect that

[140] Porter to Marcy, September 21, 1861, MS P8 [supplementary], LRAP.

[141] See below, Volume II.

[142] Item 1, Statement, " United States Government to Alexandria Gas Company, Dr." [Oct. 1, for 22,000 ft. Gas for Balloon *Saratoga*], enclosed in Third Auditor, United States Treasury, to Colonel S. H. Long, September 3, 1862, MS A632, enc., LRTE.

[143] Washington *Evening Star,* October 5, 1861; Washington *Sunday Chronicle,* October 6, 1861; Baltimore *American,* October 7, 1861; Philadelphia *Press,* October 7, 1861; New York *Herald,* October 6, 1861.

[144] Boston *Transcript,* October 10, 1861; Detroit *Free Press,* October 10, 1861; Hartford *Daily Courant,* October 10, 1861. See also description and discussion of La Mountain's free flight system, Providence *Post,* October 11, 1861.

the observer was from Beauregard's army. Some witnesses evidently possessed of the most powerful telescopes or equally powerful imagination, even claimed that the balloon bore the letters " C. S. A." painted conspicuously on its sides. Since the envelope was passing over the city at an 18,000-foot elevation, the assumption favors imagination rather than strong lenses. Other rumors were equally exciting and absurd. President Lincoln, stove-pipe hat and all, with General McClellan, were the passengers in the car, some of the sensationalists declared. A number of zealous patriots, believing the observers to be Confederates, made efforts to follow the balloon in the hope of capturing the enemy aeronauts. Several correspondents announced that one of the reconnoitering balloons had accidentally broken loose, and was lost beyond recovery. Still other reporters simply said that La Mountain was out on a joy ride. And others held out that Professor Lowe was the aerial visitor.[145]   These rumors and speculations were dispelled the following day by an announcement from the War Department that the balloon which had caused so much excitement was La Mountain's *Saratoga,* engaged in an experimental flight.[146] As has been pointed out in the discussion of La Mountain's system, the discharge of gas in order to descend at Beltsville necessitated the refilling of the envelope at the Alexandria works before operation could be resumed. On the 12th an additional supply of 22,000 cubic feet of gas was purchased to replenish that discharged in order to land after the experimental ascension.[147]

Although the demonstration of October 4 may have satisfied the ranking officers that La Mountain's theory of wind currents was essentially correct, his next operations were of the captive

[145] Philadelphia *Daily News,* October 7, 1861; Philadelphia *Ledger,* October 5, 1861; Washington *Evening Star* [second dispatch], October 5, 1861; Washington *National Republican,* October 5 and 7, 1861; New York *Times,* October 5, 1861; New York *World,* October 5, 1861; Providence *Post,* October 5, 1861; New York *Herald,* October 5, 1861.

[146] Baltimore *American* [later dispatch], October 7, 1861; New York *Tribune,* October 7, 1861; New York *World,* October 7, 1861; Philadelphia *Ledger,* October 7, 1861.

[147] Item 2, Statement, " United States to the Alexandria Gas Company, Dr.," [October 12, For 22,000 ft. gas for Balloon *Saratoga*], enclosed in Third Auditor, United States Treasury, to Colonel S. H. Long, September 3, 1862, MS A632, enc., LRTE.

variety. On October 14 he was ordered to make a captive ascension from the vicinity of Franklin's division at Camp Williams, before daylight, to observe the number and distribution of Confederate campfires. Numerous clusters of fires disclosed the enemy in considerable force near Springfield Station. Though the points of light marking the bivouacs were noticeably fewer at Fairfax Station, the volume of smoke that became visible at dawn indicated large numbers of troops in this area. The picket fires were clearly distinguished as far as the eye could see along the lines to the west and northwest of Chain Bridge.[148]

La Mountain's conclusions from his observations as shown in his report were reasonably correct. Some ten days before, a reconnaissance in force under Colonel Calvin E. Pratt from Newton's Brigade [Franklin's Division] had driven in the Confederate pickets at Springfield Station, and had heard rumors that a hostile force of 8,000 men was in the vicinity.[149] Though the size of this rumored force was greatly exaggerated, it is likely that part of this command later occupied Springfield after Colonel Pratt's troops withdrew. The fewer visible fires at Fairfax Station might also be accounted for in the plan of General Joseph E. Johnston to shelter his troops in huts rather than under canvas. He had written to the Confederate Adjutant and Inspector General a week before La Mountain's ascension, asking for an official opinion on this mode of shelter.[150] Very likely some of the troops had already begun to use huts, and the fires in these would be concealed from the air, though the smoke would be plainly visible. Such a means of shelter was widely used in the Confederate army at Manassas and Centreville later in the fall and winter,[151] and was noted from the air during a subsequent ascension.[152]

[148] E. S. Purdy [AAG, Franklin's Division] to Franklin, October 14, 1861, MS 6099, Papers of General George B. McClellan, Vol. XXVIII, Division of MSS, Library of Congress [hereinafter cited as McClellan Papers].

[149] Franklin to McClellan, October 3, 1861, 1 *O. R.*, V, 237.

[150] Johnston to General Samuel Cooper [A and IG, CSA], October 7, 1861, 1 *O. R.*, V, 891.

[151] Lieutenant General James Longstreet, *From Manassas to Appomattox* (Philadelphia, 1896), p. 61 [hereinafter cited as Longstreet].

[152] " Frank " [War Correspondent] to the editor of the Boston *Journal*, November 16, 1861, published November 22.

On the afternoon following this captive ascension, La Mountain made a free reconnaissance for the purpose of inspecting the Confederate lines opposite Camp Williams.[153] The results of this ascension are not known, but the excursion must have been regarded favorably by the various ranking officers. The next day General Smith telegraphed Colonel A. V. Colburn, of McClellan's staff, to ask if La Mountain had reported any further information,[154] and General McDowell, now commanding only a division since his severe defeat at Manassas, asked for a balloon and generator for the use of his command.[155]

On October 18 La Mountain again launched out over the Confederate lines, ascending from near Franklin's headquarters at Cloud's Mill, and took a number of observations that were reported in the press as being of great importance.[156] One correspondent enthusiastically declared that the aeronaut " doubtless saw distinctly every position of the enemy between the [Potomac] river and the Blue Ridge." [157] He maintained an elevation of 1,400 feet over the Confederate territory and reported to General Franklin on his return that he had seen a heavy battery at Aquia Creek mounting from six to ten guns. Half a mile in rear of this position he claimed to have observed a force of approximately 1,200 men. Other works were in progress along Aquia. At Fairfax Station, the force seemed to have been diminished, the place at this time being held by only a few light scouting parties. A large force was seen at Manassas and Centreville, with long trains moving in several directions. Franklin doubted the accuracy of the Aquia Creek information.

---

[153] Purdy to Franklin, October 14, 1861, MS 6099, McClellan Papers, Series 1, Vol. XXVIII. One war correspondent who evidently saw the free balloon in motion announced that another army balloon had " broken its nettings " and had escaped. Providence *Post*, October 14, 1861.

[154] Smith to Colburn, October 15, 1861, MS 6120, McClellan Papers, Series 1, Vol. XXVIII.

[155] Major A. W. Whipple to Lieutenant Colonel John N. Macomb, October 16, 1861, MS, Papers of Colonel Macomb, Records of the Bureau of Topographical Engineers, War Department Division, National Archives [hereinafter cited as Macomb Papers].

[156] Boston *Journal*, October 19, 1861; Baltimore *American*, October 21, 1861.

[157] Washington *Evening Star*, October 19, 1861.

He declared it was physically impossible to see Aquia Creek from Cloud's Mill, and suggested that La Mountain had seen a battery " higher up," if he had seen any at all.[158] Aquia Creek was thirty miles air distance from Cloud's Mill. Though the creek might be distinguished from the air at this distance, it is certain that La Mountain could not have picked up, even in a powerful glass, the battery and the camp he mentioned in his report. It is possible that he observed Occoquan Creek, fourteen miles distant from Cloud's Mill, and mistook this stream for Aquia.[159] Some strength may be given to this assumption from the fact that for several weeks prior to La Mountain's ascension, the Confederate engineers had been constructing heavy batteries near the mouth of Occoquan Creek. These, according to General E. P. Alexander, were unmasked to the Federals about the same time that La Mountain made his reconnaissance.[160] Franklin evidently had this in mind when he suggested that the aeronaut had seen a battery " higher up."

But La Mountain had not taken his observations from Cloud's Mill; he had been carried well into the Confederate territory. The number of troops he reported in the vicinity tallies reasonably well with the Confederate returns as of the end of September. General T. H. Holmes, commanding the Department of Fredericksburg, reported an aggregate strength of 869 at Aquia Creek,[161] and this force might well have been enlarged by the middle of October. Also, the estimates of troops made during these reconnaissances could not be expected to be accurate within the last hundred men. As to the battery, there is evidence that it might have been there at the time. During the conference of general officers with the Confederate President at Fairfax Court House on October 1, President Davis advised that a battery be planted near Aquia Creek where the channel of

[158] Franklin to McClellan, October 19, 1861, MS 6185, McClellan Papers, Series 1, Vol. XXVIII. See also Providence *Press*, October 19, 1861.

[159] See Maps, "Northeastern Virginia and Vicinity of Washington," *Atlas to Accompany Official Records*, I, Plates vii and viii, Sheets 1 and 2.

[160] E. Porter Alexander [C of A, Longstreet's Corps], *Military Memoirs of a Confederate* (New York, 1907), p. 58 [hereinafter cited as Alexander].

[161] "Abstract from Return of Department of Fredericksburg, for September, 1861," 1 *O. R.*, V, 884.

the Potomac was sufficiently narrow to be swept by the fire of heavy ordnance mounted near the entrance of Aquia.[162] The observation of diminished forces at Fairfax, and the concentration of large masses of troops with trains at Manassas and Centreville was entirely correct. Johnston's main force at this time was in process of withdrawal from Fairfax to Manassas and Centreville, the movement being completed by October 19, the day after La Mountain made his ascension.[163] The aeronaut thus reported the results of this movement accurately, and the trains he detected moving in " several directions " were doubtless those of Johnston's force that had just changed its base.

After taking his observations, La Mountain ascended into the upper stratum, entered the counter current and returned to his own lines, descending in the camp of Brigadier General Louis Blenker's German Brigade. Here his reception was anything but pleasant. As he neared the earth he was greeted by a volley of musketry which cut the network of the balloon and riddled the lower part of the envelope. One bullet, according to La Mountain, passed unpleasantly close to his head.[164] As though this were not enough, the balloon was surrounded as it reached the ground, by " an infuriated crowd of officers and men, who were intent upon destroying the balloon, myself included," La Mountain reported to Franklin. " My netting was cut by their knives, they refused me the privilege of alighting, dragging me by the ropes over the rough ground, causing much danger and some damage, although I showed them my passes and otherwise explained my position. The officers

[162] Jefferson Davis, *The Rise and Fall of the Confederate Government* (New York, 1881), I, 452.

[163] Joseph E. Johnston, *Narrative of Military Operations During the Late War Between the States* (New York, 1874), p. 77 [hereinafter cited as *Johnston's Narrative*]; Longstreet, p. 61; Colonel Alfred Roman [IG, Beauregard's Staff], *The Military Operations of General Beauregard* (New York, 1884), I, 154-155 [hereinafter cited as Roman].

[164] La Mountain to Franklin, October 21, 1861, MS L96, LRAP; Philadelphia *Daily News*, October 21, 1861; Philadelphia *Ledger*, October 21, 1861; Providence *Press*, October 19, 1861; New York *Herald*, October 19, 1861. A correspondent of the *Scientific American* wrote that " a German regiment, mistaking him [La Mountain] for a secession buzzard, blazed away at the aeronaut for several minutes." *Scientific American*, November 2, 1861.

seemed to be more unreasonable than the men, and paid no attention to my expostulations." La Mountain took pains to add that once in the sanctuary of General Blenker's headquarters, he was received with courtesy, and men were detailed to assist him with his damaged balloon. He respectfully requested that some precautions be taken by the commanding general to prevent a similar occurrence in the future.[165]

Franklin had telegraphed McClellan the results of La Mountain's observations and had mentioned the attack on the aeronaut even before the latter's formal complaint had been submitted.[166] When the written protest arrived, he forwarded it at once to the Commander-in-Chief, who in turn referred it to Blenker for explanation.[167] McClellan had previously sent Blenker an order to investigate the affair and to warn all men in his command that the balloons operating in the vicinity were friendly and not to be fired on.[168] The arrival of this order followed shortly by La Mountain's formal complaint, endorsed by the commanding general for explanation, drew an angry denial of the whole affair from Blenker. A rigid investigation, he declared, failed to disclose any truth in the aeronaut's charges. Furthermore, La Mountain had been cordially received at the brigade commander's quarters, and every assistance rendered him.[169] La Mountain had mentioned the brigadier's courtesy, but the attack had taken place before the general's headquarters could be reached. Why Blenker should have denied the affair is not at all clear. Probably his men, on seeing the balloon approach and descend from the direction of the Confederate lines, thought it was manned by an enemy scout, and acted accordingly. Later when the real status of things was known, they were perhaps ashamed to admit what they had

[165] La Mountain to Franklin, October 21, 1861, MS L96, LRAP.

[166] Franklin to McClellan, October 19, 1861, MS 6185, McClellan Papers, Series 1, Vol. XXVIII.

[167] Franklin's indorsement, October 22, La Mountain to Franklin, October 21, 1861, MS L96, LRAP; McClellan's indorsement, October 24, La Mountain to Franklin, October 21, 1861, MS L96, LRAP.

[168] McClellan to Blenker, October 19, 1861, MS [number omitted], Telegrams Sent, Army of the Potomac [misc. unnumbered volume], War Department Division, National Archives [hereinafter cited as TSAP].

[169] Blenker to McClellan, October 23, 1861, MS B355, LRAP.

done, hence Blenker's inquiry showed negative results. Possibly La Mountain exaggerated their ferocity; most of the officers and men of Blenker's brigade were native Germans, or from German-speaking families, many of whom understood English with difficulty, and this may have added to the confusion when La Mountain explained his identity. Or, Blenker may have been simply shielding his men from blame even though he knew the truth of the charges.

For a month after his experience with Blenker's troops, La Mountain's affairs with the army prospered well. He was assigned a detail of forty men and an officer to assist him in managing the balloon, and also a competent draftsman, N. Frank White, a fellow townsman from Troy.[170] His old technical assistant, Albert Kendrick, who had served in the operations at Fortress Monroe, was also kept in service and attached to La Mountain's headquarters.[171] The work performed had been recognized by the commanding general, and La Mountain's two balloons, *Atlantic* and *Saratoga,* were considered so satisfactory that they were purchased by the Government for the army's permanent use.[172] The amount paid, including several additional items of expense, was slightly over $3,300.[173] Toward the end of October the aeronaut requested that his assistant White, who now held a commission as lieutenant of volunteers, be assigned to recruiting duty, to raise a company of men to be especially trained for balloon service. But the request seems to have gone no farther than the office of McClellan's adjutant general where nothing more was heard of it.[174] The idea of raising such a company was evidently pro-

[170] New York *Times,* October 21, 1861.

[171] Washington *Evening Star,* December 4, 1861; Detroit *Free Press,* December 15, 1861.

[172] Colonel John N. Macomb's weekly report to Headquarters, Army of the Potomac, October 26, 1861, MS M309, LRAP [hereinafter cited as Macomb's Weekly Report].

[173] " 'Estimate of Funds on Account of Appropriations for ' Surveys and Military Defenses,' " October, 1861, enclosed with Macomb to Bache, October 23, 1861, MS M1310, enc., LRTE. The exact amount paid was $3,338.15.

[174] La Mountain to Headquarters, Army of the Potomac, October 31, 1861, MS Letterbooks of the Army of the Potomac, Vol. IX, p. 345 [hereinafter cited as LBAP].

duced by the transfer of his detail, a change which required a new assignment of untrained men.[175] The administrative system followed in assigning men for these special details provided for a new platoon every fifteen days. Hence no sooner had La Mountain given his men some rudiments of the technique in balloon management, when a fresh lot of inexperienced volunteers were turned over to him. In an effort to reduce the evils of this arrangement, La Mountain requested that ten men picked for their aptitude in the work be detailed permanently to his headquarters. This request was granted.[176] With the exception of the difficulties attending the assignment of assistants, and the rough handling he had momentarily received from Blenker's troops, La Mountain's service with the Army of the Potomac was progressing smoothly, and from all indications he was on the way to permanent success, particularly with respect to his system of free reconnaissance.

As has been mentioned, La Mountain nearly always went alone in these free flights. Colonel Colgate seems to have been the only officer to have taken part in one of them. The only

[175] La Mountain to Headquarters, Army of the Potomac, October 30, 1861, MS LBAP, IX, 345.

[176] La Mountain to Captain Wilson [Kearney's New Jersey Brigade], December 24, 1861, endorsed favorably by [Capt.?] Cannon, MS L5, Letters Received, Kearney's Brigade, Army of the Potomac, War Department Division, National Archives [hereinafter cited as LRKB].

Not one of the assignment rolls of troops attached to La Mountain's train seems to have been preserved. The names and units of the ten picked men are given in La Mountain's communication:

> 1st Regiment, New Jersey Volunteer Infantry
> Private D. W. Christian,   Company " H "
> Private T. Foster,   Company " H "
> Private John Werner,   Company " H "
>
> 2nd Regiment, New Jersey Volunteer Infantry
> Corporal William A. Lynn, Company " A "
> Private John Clampit,   Company " A "
> Private William Leonard,   Company " H "
> Private G. Squier,   Company " G "
> Private C. Wallenberg,   Company " D "
>
> 4th Regiment, New Jersey Volunteer Infantry
> Private John Austin,   Company " G "
> Private Jacob Klemmer,   Company " B "

other exception was a reporter, a correspondent of one of the Boston papers, whom La Mountain invited to accompany him in a flight over Manassas and Centreville. The correspondent accepted, and later wrote the details of his experience to his editor. So far as can be ascertained, if the tale of the reporter is to be believed, this letter is the only eye-witness description of one of these excursions ever to come to light, other than La Mountain's official military reports, and it illustrates something of La Mountain's method as well as an outsider's impression:

Being at his [La Mountain's] quarters, he kindly invited me to accompany him on a reconnaissance. Having faith with him in that ever-present upper current towards the east, I did not hesitate, though I knew that the lower current towards the west would in a very few minutes carry us far over inside the enemy's lines. Stepping into the car with him, he cut loose, and in a moment, as it were our great army lay beneath us, a sight well worth a soul to see—brown earth fortifications, white tented encampments, and black lines and squares of solid soldiery in every direction. So enchanted was I with the scene that I well nigh forgot that we were drifting enemyward, until Fairfax Court House lay beneath us, and I had my first sight of the enemy, in the roaming squads of rebel cavalry visible in that vicinity. Soon Centreville and Manassas came in full sight, and there in their bough huts lay the great army of the South. All along they stretched out southeasterly toward the Potomac, on whose banks their batteries were distinctly visible. So plain were they below, their numbers could be noted so carefully that not a regiment could escape the count. . . . I do not have time at present to give you as full description as I would like, of that great scene, a scene that would furnish material for a long letter; enough to say that it was superlatively grand and interesting.

The Professor, satisfied with his reconnaissance, as well he might be, after noting down the strength of the forces and their position, discharged ballast and started for that higher current to bring him back. Now I acknowledge I looked anxiously and (I am sure I was excusable) nervously for a backward movement, conscious that to come down where we were was death, or at least the horrors of a Richmond tobacco prison. Up, up we went, but still bearing west and south. I looked at the Professor's face. It was calm and confident, so I felt assured that all was right. That assurance became a settled thing, when in a few moments we commenced passing gently back to the east. *We had struck the Professor's current.*[177] Back, back we went, as though a magnet drew us, until our own glorious stars and stripes

---

[177] The italics in the quotation are the reporter's.

floated beneath us, and we came down gradually and smoothly into the encampments of General Franklin's division, where we were surrounded by enthusiastic soldiers who had been excitedly watching the trip.[178]

Although the accuracy in counting the Confederate troops was doubtless exaggerated, the Boston correspondent's narrative has preserved a graphic picture of what one of these free flights must have been like, a picture that cannot be found in the official military reports.

The period of La Mountain's success was unfortunately short lived. On the same day that the newspaper correspondent was writing his enthusiastic narrative to his editor, misfortune overtook the aeronaut in the loss of the balloon *Saratoga*. Early on the morning of November 16 the large envelope escaped from its moorings at Cloud's Mill during a heavy wind, and was blown off over the Confederate lines.[179] La Mountain blamed the accident on the carelessness of a soldier detailed to balloon duty from one of the New Jersey regiments of General Philip Kearney's brigade.[180] It is difficult to see how one man could have been wholly responsible, although circumstances may have made this possible. Probably the man's inexperience with the equipment, particularly during the strong wind, can be excused, but the accident spoke eloquently of the need of trained troops for this specialist duty. The accident occurred before La Mountain's request for permanent assignment of ten picked men had been made, and probably led him to suggest this means of remedying the unsatisfactory arrangements imposed by the system adopted by his superiors. Like most reforms of this kind, it was achieved only after serious damage had occurred.

The wind was also a large contributing factor to the incident. Colonel John N. Macomb, recently detailed to supervise all balloon activities in the Army of the Potomac, reported to McClellan that the wind " was blowing a gale from the north-

---

[178] " Frank " [War Correspondent of the Boston *Journal*] to Editor of the Boston *Journal,* November 16, 1861, published November 22.
[179] Macomb's Weekly Report, November 16, 1861, MS M427, LRAP.
[180] La Mountain to Franklin, December 21, 1861, MS L222, LRAP.

ward and westward, and it is doubtful if we ever recover it
[the *Saratoga*]." [181] Macomb's doubts were correct: the *Saratoga* was never recovered.

The loss of the *Saratoga* left La Mountain with only the
small balloon *Atlantic,* with which he had operated at Fortress
Monroe. This aerostat had been subjected while inflated to the
intense July and August heat on the Peninsula, and was now
exceedingly friable and otherwise in a poor state of repair. [182]
Ascensions with this inferior equipment after having used the
first-class *Saratoga* were naturally irksome to La Mountain, and
he regarded with envy the new government-built balloons that
were now under charge of his adjacent rival, Lowe. The latter
was now in the process of organizing a regular corps of balloonists for the Army of the Potomac, and had built at government order and expense four excellent balloons to be placed
in service at various points along the Federal lines. About the
first of December two of these balloons were temporarily idle
in Washington, pending assignment to aeronauts of Lowe's
corps. La Mountain was aware of Lowe's activities, and being
informed that the two balloons were idle at the Capital, he
applied to General Franklin for one of them to replace the lost
Saratoga. The application itself seems reasonable enough, but
La Mountain threw to the wind his promise of cooperation and
good will made during the interview with General Porter, and
charged Lowe with deliberately storing the new balloons, first
to prevent his rival from using them even though they were idle,
and second, with the plan of buying them, unused, at the end of
the war for a mere trifle. Lowe was inspired to these unscrupulous schemes, La Mountain directly charged, " through
motives of professional jealousy at my superior reputation as an
aeronaut." La Mountain was apparently so convinced of his
rival's wrongdoing that he begged Franklin to see General
Marcy, McClellan's chief of staff, personally and lay before him
the charges. [183] Franklin must have been perplexed at this communication. Having no personal knowledge of either of the

[181] Macomb's Weekly Report, November 16, 1861, MS M427, LRAP.
[182] La Mountain to Franklin, December 21, 1861, MS L222, LRAP.
[183] La Mountain to Franklin, December 8, 1861, MS L131, LRAP.

two aeronauts except what he had seen of La Mountain's work, he indorsed the letter on to McClellan's headquarters with the comment that La Mountain was of little use to his division without a larger and better balloon. If it should be true that two new balloons were in storage and were not being used, he recommended that one of them be assigned to La Mountain.[184]

While his request for one of Lowe's balloons went forward through the various channels of military correspondence, gathering indorsements as it went, La Mountain continued to operate as best he could with the one balloon left at his disposal. Despite his inferior equipment, he continued his free ascensions, and attracted much notice from the press. Some of the correspondents proclaimed that his experiments and aerial activities would soon " eclipse the feats of his rival, Professor Lowe," [185] publicity of a sort that did not help to bring about the cordiality and good will that Porter hoped for between the two men.

On December 10 La Mountain ascended from Cloud's Mill at half-past three in the afternoon, and soared out over the Confederate territory to accomplish one of the most detailed inspections of the enemy positions that he had yet made. Hovering for several hours over the hostile lines, he noted many items of importance, and then rose to an altitude of 17,000 feet, entered the prevailing easterly current, and descended in the camp of the 2nd Rhode Island Volunteer Infantry, four miles from the Capital. He reached his headquarters with the balloon late that evening, and reported the results of his flight to General Franklin.[186] His report on this occasion contained more information than he had previously obtained in any single ascension. Evidently the *Atlantic* was not so dilapidated as to preclude successful operations, and Franklin's statement that La Mountain was " of little use " without a larger and better balloon does not seem consistent

[184] Franklin's indorsement, December 9, 1861, La Mountain to Franklin, December 8, 1861, MS L131, LRAP.
[185] Alexandria *Local News,* December 10, 1861; Baltimore *American,* December 9, 1861; New York *World,* December 7, 1861; New York *Herald,* December 12, 1861.
[186] La Mountain to Franklin, December 11, 1861, MS L211, LRAP.

with this particular accomplishment. However, the aeronaut may have taken far greater risk in making the flight in the old aerostat; but from all indications, his personality and character would eliminate his consideration of personal danger.

The main body of Johnston's army was seen to be concentrated still at Manassas and Centreville, with a very noticeable increase in field works and entrenchments between Manassas and Fairfax Court House, five or six miles southwest of the latter village. An increase in force estimated at four regiments of infantry with some attached artillery was noted at Fairfax since the last air reconnaissance, but no large force was seen occupying this ground, as had been rumored by Union scouts and pickets. The new regiments were seen distinctly drilling in fields east of the court house and south of Little River Turnpike.[187] A body of infantry estimated at 1,500, with 200 or 300 cavalry, was observed drilling in an open field behind a large hill south of Gooding's farm half-way between Little River Turnpike and the Orange and Alexandria Railroad. Midway between Fairfax Court House and Vienna, a regiment of cavalry was seen in bivouac. From four to six regiments were discovered a mile south and southwest of Occoquan, and two more regiments were seen on the Old Ox Road between Occoquan and Fairfax Station. In addition to these specific dispositions, La Mountain reported certain generalities as a result of his reconnaissance. The scouting parties operating in close proximity to the Union lines along the outer defenses of Washington, he concluded, had been considerably increased, but there had been no material change in the disposition of the main Confederate force.[188]

General Franklin immediately telegraphed the substance of the aeronaut's report to McClellan's chief of staff, and advised that the information be checked with Major Whipple's [189] pho-

---

[187] The main turnpike between Alexandria and Fairfax Court House. See Map, *Atlas to Accompany Official Records,* I, Plate vii, Sheet 1.
[188] La Mountain to Franklin, December 11, 1861, MS L211, LRAP; Franklin to Marcy, December 11, 1861, MS 7025, McClellan Papers, Series 1, Vol. XXXII; also dispatches in New York *World,* December 12 [two items], and Philadelphia *Ledger,* December 12, 1861; Providence *Post,* December 12, 1861.
[189] A. W. Whipple, who supervised Wise's activities with the army earlier

tographic map of the region to verify La Mountain's identification of localities and landmarks. Franklin did not regard all the data obtained as free from error, and suggested that " the information be taken with a good deal of allowance," but qualified his doubt by assuring Marcy that La Mountain had declared the visibility and observation " uncommonly good," and that he had been " all over the enemy's lines." [190]

In fact, La Mountain's identification of places and his estimates of distance were remarkably correct. He had identified Gooding's farm and the Old Ox Road perfectly with reference to his points of orientation. And his general sense of direction, as disclosed in the details of his report, show a familiarity with the region and a knowledge of its topography.[191] While he undoubtedly had a map of the region to help him, orientation and ground identification from a moving balloon is not a simple task, even for a trained topographical officer. Also, the maps used in this period were notoriously poor. The main point of doubt might be whether he actually saw the troops mentioned at the places indicated in his report.

The statement of the main force of Confederate troops being still at Manassas and Centreville proved entirely accurate. This area was occupied by Beauregard's First Corps,[192] the largest unit of the Confederate army in the Potomac District.[193] The units La Mountain reported drilling south of Gooding's farm might well have been three regiments that Beauregard had ordered to encamp " as near the Orange and Alexandria Railroad as the ground will permit in the direction of Manassas," [194] and may have been maneuvering on this open ground some

in the summer. He was promoted to major on September 9, 1861 [Francis B. Heitman, *Historical Register and Dictionary of the United States Army*, I, 1025].

[190] Franklin to Marcy, December 11, 1861, MS 7025, McClellan Papers, Series 1, Vol. XXXII.

[191] Compare the information and points of reference in La Mountain's report with McDowell's map, *Atlas to Accompany Official Records*, I, Plates vii and viii, Sheets 1 and 2.

[192] Special Orders, No. 513, Headquarters, 1st Corps, Army of the Potomac [Confederate], November 30, 1861, 1 *O. R.*, V, 973.

[193] "Abstract of Return, Department of Northern Virginia," November, 1861, 1 *O. R.*, V, 974. The aggregate strength of the 1st Corps was 29,885.

[194] Par. I, Special Orders, No. 513, Headquarters, 1st Corps, Army of the Potomac [Confederate], November 30, 1861, 1 *O. R.*, V, 973.

distance east of their camp areas. The increased fortifications and earthworks reported five or six miles back from Fairfax Court House were undoubtedly the works erected by Confederate engineers on the high ground near Centreville, mentioned by Johnston in his narrative of this period.[195] This high ground plots an even six miles west and southwest from the court house on General McDowell's map, thus showing that La Mountain's judgment of the distance and direction were essentially accurate.[196] Since he was correct on these points, it is reasonable to assume that the other information submitted may have been equally trustworthy.

The promiscuous intrusion over their positions by the daring Federal aeronaut doubtless annoyed the Confederate commanders considerably. General Johnston, in giving instructions to General W. H. C. Whiting in case it should prove advisable to attack the Federal forces in the vicinity of Dumfries, warned his brigadier that the " infernal balloon " might interfere with the success of such a movement.[197] In a similar vein General Longstreet later wrote that the security of the Confederate forces at Munson's Hill, and the deception of the Federals by the use of dummy guns could no longer be assured after the Unionists began " to realize all their advantages by floating balloons over our heads." [198] These " quaker " guns, fashioned of blackened logs and " stove pipes of different caliber," were later specially camouflaged with sheds and covers of brush and foliage in order to conceal their true nature from the observation of the Federal balloonists,[199] of whom La Mountain, because of his free ascents, evidently proved particularly troublesome. This camouflaging was effected in compliance with a specific order from Beauregard to his division commander, Longstreet:

As it may become suddenly important to prevent the enemy's balloon observations from discovering whether or not we have guns in our

---

[195] *Johnston's Narrative*, p. 78.
[196] Cf. McDowell's Map, " Northeastern Virginia and the Vicinity of Washington," *Atlas to Accompany Official Records*, I, Plate vii, Sheet 1.
[197] Johnston to Whiting, December 5, 1861, 1 *O. R.*, V, 982.
[198] Longstreet, p. 60.   [199] Roman, I, 155.

Batteries, or more properly to let them believe that *we have,* you will have at once the position of each gun protected from aerial vision, by a rough shed of leaves and brushwood, elevated about 6 feet from the ground or the height of the crest, putting in each embrasure a piece of wood of the proper size (blackened) to represent a gun.

Endeavor to have this done as soon as practicable.[200]

It does not appear, however, that La Mountain took the bait thus prepared for him at Beauregard's orders. His dispatches of the period are conspicuously silent as to emplaced cannon along the Confederate lines. Possibly the camouflaging was so well done as to conceal the dummy guns themselves from aerial inspection.

La Mountain's ascension of December 10 drew wide comment from the northern press. Accounts of his flight, with details of the information secured, appeared in newspapers from Alexandria to Boston, and as far west as Detroit. One correspondent, with humorous speculations over the Confederate discomfiture at being thus freely observed, gave a detailed description of La Mountain's system of operation, with explanations of his proven theory of wind currents.[201] This widespread publicity extolling the work and ability of La Mountain had serious repercussions in Lowe's camp.

By December Lowe's efforts to secure the adoption of aeronautics as a real branch of the service had produced positive results. He was now supervising the operations of a loosely organized balloon corps, with the accepted title of chief balloonist in the Army of the Potomac. The activities of his organization extended from Hooker's division at Budd's Ferry on the Lower Potomac to Stone's Corps of Observation on the Upper Potomac at Edwards Ferry, as well as at intermediary

[200] Beauregard to Longstreet, December 5, 1861, MS, Papers of General Pierre Gustave Toutant Beauregard, Division of MSS, Library of Congress [hereinafter cited as Beauregard Papers].

The document cited is in a MS Letterbook, "Private and Official Letters, from March 1, 1861 to March 31, 1863," p. 38.

[201] Alexandria *Local News,* December 12, 1861; Baltimore *American,* December 12, 1861; Philadelphia *Inquirer,* December 12, 1861; Philadelphia *Press,* December 12, 1861; New York *Tribune,* December 12, 1861; Hartford *Daily Courant,* December 12, 1861; Boston *Journal,* December 12, 1861; Boston *Transcript,* December 12, 1861; Detroit *Free Press,* December 15, 1861.

sectors along this line. His plans and methods had received the technical approval of Professor Joseph Henry of the Smithsonian Institution, and he had gained the confidence and respect of McClellan, Porter, and other ranking officers. Although he had been one of the several professional balloonists who believed in the existence of the upper easterly current, and had demonstrated its presence in his flight to South Carolina the previous April, he had no belief in its practical use for free reconnaissance over the enemy lines.   It is evident that he regarded La Mountain's activities as the product of showmanship and exhibitionism, that could lead to no permanent results in military science, and that might, in the event of a spectacular failure, retard the progress of military aeronautics. La Mountain's presence and sensational activities within the sphere of the balloon corps' operations, Lowe feared, might soon become known as associated with the work of his assistants, a possibility that was not wholly unlikely, and entirely repugnant to him. He evidently believed that La Mountain's reports of observations were not only exaggerated and valueless from the military standpoint, but also at times wholly ridiculous.   The continuation of such activities within the immediate theater of his own endeavor, he feared, would bring eventual distrust of all military aeronautics, and thus impede the progress of his corps.   Added to these apprehensions there was also the possible feeling of resentment towards the presence of one whom he evidently regarded as a free lance over whom he had no control, acting within the vicinity of his own operations.[202]

When the northern press published *en masse* accounts of La Mountain's latest ascension, and in some cases mentioned Lowe as active in the same field of operations, the latter seems to have felt that the showmanship of his rival had gone too far. Whether Lowe had any knowledge of La Mountain's unfounded charges against his character, and also the latter's attempt to secure possession of one of the corps' balloons, is a

[202] The statement of Lowe's attitude is based on the sentiments expressed in several drafts and copies of letters addressed to McClellan and members of his staff, found among the MSS in the Lowe Papers. Also in other miscellaneous letters in this collection.

matter of speculation. Such knowledge would doubtless have reflected strongly in his attitude toward La Mountain. In any case, Lowe seems to have determined to check the activities of his publicized rival, or at least to secure his removal to another theater of the war. He clipped several notices concerning La Mountain's work from papers that had circulated in his camp, and sent them with a letter of protest to General McClellan. He had no desire, he declared, to question the policy of the commanding general in regard to aeronautics with the army, " but when my name and reputation as an aeronaut becomes mixed up with such extravagant statements, it places me in a very unpleasant light before the public." Pointing to his long efforts and success in creating an efficient branch of the service in the establishment of the present balloon corps, he requested that if the General-in-Chief desired to create two branches of aeronautics within the army, then these organizations should " be as far apart from each other as possible." In addition to these considerations, Lowe explained his objections to La Mountain's operations so close to his own. With two branches of aerial service, widely different in theory and practice, active in the same general area, confusion might easily arise as to the responsibility for the general results. A spectacular failure in one might well cause the discredit of the other. Lowe naturally anticipated no failure of his own corps; the danger was, in his mind, blame attaching itself to ballooning in general should his rival blunder. " My efforts to further the interests of ballooning may be rendered nugatory," he wrote McClellan, " or become identified with experimental efforts of those whose theories are widely different from my own. Besides, if such exaggerated statements as the enclosed [newspaper clippings] obtain publicity, the use of balloons for war purposes will soon be treated with ridicule and soon become very unpopular." [203]

[203] Lowe to McClellan, December 18, 1861, MS [copy], Lowe Papers.

War correspondents had already poked fun at the activities of La Mountain. In October the Cincinnati *Gazette* had published the following item ridiculing the aeronaut's operations:

" La Mountain has been up in his balloon, and went so high that he could see all the way to the Gulf of Mexico and observe what they had for dinner at Fort Pickens. He made discoveries of an important character, my Boy, and says

Not content with having addressed this letter to McClellan, Lowe wrote a similar communication to Colonel A. V. Colburn, one of the assistant adjutants general, expressing the same ideas. " These outside efforts have in a great degree interfered with my management, and identified my movements with those of whom I have had no knowledge," he declared. And further,

I do not wish to find fault with any aeronauts the general desires to employ, but I think movements independently conducted, without head or direction, confuse and retard the general business, leading to a complication of the aeronautic department. . . . The men under my direction, as well as myself, are perfectly willing to perform any duty, however hazardous, in the prosecution of the aeronautic service, but would prefer to see a unity of action, and have this branch of the service placed under an acknowledged head, who should control *all* balloon movements, under the General's direction, and not have the whole Corps responsible for the experiments of those who are not acting in concert with them.[204]

It is clear as chief balloonist, Lowe felt that his jurisdiction over the corps should extend to all aeronautic operations. And in this connection he asked that his status in his present acknowledged rating be more clearly defined.

To what degree these statements were prompted by professional jealousy, intensified by the volume of publicity that attended La Mountain's spectacular achievements, and how much of them was the expression of a sincere belief that the man's work was really humbug, and might later bring discredit on ballooning as a military science; these are judgments impossible to determine. Equally impossible is it to determine whether Lowe desired to extend his authority over all balloon operations from motives of self-aggrandisement or from a genuine desire to advance the new corps as a permanent and successful branch of the army. It is true that Lowe had a stake in the matter. He had spent his own money freely in experi-

that the rebels have concentrated several troops at Manassas. A reporter from the *Tribune* asked if he could see any negro insurrections, and he said that he did see some black spots moving around near South Carolina, but found out afterwards that they were ants which had got into his telescope."
    Cincinnati *Gazette*, October 22, 1861; published also in Frank Moore (ed.), *Rebellion Record,* III [Incidents], 28.
    [204] Lowe to Colburn, December 18, 1861, MS [copy], Lowe Papers.

ments and practical demonstrations that eventually led to the creation and operation of the balloon corps under his supervision. He had consulted with scientific men, such as Professor Henry, and his tests had been conducted before a board of the Smithsonian, whose members pronounced them to be sound.[205] La Mountain, he felt, in securing his original employment, had simply been fortunate enough to gain the favor of General Butler, through no legitimate tests nor scientific demonstrations. And he had later happened to appear in the Potomac region after Lowe's efforts and practical applications at his own expense led the ranking generals to recognize the military value of the balloon. Added to these views was the fact that La Mountain had more than once ridiculed Lowe's civilian activities prior to the war.[206]

Whether La Mountain really had interfered with the concerted action of the balloon corps as a whole, and whether his widely publicized excursions would have resulted in discredit to ballooning in general should they later fail during combat, are also considerations that cannot be judged. There is little evidence that there was any outward interference from La Mountain's activities. The second possibility belongs to the realm of "might-have-beens." It is obvious that after the army once took the field in a moving situation, and in one which might not conform geographically to the requirements of the easterly current, La Mountain's system of free reconnaissance could not have been feasible. But this seems hardly enough to justify all the statements that Lowe offered to McClellan and Colburn, even though he was acting in good faith. There is much room for argument on his side; but in the main, it appears that he was totally unaware of, or arbitrarily denied, the accuracy and value of La Mountain's reports. Also, his sincerity might be suspected because of his expressed desire to have all balloon operations placed "under one acknowledged head," since he obviously expected to be that head. All things considered,

[205] See Chapter v, *infra,* pp. 176, 180-181.

[206] La Mountain's derision was largely directed against Lowe's trans-Atlantic project, and its first reverses drew more derogatory comments in the public press from Lowe's rival. See Chapter v, *infra,* p. 158, n. 10.

however, it seems doubtful under the same circumstances, whether any other individual would have acted otherwise. Had La Mountain occupied Lowe's position, he probably would have launched a more violent protest. His ridicule of Lowe's civilian work before the war; his contempt of Lowe's aerial telegraph; and his unfounded charges against Lowe's personal integrity in the matter of planning to buy the balloons after the war for almost nothing: all these incidents indicate that La Mountain entertained personal malice towards his rival, and that he was motivated by jealousy, more so than Lowe. In fact the case was one of complete incompatibility and irreconcilable opposition, and bad blood was bound to result between the two as long as they were operating in the same vicinity.

La Mountain's letter to Franklin on December 8, asking for one of Lowe's balloons and charging the chief aeronaut with fraudulent intentions, had received no attention from higher authority despite Franklin's indorsement. No notice was taken of the charges, and no action was taken on the request for the alleged idle balloon. On the 21st, La Mountain again addressed Franklin, this time asking for the authority to build two new balloons at government expense, and enclosed two duly executed requisition forms. In this communication La Mountain in turn sought to discredit Lowe's operations, and declared that captive ascensions were of no value compared to the advantage of his free flights.[207] Since his own beginning in army ballooning had been confined to captive operations, he was hardly consistent in theories and practice. In dismissing the work of Lowe as inferior, he also dismissed all his own operations at Fortress Monroe as similarly ineffective.

Regardless of this element of inconsistency, La Mountain pressed his point. In order that the army might derive the benefit of his superior work, it became necessary to secure the proper means of carrying it on. The old balloon *Atlantic* was now in such poor condition as to render it almost useless. " A soldier might as well be expected to render efficient service with a flintlock musket as myself to make complete and frequent

[207] La Mountain to Franklin, December 21, 1861, MS L222, LRAP.

aerial reconnaissances with the means at my command," was his protest to General Franklin. Since the high command seemed not inclined to allow him the use of one of Lowe's balloons, La Mountain therefore asked that orders be issued to furnish him with materials for two new balloons of his own.[208] General Franklin forwarded the application to McClellan with the comment that La Mountain's equipment was now practically useless, and if La Mountain were to be retained in service, he ought to have at least one new balloon. Franklin also indorsed La Mountain's system of free reconnaissance, and declared that "the proposition is a good one." [209] Similarly he approved the requisition forms La Mountain had submitted for materials.[210]

Unfortunately for La Mountain, this letter was written three days after Lowe's communications of protest reached McClellan's headquarters. The commanding general replied to the favorable indorsements by inquiring whether Franklin desired to retain La Mountain in service, or would rather prefer one of Lowe's assistants from the balloon corps.[211] McClellan's action thus indicated that Lowe's protests had begun to take effect. Franklin answered that he did not know Lowe personally, and had had no experience with his ability. La Mountain, he added, "appears to work energetically, but I doubt whether the results obtained from his employment justify its cost." [212] This statement was scarcely consistent with Franklin's expressed approval of La Mountain's work, and his favorable indorsements on the two requisition forms for new balloon materials. If he doubted that the results were worth the cost, his action in approving the construction of two additional balloons seems

---

[208] *Ibid.*

[209] Franklin's indorsement, December 22, 1861, MS L222, LRAP.

[210] Quartermaster Requisition Forms, dated December 21, 1861, MS L222, enc., LRAP. La Mountain seems to have been confident that he would be allowed to build the two new balloons. A correspondent of the New York *Herald* visited his headquarters two days after he sent in his requisition and found him carefully examining samples of silk. La Mountain then told his visitor he was about to order several thousand yards of silk for two new balloons the government was about to build for him. Cf. Special Correspondence of the New York *Herald*, December 26, 1861.

[211] McClellan's indorsement, by Lieutenant Colonel James Hinds [ADC], December 23, 1861, MS L222, LRAP.

[212] Franklin's [second] indorsement, December 27, 1861, MS L222, LRAP.

hard to explain. Franklin conceded, however, that La Mountain had " done as much and as intelligently as any balloonist could." [213] Evidently the General did not regard balloonists in general as a very intelligent lot. Still, he advised McClellan that he would as soon have La Mountain as any one else, and re-forwarded the requisitions for materials asked for by the aeronaut. [214]

Despite the approval of Franklin, La Mountain's request was denied. On December 30, having received no reply, he bolted the regulation channels of military correspondence, and renewed his application to McClellan directly. [215] By this time Lowe's letters of protest had done their intended work. On December 27 the general commanding had reached a decision on the original application of the 21st, a decision entirely in Lowe's favor. He instructed an officer of his staff to inform La Mountain that,

It is his [General McClellan's] wish that all balloons shall be under the superintendence of Mr. Lowe. Upon this basis if you can come to an understanding with Mr. Lowe, it may be of interest to yourself and the service. [216]

McClellan might as well have suggested that La Mountain come to an understanding with General Beauregard or President Davis. At this stage of affairs, understanding between the two aeronauts was simply out of the question. The matter thus having been disposed of, McClellan ignored La Mountain's letter of the 30th. For the time being, the question appeared to be settled. La Mountain still continued to carry on with the dilapidated *Atlantic*. [217]

[213] *Ibid.*

[214] *Ibid.*; Quartermaster Requisition Forms, dated December 21, 1861, with Franklin's indorsement, MS L222, enc., LRAP.

[215] La Mountain to McClellan, December 30, 1861, MS L243, LRAP.

[216] Lieutenant Colonel, ADC [signature omitted, unsigned entry in outgoing letterbook], to La Mountain, December 27, 1861, MS LBAP, I, 603, entry 1279.

[217] The extent of La Mountain's operations with the old *Atlantic* during this period cannot be determined. There are no records of ascensions after December 10. But since he was retained in the service and paid a regular salary during this period, it would seem that he performed some service. It is not likely he would have been carried on the active payrolls without rendering duty of some kind.

His persistence, however, was not broken. In February he renewed his efforts to secure possession of one of the balloons attached to Lowe's corps. This time he approached Colonel Macomb, who among other staff duties as Chief Topographical Engineer of McClellan's army, was in general supervision of all balloon operations for the purpose of coordination and administration.

Towards the middle of February, two of Lowe's balloons had been taken from the Potomac lines and sent to Washington. La Mountain evidently had knowledge of this transfer of equipment, and decided that again the time might be propitious for obtaining the use of one of these aerostats, since at the moment they were not being used in the field. Having made no progress in his previous approaches to McClellan's headquarters, even with the support of one of the division commanders, La Mountain decided to try Macomb, who was technically Lowe's immediate commanding officer. What were his persuasions, and how he won Macomb over to his suggestion after it had once been denied by the General-in-Chief, remain matters of speculation.[218] Despite McClellan's expressed policy, Macomb approved La Mountain's new application and on February 15 ordered Lowe to turn over to La Mountain one of the large balloons in Washington, together with all necessary appendages.[219] La Mountain was sent as bearer of the order, and returned to report that Lowe positively refused to comply with the instructions he had delivered.[220] Macomb thereupon issued a second order directing Lowe to comply immediately with his instructions of February 15, closing his communication with the curt comment that " this matter admits of no further delay." [221]

[218] The correspondence relating to this part of the affair is missing from the files of the War Department. Neither the Lowe Papers nor the Macomb Papers contain items relating to it. Possibly La Mountain's persuasions were all verbal in personal interviews with Macomb.

[219] Macomb to Lowe, February 15, 1862 [holograph], MS M243, enc., LRAP; also MS [copy in Lowe's hand], Lowe Papers.

[220] Macomb's Note, *ibid.* [holograph], February 17, 1862, MS M243, enc., LRAP; also MS [copy in Lowe's hand], Lowe Papers.

[221] Macomb to Lowe, February 17, 1862 [holograph], MS M243, enc., LRAP; also MS [copy in Lowe's hand], Lowe Papers.

Again La Mountain was sent to deliver the new order, and again he returned to report to Macomb that "Mr. Lowe refused to recognize [Macomb's] authority and would not obey." [222]

Colonel Macomb, as supervising officer of the balloon service, was now faced with a serious problem of discipline. Under ordinary circumstances, he could have ordered Lowe's arrest, and sent a detail to enforce his order and take over the government property he demanded. But Lowe was a civilian, and not entirely subject to military law. It is not likely, however, that this consideration would have influenced Macomb in such a matter. But the Chief Aeronaut was a valuable expert, and his services were needed to supervise the technical details of the corps, a task that neither Macomb nor the officers of his staff could undertake of themselves. Consequently Macomb simply forwarded the disregarded orders and his own notations concerning them to Brigadier General Williams, McClellan's chief assistant adjutant general, with the statement that "Mr. T. S. C. Lowe, aeronaut in the service of the United States, and on duty under my command, has positively refused obedience to my orders." [223] General Williams then referred the papers to Lowe himself, "for such explanation as he may have to offer." [224]

Lowe in the meanwhile sensed that trouble would result from the series of orders he had received. On the 18th, the day that Macomb reported his insubordination to Williams, he called at the former's headquarters to explain why he had not complied with the two sets of instructions. Macomb refused to see him, although he waited several hours. Consequently Lowe reduced his explanations to writing for his chief's information. The two balloons, he declared, were neither idle nor available for La Mountain's use. One of them, the *Intrepid,* had been

[222] Macomb's Note, *ibid.* [holograph], February 18, 1862, MS M243, enc., LRAP; also MS [copy in Lowe's hand], Lowe Papers.

[223] Macomb to Williams, February 18, 1862 [holograph], MS M243, LRAP; also MS [copy in Lowe's hand], Lowe Papers.

[224] Williams' indorsement, February 19, 1862, *ibid.,* MS M243, LRAP; also MS [copy in Lowe's hand], Lowe Papers.

brought from its station with General Stone's command at Poolesville for the purpose of recoating, severe winter weather having injured its varnish.[225] There was entire truth in this statement. On January 20, and again on January 25, Stone had written to Lowe, asking that a smaller balloon be sent to replace the *Intrepid*. The larger aerostat, he informed the Chief Aeronaut, "has suffered in varnish from the excessively bad weather." [226] Five days later, before Lowe could make the desired substitution and repair the large envelope, Stone repeated his request: " The balloon *Intrepid* got an inch of ice on it last night, and is reported much injured. Hurry up the smaller one." [227]

The second balloon that La Mountain had alleged to be idle in Washington, had originally been brought in for shipment to General Halleck's army in Illinois.[228] It could not be assigned to La Mountain for the simple reason that it had been temporarily sent to Stone at Poolesville to replace the *Intrepid* while repairs to the latter were being effected. These facts, Lowe assured Macomb, had been made clear to La Mountain when he presented the orders. La Mountain's statement that Macomb's instructions were summarily ignored, Lowe declared, was a gross and deliberate misrepresentation.[229] This written explanation Macomb also refused to receive, whereupon Lowe took it in person to McClellan's headquarters, and handed it to Assistant Adjutant General Colburn, in order that his statements could become a matter of record.[230]

Having filed his explanation to Macomb with the adjutant general's office, Lowe then prepared a reply to General Wil-

[225] Lowe to Macomb, February 18, 1862, MS [copy in Lowe's hand], Lowe Papers; also MS LRAP, X, 551; also MS L83, LRAP [original].

[226] Brigadier General Charles P. Stone to Lowe, January 20, 1862 [telegram], MS, Lowe Papers; also printed in 3 *O. R.*, III, 269.

[227] Stone to Lowe, January 25, 1862 [telegram], MS Lowe Papers; also printed in 3 *O. R.*, III, 269.

[228] Lowe to Headquarters, Army of the Potomac, February 12, 1862, MS L73, LRAP; also MS LBAP, X, 550.

[229] Lowe to Macomb, February 18, 1862, MS L83, LRAP [original]; also MS [copy in Lowe's hand], Lowe Papers; also MS LBAP, X, 551.

[230] Lowe to Williams, February 19, 1862 MS [copy in Lowe's hand], Lowe Papers.

liams' indorsement, which had demanded the reasons for his insubordination. He reiterated the statements made to Macomb, and pointed out that this officer had refused either to see him or accept his written statements explaining that compliance with the instructions had been physically impossible. He then embarked on a detailed narrative of his prior service, calling attention to his status as chief aeronaut of the army. He then related that much of his equipment, although built at government expense, was of his own design and invention, and that he did not wish this equipment to be used by others without instructions from him. Also, he asserted, it did not seem right that his improvements and inventions should be freely given to a rival aeronaut who was admittedly hostile to him in their common profession. Furthermore, the portable generators were likewise of his own invention, on which he had applied for patents. " This machinery is entirely new," Lowe wrote, " and the designs are original with me. I have taken pains to instruct all the aeronauts in the service employed by me, in its use, and without such instruction, I am unwilling to risk this machinery in their hands [even] knowing that they are honest men and skillful in their profession. I feel more confidence in their use of it than I can in a person who has no reputation in his profession in the minds of scientific men."

Having thus thrown down the gauntlet with respect to La Mountain as an aeronaut, Lowe then proceeded to list a series of charges, which if true, stamped La Mountain as a rascal of the first water. This independent balloonist was, he asserted,

. . . a man who is known to be unscrupulous, and prompted by jealousy or some other motive, has assailed me without cause through the press and otherwise for several years. . . . He has tampered with my men, tending to a demoralization of them, and in short, has stopped at nothing to injure me . . . so much so that it is impossible for me to have any contact with him, as an equal in my profession, with any degree of self respect.

I did not have time to write to Lt. Col. Macomb on receipt of his first order, sent by this man, but sent word of explanation by him, who no doubt perverted what I did say, or told the Colonel a story of his own manufacture. . . .

This man La Mountain has told my men that he is my superior, and is considered so by the Commanding General. . . . He says that he is paid by Lieutenant Colonel Macomb two hundred dollars more per month than I am paid. . . .

I do not think that I should serve this man by giving to him possession of my improved balloons and portable gas generator for his examination . . . yet if it is the desire of the General that I should do so, I will most cheerfully comply. . . .

Without the improvements that I have made in the manufacture and management of balloons, they could be of little service to the Government; add to that my invention of the portable gas generator, which I am using for the benefit of the Government Service. I submit that I should not be interfered with in the management of this matter, at least until I have instructed men in the use of my invention.[231]

These were fighting words. Proof of the charges against La Mountain was not submitted, but Lowe's statements concerning his own inventions, and the attacks on his civilian work were substantially true. If his letters of complaint on December 18 were successful in thwarting La Mountain's plans, then the attack that Lowe hurled at his rival in this reply to General Williams was even more than successful. The ranking officers and their staffs by this time were doubtless weary of bickerings about balloons and who should use them. They had taken Lowe's part in the previous squabble. The best solution, they probably considered, would be to remove the source of dissension in the aeronautic service. After all, Lowe had been the first aeronaut to receive permanent employment in the Army of the Potomac, and now had built up an organized corps of balloonists. La Mountain, therefore, must go, and then there might be an end to controversy, charges, and counter-charges. On February 19, the day after Lowe wrote his explanatory letter to General Williams, McClellan issued an order directing Macomb to dismiss La Mountain from the service.[232] With the execution of this order, La Mountain's military career came to an end, and the fact was announced to the public the same day, in the same columns that once had lauded his daring feats of aerial reconnaissance: " Notwithstanding attempts of in-

[231] *Ibid.*

[232] Assistant Adjutant General [signature omitted, unsigned entry in outgoing letterbook] to Macomb, February 19, 1862, MS LBAP, I, pp. 671-672, entry 1408.

terested parties to create a contrary impression, we learn from the best authority that Professor Lowe is head of the Aeronautic Department of the entire army. Professor La Mountain, who once had a position of some sort, is no longer connected with the army.[233]

The dismissal of La Mountain for reasons apparently not connected with his work as an aerial observer, after he had been the first aeronaut to achieve success in this new field, raises the question of his value to the army. What was the opinion of his work among the several general officers; how did they regard the information he obtained during his captive ascensions and free flights; and why was he summarily dismissed simply because of his difficulties with Lowe, if he had made substantial contributions to the stock of military intelligence acquired by the Federal generals? These questions require consideration, even though conclusive answers cannot be formulated.

The evidence is conflicting and often misleading. Certainly in Butler's opinion, La Mountain's work at Fortress Monroe had been entirely successful. Butler's wholehearted support and encouragement of the aeronaut's efforts, even though the latter had not carried out his original promises; his approval of La Mountain's accounts although they were several times the amount authorized; his approval of La Mountain's plans for the employment of a larger balloon; and his recommendations that La Mountain's scheme for offensive tactics from the air be tried: all these acts indicate that Butler was strongly in favor of a continuation of the work, and believed the results to be worth the cost. Furthermore, Butler had declared that the information obtained was " of great advantage," and had considered La Mountain's dispatches of sufficient importance to forward to the General-in-Chief and to the War Department. Thus in Butler's case, there can be no doubt that he was satisfied with La Mountain's work, and intended to have it continued.

If Butler's estimate was sound, then why did his successor,

[233] Philadelphia *Inquirer*, February 19, 1861.

General Wool, wash his hands of the matter and refer La Mountain to the Secretary of War rather than keep him in service in the Department where he had proven his worth to the satisfaction of the former commander? To this question no satisfactory answer can be made. There is no evidence that Butler and Wool were on unfriendly terms, which might have led the latter to cast out an innovation that his predecessor had approved. If the aeronaut's work had been valuable to Butler, then it should have been likewise valuable to Wool. The situation in the Department had not changed when Wool took command. These considerations lead to the assumption that the matter was probably one of opinion; on Butler's part, favorable; on Wool's, perhaps skeptical.

A final consideration lies in the possibility that Wool, as a newly appointed commander of the Department, might have been cautious about innovations of which he had no knowledge or experience, and consequently sought the advice and guidance of the Secretary of War before taking the matter up. But if this is true, why did not Butler, if he so heartily approved the use of balloons, make some arrangements with Wool for this purpose when the latter succeeded him? It is perfectly clear that Butler intended that La Mountain should continue his work in the Department, even though he knew that Wool was to replace him as commanding officer. Notice of the coming change of command was sent him on August 11, and some twelve days later, even after Wool had superseded him, Butler had ordered the steamboat company officials to transport whatever men and materials that La Mountain should wish brought to the Fortress. Yet when Wool took up his duties at Fortress Monroe, Butler had made no arrangements with him concerning La Mountain, and if Wool is to be believed, never mentioned the aeronaut at all.

The situation on the Potomac lines was somewhat different. By the time La Mountain arrived in Washington with Wool's letter to Secretary Cameron, Lowe had already convinced several of the ranking officers of the value of balloon reconnaissance. As early as September 9, Fitz John Porter had recommended the construction of two additional balloons for Lowe's use,[234] and several days later had cordially assured Lowe of his value to the army.[235]   Smith's request for a balloon to be permanently assigned to his division was undoubtedly the result of his witnessing Lowe's operations before La Mountain arrived on the scene.   Yet Smith attached importance to the information obtained during La Mountain's free ascensions, having telegraphed McClellan's headquarters the day after his flight of October 14, to ask what had been reported.   McDowell's similar request for a balloon was made after La Mountain had begun his free flights in the vicinity, and may have been prompted by the work of both aeronauts.

The evidence from the statements and action of those under whom La Mountain served, however, is as conflicting and misleading as that which relates to the operations at Fortress Monroe.   General Franklin, his commanding officer, under whose command nearly all the ascensions were made, was far from consistent in his expressed opinions and action.   When La Mountain asked for one of Lowe's balloons and attempted to impeach his character, Franklin supported him, and approved the request.   Later, on the same issue, he approved La Mountain's demand for two new balloons, signed the requisitions for materials, and forwarded them twice to McClellan's headquarters, once after the commanding general had rejected them. Franklin also indorsed La Mountain's system of free flights, and declared that he believed the aeronaut's proposals for its continuation with new equipment reasonable and sound.   Yet despite these statements and acts, Franklin had declared in an indorsement to McClellan that he doubted if the results of La Mountain's service justified its cost.   In view of these conflicting statements, no sound basis for determining Franklin's true opinion can be established.

The same problem arises concerning the General's estimate of specific information obtained.   Franklin doubted the accuracy of some of La Mountain's reports, such as the observation of

[234] Porter to Lowe, September 9, 1861, MS, Lowe Papers; also printed in 3 O. R., III, 261.

[235] Porter to Lowe, September 11, 1861, MS, Lowe Papers; also printed in 3 O. R., III, 261.

the battery at Aquia Creek. He also advised that some of the information in another report be taken with " a good deal of allowance." Yet he believed the information important enough to telegraph immediately to McClellan's headquarters, and also qualified his skepticism by reminding McClellan of the extent of La Mountain's excursion over the enemy lines, and the good state of visibility at the time.

Franklin's opinions, whatever they may have been, did not alter the fact that La Mountain's activities were a source of annoyance and apprehension to the Confederate commanders. Johnston's warning to Whiting with reference to the " infernal balloon," and Beauregard's specific order to camouflage the dummy guns from aerial reconnaissance are clear indications that the Federal balloonist was regarded as a nuisance and a menace to Confederate security. It is equally apparent from Longstreet's mention of " floating balloons over our heads," that La Mountain was the source of annoyance. Such a statement could hardly have referred to Lowe's captive ascensions, nor was Lowe's range of observation close enough to have required Beauregard's precaution of camouflaging the " quaker " guns from his view.

The question of La Mountain's dismissal because of the difficulties with Lowe is perhaps easier to explain than the inconsistencies of the general officers. Lowe had already established his reputation in the Potomac army. He had won the confidence of several of the ranking officers. And he was working diligently to develop aeronautics as an organized branch of the service. La Mountain was a newcomer, who was, according to Lowe, a troublemaker and a rascal. It is not unusual that the commanding general should have taken Lowe's part. He was well known and liked by McClellan and Porter, and Smith and the topographical officers were familiar with his work and reputation. Also, Lowe's idea of organizing aeronautics into a regulation hierarchy, as a corps with commander, assistants, and lesser subordinates, subject to the control of the general officers, naturally fitted the traditional scheme of army organization better than did the independent ways and methods of his rival. Hence Lowe was supported, and La Mountain discarded, though his dismissal under these conditions does not reflect discredit on his actual work. The final question of the propriety and wisdom in dismissing La Mountain for such reasons if his work had produced valuable results, must remain unanswered. Evidence of the generals' opinions, as has been demonstrated, is too scant and conflicting to form a basis of conclusion. Furthermore, the reasons and motives that frequently directed various phases of army administration, particularly under the present circumstances, are beyond speculation.

In conclusion, it may be said that John La Mountain was an able aeronaut, courageous and energetic in the prosecution of his duty. He accomplished results of importance, and was the first aeronaut to perform successful operations in the United States service. His character and personality are not important to this narrative, and his relations with Lowe, which led to his removal, were the misfortune of long animosity and complete misunderstanding between the two men. In the main, it seems that he deserved a better reward than the curt dismissal that ended his army career.[236]

---

[236] After his dismissal from the army, La Mountain continued as a professional civilian aeronaut, until his death at South Bend, Indiana, on February 14, 1870. A contemporary obituary notice states that his health was seriously undermined from exposure and an accident that occurred sometime before his last illness, from which he died at the premature age of 41. Appleton's *American Annual Cyclopaedia, 1870* (New York, 1871), X, 559.

CHAPTER V

THE EARLY CAREER AND WORK OF T. S. C. LOWE, CHIEF AERONAUT OF THE ARMY OF THE POTOMAC

The dismissal of La Mountain from the army in February, 1862, left his rival, Lowe, the dominant figure in the scene of military aeronautics. James Allen had made his initial attempts at independent service, and because of failure for which he was not alone responsible, had been considered unsuccessful, and withdrew until called for service with the Federal balloon corps. Similarly the veteran balloonist John Wise had experienced misfortune and had permanently retired from the scene. And finally La Mountain, though he had done good work and had accomplished results of importance, was dismissed for reasons extraneous to his aerial operations. Lowe, with his steadily developing group of military aeronauts, was now the undisputed master of balloon operations of the Federal army. Around him the affairs of the first air corps in American history were destined to center for the next two years.

Thaddeus Sobieski Constantine Lowe [1] was born at Jeffer-

---

[1] All available references in print which include mention of Lowe's full name invariably give his second middle name as " Coulincourt." Even George Wharton James, an intimate friend of Lowe, and a frequent sojourner in the Lowe household, gives the name incorrectly [George Wharton James, " Thaddeus S. C. Lowe: One of America's Greatest Inventive Geniuses and Benefactors," *The Arena,* XXXVIII (1907), 388], and the error has been perpetuated in the *Dictionary of American Biography* [XI, 452].

This error has been caused by Lowe's temporary adoption during the period of his trans-Atlantic project, of the assumed name of " Carlincourt " [not Coulincourt], which seemed to him more euphonious and distinguished than the baptismal " Constantine." In this change he was encouraged by his young French wife [Leontine Gachon, daughter of an officer of Louis Philippe's Royal Guard], who preferred the assumed French form. When press notices concerning Lowe appeared referring to him as " Mr. Carlincourt Lowe," his brother Pembroke G. Lowe wrote several letters of protest to his much publicized kinsman, declaring that the change was sheer affectation and pretense, and demanded that he discard the high-sounding French name. Later Lowe dropped the assumed name and adopted his habitual signature of " T. S. C. Lowe."

The sources for Lowe's correct name are: Pembroke G. Lowe to T. S. C. Lowe,

son Mills, Coos County, New Hampshire, on August 20, 1832. He was the son of Clovis and Alpha Green Lowe, whose forebears had come from England and settled in Massachusetts toward the middle of the seventeenth century. The elder Lowe served in the legislature of New Hampshire at intervals from 1832, and appears to have been a man of strong pioneer spirit and rugged vitality.[2] Young Thaddeus, destined to become a nationally prominent scientist and inventor [3] as well as an aeronaut, received only an ordinary grammar school education, but at an early age developed a marked aptitude for applied science. While still a boy he became interested in air currents and theories of aerostatics. Like his later professional contemporary, Wise, his first experiments in this field were made with kites and the family cat for a passenger.[4] In his later 'teens Lowe became an assistant to a travelling prestidigitator, whose exhi-

April 2 and 28, 1860, MSS, Lowe Papers; the Lowe family Bible, in which the name is inscribed; and the various patents issued on his several inventions by the governments of the United States, Great Britain, France, Germany, Finland, Norway, Portugal, Canada, Brazil, Argentina, and Australia, MSS, Lowe Papers.

Articles giving the name " Carlincourt" appear in: New York *Tribune,* September 16, and October 24, 1859; New York *Sun,* October 28, 1859; *Harper's Weekly,* September 24, 1859; New York *Herald,* September 29, 1859; and *Courrier des États-Unis,* September 24, 1859. The assumed name is also thus spelled in Pembroke G. Lowe's letters to his brother, cited above.

[2] Clovis Lowe later became an assistant to his son in the Balloon Corps of the Army of the Potomac, and served for some eighteen months, until arbitrarily discharged by Captain Cyrus B. Comstock.

When Lowe was taken ill on the Peninsula and was invalided home, the elder Lowe assumed charge of the administration of the corps, and at the orders of Brigadier General A. A. Humphreys, took charge of all the equipment and materiel and supervised its transportation back to Washington from Harrison's Landing [Humphreys to Clovis Lowe, August 13, 1862, MS, Lowe Papers; also printed in 3 *O. R.,* III, 291].

Although well over sixty years of age, he served in all weather and under all conditions in the field and, according to his son, in fifteen months' campaigning he was not ill a single day. He sent four sons to the war, either as officers of volunteers or in various civilian capacities, such as the Chief Aeronaut. [T. S. C. Lowe to Captain John B. Howard (AQMG), January 31 1863, MS L100, enc., LRSW; T. S. C. Lowe to Canby, January 22, 1864, MS L100, enc., LRSW.]

[3] For a brief statement of Lowe's various inventions and scientific attainments after the war, see below, Volume II.

[4] Cf. Wise's youthful experiments with kites and parachutes, in which he used the Wise family cat and her kittens as passengers. John Wise, "Autobiographical Sketch," *Through the Air: A Narrative of Forty Years' Experience as an Aeronaut* (Philadelphia, 1873), pp. 27-31.

bitions in small rural communities frequently involved the demonstration of simple chemical experiments. It was through association with this free-lance showman that Lowe first acquired an interest in chemistry, which led to his serious study of this science. As humble as was the early influence that led to his quickening interest in this sphere of inquiry, it was nevertheless the impelling force that started Lowe on the road to real scientific endeavor.[5]

In 1854, at the age of twenty-two, he determined to adopt the study and practice of aeronautics as a profession. Being without sufficient resources to pursue a formal education in the sciences, he continued to give quasi-scientific demonstrations and exhibitions of his own,[6] and thus earned enough to permit private study along his chosen lines. The path was thorny enough, and four years elapsed before he had accumulated sufficient information and funds to begin active practice as an aeronaut. In 1858 he built his first balloon and in it made his first ascension at Ottawa, Canada, in connection with a celebration in honor of Cyrus Field's trans-Atlantic cable. The occasion was appropriate. The following year was to witness his spectacular attempt to span the Atlantic by air travel, even as Field had spanned it by wire communication. From this point on, his progress was steady, and in a year's time he had earned a wide reputation as a capable and successful balloonist.[7]

[5] Lowe had once considered the study of medicine, but had abandoned the idea. The above sketch of his early life is based largely on his own account in the first chapter of his unpublished memoirs, " My Balloons in Peace and War," the MS of which has been made available to the writer through the courtesy of his daughter, Mrs. Henry M. Brownback. This source will be cited hereinafter as Lowe's MS Memoirs. A description of the work will be found in the Bibliographical Essay, Volume II.

[6] Because of this earlier activity as a travelling entertainer and lecturer on pseudo-scientific subjects, Wise later attacked Lowe in the press during the period of excitement attending the proposed trans-oceanic voyage, calling him " a magician by profession," and " by practice in balloon progress an unscrupulous plagiarist." Wise to the editor of the New York *Sun,* November 12, 1859, published November 15.

[7] Lowe had met and discussed aerostatics with the distinguished English aeronaut Charles Green, and also the French civil and military aeronaut Eugène Godard. He had also become acquainted with the American balloonist W. D. Bannister.

The details of Lowe's earlier aeronautic career are based on Lowe's MS

In the close of the decade of the fifties that witnessed a rapid increase in popular interest in aeronautics, one idea stood out in the minds of several of the more prominent aeronauts: the dream of crossing the Atlantic by air.[8] This idea, long cherished and considered by air-minded men, had become the aim and purpose of Wise, La Mountain, and Lowe alike.

The idea of utilizing the prevailing easterly current, exploited by La Mountain in his military operations later, for motive power to carry a free balloon across the ocean, had impressed Lowe as well as his contemporaries. In 1859 he built at Hoboken, New Jersey, a huge "airship," the *City of New York*, in which he proposed to make the trans-Atlantic voyage.[9] The

Memoirs; a sketch of his life in *Balloon Pictorial* [a periodical publication edited by Lowe], July, 1866; a sketch of his early aeronautic activities published in the New York *Sun*, October 28, 1859; and numerous letters and other MS material in the Lowe Papers.

The statement of his reputation is based on the mass of press notices from all over the country preserved in his personal scrapbook. This source will be cited hereinafter as Lowe's Scrapbook. All newspaper citations from it will be designated with an asterisk [*].

[8] The idea of a trans-Atlantic air voyage had long appealed to the minds of American aeronauts. Both Wise and La Mountain had made active preparations to make the attempt [see chapters i, iii, and iv, *supra*]. Nearly sixty years after Wise had first petitioned Congress for funds to carry out his trans-Atlantic project, the aged aeronaut King was still urging the feasibility of the scheme. S. A. King, "How to Cross the Atlantic in a Balloon," *Century Magazine* LXII (1901), 844-859.

[9] T. S. C. Lowe, *The Airship City of New York: A Full Description of the Airship and Apparatus to be Employed in the Aerial Voyage to Europe* (New York, 1859), a pamphlet of 24 pages; *Harper's Weekly*, September 24, 1859; *Leslie's Illustrated Newspaper*, November 12 and 19, 1859; New York *Herald*, September 24 and September 29, 1859; New York *Times*, September 16, 1859; *Courrier des États-Unis*, September 24, 1859; [*] New York *Tribune*, September 16, 1859; *American Gas Light Journal*, October 1, 1859; [*] Lowe's MS Memoirs, Chapter ii.

The *City of New York* [later called *Great Western*] was an enormous affair, and would appear large even when compared with modern lighter-than-air craft. The gas bag at its widest point of circumference was 130 feet in diameter. Its height from boat to valve was 200 feet. The gas capacity of the envelope was 725,000 cubic feet. Over 6,000 yards of twilled muslin went into its construction. The gross weight of the envelope, machinery, and appendages was 2½ tons. The lifting power of the balloon when inflated was 22½ tons. The car was designed to accommodate comfortably six persons. A Francis metallic lifeboat, named *Leontine* in honor of Mrs. Lowe, was suspended beneath, connected with the car by two fixed ladders, and carried specially designed hand-driven propeller and sails. Other equipment included parachutes for dropping

project was widely publicized, and drew wide attention and interest from all sections of the country. Lowe, hailed as a scientist and a humbug alike, became a headline figure as a result.[10] A series of misfortunes and unforeseen difficulties, however, prevented the initial attempt from materializing. Damage to the giant envelope occurred during the first inflation in New York; unavoidable delays and unfavorable weather also

mail and messages, drags, anchors, buoys, sounding line and windlass, meteorological, nautical, and navigating instruments, heating apparatus, lanterns, signal lights, and a large supply of provisions. For a full technical description and explanation of the ship and its machinery, see Lowe, *Airship City of New York*, pp. 11-16.

A piece of the original envelope fabric is preserved in the Lowe Papers; the valve is in the Lowe Collection in the Aircraft Building, United States National Museum [Smithsonian Institution].

The huge envelope was later cut up and used for ground cloths at the balloon stations of the Federal army during the war. See Chapter vii, *infra*, p. 243.

[10] New York *Times*, October 25, 1859; New York *Morning Express*, November 2 and 3, 1859; [*] Boston *Transcript*, November 8, 1859; New York *Evening Post*, November 11, 1859; New York *Sun*, October 28, 1859; Philadelphia *Ledger*, October 5, 1859; Petersburg [Va.] *Express*, October 4, 1859; [*] *Courrier des États-Unis*, November 16, 1859; [*] *Nassau Guardian and Bahamas Islands Advocate and Intelligencer*, January 18, 1860; [*] also large number of unidentified clippings in Lowe's Scrapbook, including the following with dates omitted: New York *Tribune*; New York *World*; Philadelphia *Press*; Utica *Telegraph*; Utica *Observer*; Cincinnati *Commercial*; Boston *Journal*; Baltimore *American*; Washington *Evening Star*; Buffalo *Courier*; Cleveland *Herald*; St. Louis *Missouri Democrat*.

Wise and La Mountain scoffed and derided the project in letters and articles addressed to the editors of various newspapers. Samples of their attacks appear in "The Big Balloon a Humbug," letter of John Wise to the New York *Express*, November 12, 1859, published November 15; La Mountain to the New York *Express*, undated clipping in Lowe's Scrapbook; Wise to the New York *Tribune*, September 16, 1859, published September 17; "Nous Verrons" [La Mountain?] to the New York *Express*, November 11, 1859; and other unidentified clippings in Lowe's Scrapbook. Wise, however, changed his opinions the following year, and wrote to Lowe, offering to join him in the project. Wise to Lowe, June 16, 1860, MS, Lowe Papers.

The humorously inclined critics of the intended voyage loosed their talents in a series of comic verses, and others penned doggerel in praise of Lowe as a pioneer and scientist. Among such poetic efforts appeared: Jacques Maurice, "Jonothan to Lowe," in New York *Tribune*, November 29, 1859; Hans Lubeck, "The Pig Palloon Schip," a broadsheet published by H. De Marsan, 38 Chatham Street, New York City [Lowe Papers]; Anon., "The Mammoth Balloon," in Boston *Herald*, November 5, 1859; Anon., "The Song of the Flying Machine," *Comic Monthly*, October, 1859; Anon. [J. B.], "The Big Balloon," New York *Mercury*, clipping with date omitted in Lowe's Scrapbook; Anon. [J. A. A.], "A Colloquial Epigram," New York *Evening Post*, clipping with date omitted

interfered; and finally the approach of winter caused Lowe to postpone his projected flight until the following year.[11]

Shortly after this temporary failure, Dr. John C. Cresson of the Franklin Institute, also president of the Point Breeze Gas Works, invited Lowe to bring his aerial leviathan to Philadelphia for the next attempt. Lowe readily accepted, and arrived in the city with his airship in the winter of 1859.[12] At the suggestion of Horace Greeley's *Tribune,* the great balloon's name was changed to *Great Western.*[13] The winter months Lowe spent in experiments with smaller balloons and in enlisting the aid of scientific and wealthy men to support his project.[14] In the interim before the summer season he was invited to entertain the newly arrived first embassy from Japan with a demonstration of aeronautics. A fete was tendered the Japanese dignitaries, in which Lowe, with the assistance of William Paullin,

in Lowe's Scrapbook; Anon. [J.], " To Professor Lowe," and Anon. [I.], " Professor Lowe," unidentified clippings in Lowe's Scrapbook.

The comic elements were also manifest in a series of amusing caricatures portraying the coming experiences of Lowe after he should embark on the flight. Cf. " Professor High's Adventures," *Comic Monthly,* November, 1859.

One of the most extraordinary comments in the press came from a paper evidently Southern in locality, which published a humorous and fantastic story that the real purpose of Lowe's activities was to sail over Charlestown [West] Virginia, and effect the rescue of John Brown. This " Bloody and Brutal Plot," as the heading ran, involved the terrorizing of Charlestown by sailing over it and dropping bombs. When the populace should be thus put to flight, the balloon would be made fast to the jail where Brown was incarcerated, the doors forced, and the would-be negro insurrectionist would be rescued. The airship would then cast off and sail on to England, where Brown would be released, and where " the abolitionists of that country will lionize the old sinner." Unidentified clipping in Lowe's Scrapbook.

Perhaps the best example of confidence in the success of Lowe's flight was in the request of a linen manufacturer who sent Lowe a package of his best linen collars, which he asked the aeronaut to present with his compliments to Chancellor of the Exchequer William Gladstone, with the hope that the distinguished cabinet minister would wear them and thus set the fashion in England. William E. Lockwood to Lowe [date omitted], 1859, MS, Lowe Papers.

[11] MS Log of Inflation, *City of New York,* October 22 to November 26, 1859, Lowe Papers. Boston *Journal,* November 12, 1859; New York *Herald,* November 12, 1859; New York *Tribune,* November 12, 1859; Rochester *Daily Express,* November 12, 1859; [*] Lowe to the Editor of the New York *Times,* June 8, 1860, published June 9.

[12] Lowe's MS Memoirs, p. 16.

[13] *Ibid.*; also unidentified clipping in Lowe's Scrapbook.

[14] Lowe's MS Memoirs, p. 19.

who was later to serve with him in the army balloon corps, sent up several small balloons and then made an ascension before the visiting embassy. The *Great Western* was on exhibition and elicited much interest from the Mikado's emissaries.[15]

The early summer of 1860 saw the preparations for the trans-Atlantic flight again complete. A trial voyage on June 28 demonstrated that the giant airship was capable of ascension and could be maneuvered successfully.[16] Damage to the fabric of the envelope after the balloon was brought safely to earth on the New Jersey sand flats, however, was later to prevent a successful taking off for the long-heralded flight to Europe.[17] When the final inflation for the voyage took place on September 8, the weakened envelope was unequal to the strain of gas pressure and the effects of a gust that sprang up during inflation: it burst half an hour before the intended departure.[18]

[15] *Ibid.,* pp. 19-20; Boston *Transcript,* September 9, 1860; unidentified clippings in Lowe's Scrapbook; also card of admission to Japanese embassy fete, showing a figure of Lowe's balloon *Constitution,* and the words, " Japanese Embassy, Admit Bearer," Lowe Papers.

Thirty-eight years later Lowe met the sons of one of the Japanese dignitaries, who recalled their father's interest in the demonstration Lowe gave in 1860. T. Nakahaina to Lowe [Tokio], December 9, 1898, MS, Lowe Papers.

[16] Garrick Mallery, "A Ride in the Mammoth of the Air," in Philadelphia *Inquirer,* June 30, 1860; Philadelphia *Press,* June 29, 1860; Philadelphia *Ledger,* June 29, 1860.

Mallery was an associate editor of the *Inquirer,* and was one of the five passengers who accompanied Lowe in the trial flight. His enthusiasm over the success of the test and his confidence in Lowe and the *Great Western* were plainly expressed in his article. " The experiment about the balloon itself succeeded to the most sanguine hope. The monster in both ascent and descent, and in that generally most difficult matter of landing, was thoroughly under the control of its skillful builder as ever was a horse in harness." Lowe, he added, managed " the great creature like a charm. . . . In regard to the latter project [trans-Atlantic voyage] the experiment was also signally successful."

Mallery later distinguished himself as lieutenant colonel commanding the 13th Pennsylvania Cavalry, Gregg's Division, Pleasonton's Cavalry Corps, Army of the Potomac. Cf. " Organization of the Army of the Potomac, October 10, 1863," 1 *O. R.,* XXIX, i, 224.

[17] Philadelphia *Inquirer,* June 29, 1860; Lowe to the Editor of the *Inquirer,* October 2, 1860, published October 3.

[18] Philadelphia *Bulletin,* September 7 and 8, 1860; [*] Philadelphia *Inquirer,* September 8, 9, 10 and 11, 1860; Philadelphia *Pennsylvanian,* September 8, 1860; [*] Philadelphia *Press,* September 8 and 9, 1860; Philadelphia *Ledger,* September 10, 1860; Philadelphia *Sunday Mercury,* September 9, 1860; [*] also numerous clippings in Lowe's Scrapbook.

Another attempt on September 29 resulted in similar disappointment.[19] Lowe now realized that the *Great Western*, at least with its old envelope, was not destined to make the transoceanic crossing.

Despite this series of failures, Lowe was more determined than ever to prove the feasibility of his idea and carry his plans through to success. He exhausted the last of his resources in making preparations for a new envelope, and in the manufacture of a smaller balloon for additional experiments.[20] He then made a further appeal to the wealthier and more prominent citizens of Philadelphia for assistance in renewing his attempted European flight.[21] Several thousand dollars were raised in this manner, but the sum was insufficient to carry the project to completion. Then a committee of prominent Philadelphians, including Dr. Cresson, George Harding of the *Inquirer,* and others, addressed a memorial on Lowe's behalf to Professor Joseph Henry, Secretary of the Smithsonian Institution, stating that they had full confidence in Lowe and his project, and recommended that the Smithsonian furnish aid and advice " as may assist in the success of the attempt." [22]

The memorial was read before a meeting of the Board of Regents of the Smithsonian on February 16, 1861, and a reso-

---

[19] Philadelphia *Inquirer,* October 1, 1860; Philadelphia *Mercury,* October 1, 1860; [*] Lowe to the Editor of the *Inquirer,* October 2, 1860, published October 3.
    Even after the previous failures, one of Lowe's backers gave him a personal letter of introduction to be given to Ambassador Dallas when he landed in England, in which he was presented as the aeronaut " who has accomplished one of the greatest feats on record." William N. Reilly to George M. Dallas, September 29, 1860, Lowe Papers.
[20] Philadelphia *Evening Bulletin,* October 27, 1860; [*] Philadelphia *Press,* October 20, 1860.
[21] Lowe to John C. Cresson, November 10, 1860, MS, Lowe Papers; also unidentified clippings in Lowe's Scrapbook.
[22] Philadelphia Committee to Joseph Henry, December [day omitted], 1860, MS [copy], Lowe Papers; also printed in *Annual Report of the Regents of the Smithsonian Institution* (Washington [*HR Misc. Doc.,* 36 Cong. 2 Sess.], 1861), p. 113; and in 3 *O. R.,* III, 253-254.
    The committee in addition to Dr. Cresson and George Harding, included J. H. Lippincott, Dr. Cheston Morris, William Hamilton, W. H. Harrison, Henry Seybert, Isaac Lea, Fairman Rogers, Dr. James C. Fisher, Dr. Thomas Stewardson, George W. Childs, John Grigg, S. S. Haldeman, John F. Frazer, and M. McMichael.

lution was adopted recommending that Professor Henry furnish Lowe with technical information and advice.[23] At a second meeting of the Regents held a week later, suitable replies to the requests of Lowe and the Philadelphia Committee were framed.[24] The latter was informed that the endowed income of the Smithsonian could not be made available for the purpose desired, but that every assistance in the form of technical service and advice would be rendered.[25] Professor Henry then wrote to Lowe that he concurred fully in the theory of the upper easterly current, and expressed his belief that it could be used successfully in making an aerial passage to Europe. Henry advised, however, that a trial voyage be made over land, from one of the interior cities to test the current more fully and to gain more experience in navigating a balloon in it over long distances.[26] Lowe thereupon went to Washington at once and discussed the matter with Henry. The action of the Philadelphia committee in bringing about this acquaintance between Henry and Lowe was to prove a greater work of assistance than was at first apparent. Henry became one of Lowe's firmest friends, and was soon to become one of his supporters and a constant adviser in his operations with the army.[27]

Acting on Henry's advice, Lowe packed up his smaller balloon, which he named *Enterprise,* and proceeded early in April to Cincinnati, the interior city selected for the point of departure for the test flight. Here he enlisted the aid of Murat Halstead, editor of the Cincinnati *Commercial,* in backing the proposed voyage. Lowe was well received, and through Halstead's influence he was invited by a number of locally distinguished business and professional men to present several lectures on his proposed test flight and trans-Atlantic project.

[23] Journal of Proceedings of the Board of Regents of the Smithsonian Institution, Minutes of Meeting of February 16, 1861, *Annual Report,* 1861, p. 113.

[24] Journal of Proceedings, Board of Regents, Minutes of Meeting of February 22, 1861, *Annual Report,* 1861, p. 117.

[25] Henry to Cresson et al., March 8, 1861, MS [copy], Lowe Papers; also printed in *Annual Report,* 1861, p. 117, and in *Scientific American,* December 14, 1861.

[26] Henry to Lowe, March 11, 1861, MS, Lowe Papers; also printed in *Annual Report,* 1861, pp. 118-119.

[27] Lowe's MS Memoirs, pp. 37-38.

With funds raised in this manner, and with the special assistance of Halstead, he was ready to start on his trial trip by April 19.[28]

Preparations for the ascent were made in a boarded off lot attached to the grounds of the Commercial Hospital.[29] Inflation was accomplished about two o'clock on the morning of the 20th, and at half past three in a dead calm, in brilliant moonlight under a cloudless sky, Lowe ascended to a height of 12,000 feet and entered the prevailing upper current which rapidly carried the balloon eastward. At the last moment, a sheaf of *Commercials*, the early morning edition still damp from the press, bearing notices of the ascension, were taken into the car as evidence of the hour and point of departure. Shortly before one o'clock on the afternoon of the same day, Lowe landed at a point nine miles west of Unionville, South Carolina. He had accomplished a flight estimated from 900 to 1,200 miles in approximately nine hours.[30]

Lowe's initial reception upon reaching earth was unusual and unpleasant. The illiterate whites and negroes of the neighborhood insisted that he was some kind of " devil " that had

[28] Lowe's MS Memoirs, p. 39.

[29] Cincinnati *Commercial,* April 20 and 22, 1861.

[30] Lowe had intended to travel due east, with the Chesapeake Bay region as his ultimate destination. He descended close to the earth after crossing the Cumberland Mountains in order to determine his approximate position. In doing so he entered a southerly current flowing between the mountain ranges which deflected his course, eventually carrying him to South Carolina. To have risen back into the upper easterly current would have necessitated the discharge of nearly all his ballast, so he decided to remain in the southerly current, considering that his experiment to test the upper current had been successful enough.

There are several accounts of the voyage by Lowe, of which the best is the journal of the flight, embellished by details of his experience in South Carolina, first published in the Cincinnati *Commercial* of April 30, 1861, an article occupying several columns. This journal and extracts from it were reprinted in the Portland [Maine] *Transcript,* May 18, 1861; Baltimore *American,* May 6, 1861; Providence *Daily Post,* May 9, 1861; New York *Tribune,* May 7, 1861.

Details of the flight also appear in articles in Charleston *Mercury* [triweekly], April 23, 1861; Charleston *Courier,* April 24, 1861; Providence *Evening Press,* May 1, 1861; Providence *Journal,* May 1, 1861; Philadelphia *Daily News,* May 3, 1861; Cincinnati *Commercial,* April 27, 1861.

The voyage is also described in full detail in Chapter vi of Lowe's MS Memoirs, and another account by Lowe appears in *Navigating the Air* (New York [The Aero Club of America], 1907), pp. 127-156.

descended upon them from the wrong direction. Being convinced at length that he was at least human, they hustled him off to the nearest jail, from which he was rescued by the local dignitaries of Unionville and offered all the courtesies appropriate to a distinguished visitor.[31] Especially kind in this respect was A. W. Thomson, editor of the local paper, also a member of the South Carolina legislature, who furnished Lowe with a letter of introduction to influential friends in Columbia, to insure his courteous reception there.[32] A local committee also drew up a certificate stating the time and place of Lowe's arrival, in order that documentary evidence of the time of flight and point of landing might be preserved.[33] Lowe thereupon prepared to return North. On April 22 he placed his balloon and instruments on a train for Columbia and proceeded to the State capital intending to continue on to Washington.[34]

But the time was unpropitious for the arrival of a northern visitor. Only a week before, Beauregard's cannon had thundered defiance to the Federal government, inaugurating formal hostilities at adjacent Charleston. The Confederate seizure of Harper's Ferry and its arsenal had pre-dated Lowe's flight by less than two days. Only twenty-four hours before the *Enterprise* ascended from Cincinnati, a Baltimore mob had attacked the Sixth Massachusetts Regiment, and on the same day that this outburst took place, President Lincoln had declared the blockade of Southern ports. And even while Lowe was on his voyage, Colonel Robert E. Lee was tendering his resignation from the United States Army. A serious time it was for a Yankee balloonist, with a bundle of abolition newspapers,[35] to be land-

[31] Journal of the Voyage, in Cincinnati *Commercial*, April 30, 1861; Lowe's MS Memoirs, Chapter vi.

[32] A. W. Thomson to John Caldwell and Colonel Richard Anderson, April 22, 1861, MS, Lowe Papers.

[33] Certificate signed by P. M. Wallace, A. W. Thomson, Dr. Jesse Lamb, Joseph Fant, Thomas McNally, and A. Powell, April 21, 1861, MS, Lowe Papers; also published in Lowe's journal of the voyage, in the Cincinnati *Commercial*, April 30, 1861.

[34] Lowe's MS Memoirs, p. 58.

[35] The Cincinnati *Commercial* was regarded by the South Carolinians as an abolition paper. Thomson in his letter of introduction to Caldwell and Colonel Anderson admitted the hostile political opinions expressed in the papers Lowe

ing in South Carolina, the stronghold of secession and Southern nationalism.

This fact was only too apparent when Lowe reached Columbia. When it became known that a Northern aeronaut had arrived in the State with abolition papers only nine hours old, the authorities immediately concluded that he was either a bearer of dispatches or a spy.[36] At the station platform Lowe was arrested by the sheriff of Columbia and conducted to jail for a second time, followed by a lusty throng of ardent secessionists who urged that he be dealt with summarily. Some of the less violent spectators advised their colleagues to "tar and feather the damn Yankee" without further ado; the more sanguine demanded that he grace the nearest tree of suitable size.[37]

Safely incarcerated, Lowe was soon visited by W. H. Boatwright, Mayor of Columbia, and several other officials. Fortunately the publicity attending the proposed trans-oceanic flight had reached Columbia, and this fact was helpful in establishing his identity. Lowe was known by reputation to several prominent citizens of the town, including members of the faculty of South Carolina College, some of whom knew of his connections with Professor Henry. The assurance of these gentlemen that Lowe was not connected with politics or the Federal service was further strengthened by the letter of introduction from A. W. Thomson, of Unionville. These evidences of integrity, along with the certificate of the Unionville committee, secured Lowe's release almost at once.[38] Mayor Boatwright then drew up a formal passport designed to conduct the aeronaut safely through Confederate territory on his return North. All authorities and private citizens were assured of

---

had with him, but added that Lowe did not know what the papers contained other than the notice of his ascent, and that "Professor Lowe says he is no politician, but only engaged in his profession as an aeronaut." Thomson to Caldwell and Colonel Anderson, April 22, 1861, MS, Lowe Papers.

One of the original *Commercials* which Lowe carried with him in the car of the balloon is preserved in the Lowe Papers.

[36] Lowe's Journal of the South Carolina Voyage, in the Cincinnati *Commercial*, April 30, 1861.

[37] Lowe's MS Memoirs, p. 59.   [38] *Ibid.*, pp. 60-61.

Lowe's integrity and purely scientific purpose in coming South, and were admonished to render him assistance and courtesy on his return journey. The seal of Columbia was affixed to make the document official.[39]

Because of the chaotic condition of railway service caused by troop movements and the severance of rail communication between Washington and the South, Lowe found the only way open for his return was to Cincinnati by way of Louisville, Kentucky.[40] He packed up his balloon and equipment for shipment by express, and telegraphed the offices of the *Commercial* that he was on his way to Louisville.[41] The same day he boarded a train for his return journey.

The route North was marked with numerous indications of the coming conflict. Encampments were clustered thickly about the various rail junctions, and Lowe's train overtook several troop-trains loaded with Confederate regiments on their way to the Virginia frontier. Recruiting posts were conspicuous near the depots in the villages and towns. Occasionally a glittering regimental band serenaded the train as it stopped to discharge and take on passengers. The martial scenes encountered at every turn along the way impressed Lowe with the magnitude of the national crisis. There was no denying that the Southerners were in deadly earnest, all the units of troops he observed seemed to be exceptionally well armed and equipped, and their enthusiasm ran high. As the train gradually neared the Ohio River, the trans-Atlantic project seemed to fade from Lowe's mind in contrast with the grim realities of the present. He was in the midst of civil war, and crossing the ocean by air seemed trivial in the face of the national catastrophe. In the North men were enrolling companies and regiments for the defense of the Union. He would offer his services as a military aeronaut and organize an army balloon corps.[42]

Lowe arrived in Cincinnati on April 26,[43] soberly impressed

[39] Safe Conduct issued by the City of Columbia, S. C., April 22, 1861, signed by W. H. Boatwright, Mayor, MS, Lowe Papers.

[40] Lowe's MS Memoirs, pp. 61-62.

[41] Lowe to the Cincinnati *Commercial*, April 22, 1861, published April 24.

[42] Lowe's MS Memoirs, pp. 63-65, 68.

[43] Cincinnati *Commercial*, April 27, 1861.

with what he had seen during his brief visit in the Confederacy. " I was fully convinced that the country was facing a severe struggle," he later wrote, and added that a feeling of patriotism had now transcended his private ambitions to span the Atlantic by air. He immediately called on Halstead to discuss plans for entering the government service as a military aeronaut.[44] Exactly what was decided in this conference is not clear, but it is obvious from later developments that the Cincinnati editor pledged himself to assist Lowe through his influence with his friend, Salmon P. Chase, Secretary of the Treasury in Lincoln's cabinet. It does not appear that Halstead offered financial assistance as he had done in backing Lowe's trial flight to South Carolina.

Lowe evidently realized that gaining a post with the army as a balloonist would not be an easy task, and doubtless foresaw that tests and demonstrations of some sort would be required before the military authorities would recognize the possibilities of balloons for observation. His foresight also discerned that such demonstrations might also be expensive. The South Carolina flight, even with the assistance of Halstead, must have drained away much of his ready money, and it now became necessary to replenish his finances to some extent before making his application to the officials in Washington.

Consequently he decided to make another public ascension in the *Enterprise.* The publicity attending the recent flight would aid in drawing spectators, and even modest admission charges to witness the new ascension would net a sizeable sum to cover expenses in Washington. On May 8 he again ascended from the Commercial Hospital lot before several thousand spectators who jammed the enclosures to see the now famous *Enterprise* go aloft.[45] Accompanying Lowe in this ascent were Junius Brown, of the Cincinnati *Daily Press,* another newspaper man, W. W. Ware,[46] and a Philadelphia lawyer named Jacob C. Freno, who later served in the army balloon corps, much to

---

[44] Lowe's MS Memoirs, p. 68.   [45] Cincinnati *Commercial,* May 9, 1861.
[46] The name is given as " West " in some of the other editions. The writer has no way of telling which is correct. The man is not mentioned in Lowe's papers.

Lowe's lasting regret.[47] The balloon landed at Bethel, Clermont County, Ohio, where Freno and Brown left the party and returned to Cincinnati. Lowe and Ware again cast off, intending to sail eastward towards Washington.[48] Strong winds, however, carried the voyagers out of their intended course, and as though to repeat the experience of La Mountain two years before, landed them in Canada at some distance from Hamilton, Ontario. Arriving in Hamilton, Lowe learned of preparations for the celebration on the 24th, of the forty-second birthday of Queen Victoria, and was engaged to make an ascension in honor of the occasion.[49] The financial return from admissions to this aerial exhibition evidently induced Lowe to repeat the demonstration. On the 29th he again went aloft, landing at the village of Saint Anne's Gainsboro, in the county of Lincoln.[50] From this point he proceeded with his balloon to the nearest railway and started for his home in Philadelphia, and thence on to the Capital.

In the meanwhile, Halstead had been as good as his word. He had written to his friend Secretary Chase in Lowe's behalf, warmly urging that Lowe's plans for aeronautic service in the army be adopted, and recommended Lowe's ability and scientific knowledge in the field.[51] Chase received the suggestions and recommendations with favor.

My dear Halstead [he wrote on May 20, while Lowe was in Hamilton], I think very well indeed of your ideas about Professor Lowe, and have spoken of them to the Secretary of War and to distinguished officials. They are well received, but there is some difference of opinion as to the balloonists to be employed. I shall urge your man.[52]

---

[47] See Chapter vii *infra*, pp. 265-267, and Volume II.

[48] Cincinnati *Commercial*, May 10, 1861. The *Commercial* of the 9th and 10th both give Freno's name incorrectly, first as " Freeman," and then as " Frenau."

[49] Broadsheet: " Queen's Birthday, Friday, May 24, 1861—Programme," in which was announced after the list of dignitaries and participants, " Professor Lowe, of Philadelphia, will make an ascent from his splendid air ship *Enterprise*." Signed by W. Gillespy, Secretary, and C. Magill, Chairman, Hamilton, Ontario, May 22, 1861. Lowe papers.

[50] New York *Commercial Advertiser*, June 4, 1861.

[51] Halstead to Lowe, May 14, 1861, MS, Lowe Papers.

[52] Chase to Halstead, May 20, 1861, Lowe's MS Memoirs, p. 69.

Having received this reply from Chase, Halstead wrote Lowe at once in care of his home in Philadelphia, enclosing a copy of the Secretary's letter, and urged the aeronaut to go at once to Washington to press his application for service. Halstead had developed a keen interest in Lowe, as well as a sincere belief in him as an able aeronaut, and desired to see him succeed in government service:

You will, I am sure [he wrote], pardon the Secretary for speaking of you as " my man " as I certainly addressed him so warmly on the subject as to become in some sort your champion or advocate in the matter.

I am fully convinced that you are the man and that you could at once render your country and your science very essential service. I will again write to the Secretary of the Treasury. . . . If you have not formed other views, I would respectfully urge you to proceed to Washington and take your balloon there. Nothing will be so convincing as a demonstration of what could be done. . . .

There is no doubt but what you could make your way with the powers that be at Washington, of whom Secretary Chase is one of the most important. He would warmly second you, and would at once know that you are a man of science and a gentleman, and therefore *not a showman*.[53]

Halstead was not alone in advising Lowe to turn his aeronautic knowledge and experience to military advantage for the benefit of the government. In reply to a letter Lowe had sent from Cincinnati concerning the results of the South Carolina flight,[54] Professor Henry also wrote on May 28 suggesting that the balloon would be " of advantage to the Government in assisting their reconnaissances of the district of country around Washington." [55] Lowe in the meanwhile was making his second ascension at Hamilton, and Henry's letter, along with Halstead's communications, awaited him on his return to his home in Philadelphia shortly after June 1.[56]

While Lowe was in Canada, Halstead had gone to Washington, and having received no reply to his first letters, telegraphed

[53] Halstead to Lowe, May 23, 1861, MS, Lowe Papers; also copy in Lowe's MS Memoirs, pp. 69-70.
[54] Lowe to Henry, May 1, 1861, MS [copy], Lowe Papers.
[55] Henry to Lowe, May 28, 1861, MS, Lowe Papers.
[56] Philadelphia *Press*, June 4, 1861; Philadelphia *Ledger*, June 4, 1861.

the aeronaut on June 5 urging that he come to Washington at once if he still intended to carry out his plan for military balloon service. "Can't you prepare balloon for transportation and come here? See Chase if you come," was his terse message.[57] The following day Lowe arrived in Washington with the *Enterprise* and took up quarters at the National Hotel.[58] He first called on his technical adviser, Professor Henry, and outlined tentative plans for military balloon service and demonstrations to establish the value of aerial observation. Henry expressed entire approval and assured Lowe of his cordial support. Lowe then sought an interview with Secretary Chase. The Secretary promised to support Lowe's plans with the other officials of the administration, and further agreed to bring the matter to the attention of President Lincoln.[59]

These were not idle promises. Chase immediately pressed the matter with Secretary of War Cameron, with the result that Cameron interviewed Professor Henry the following day and requested that he examine Lowe's equipment and witness the experiments that Lowe intended to make in demonstrating the military value of balloon observation.[60]

Lowe's plans involved a simple demonstration, before governmental officials and high military officers, of aerial observation from various altitudes with the assistance of strong field glasses and telescopes. He also proposed to carry up a small magnetic telegraph set including transmitter and receiver, to be operated by a trained telegrapher, to show that information could be readily communicated from the balloon car to a ground station by ordinary telegraph, thus solving the problem of rapid transmission of reports from the air to distant headquarters. Since the ground operator could plug into lines extended to any distant telegraph station, intelligence from the aerial station could be instantaneously transmitted to any point located in the ground telegraph net. The advantages of this system over com-

[57] Halstead to Lowe [telegram], June 5, 1861, MS, Lowe Papers.
[58] Washington *Evening Star,* June 6, 1861.
[59] Lowe's MS Memoirs, p. 70.
[60] Henry to Cameron, June 21, 1861, MS [copy], Lowe Papers; also copy in Lowe's MS Memoirs, pp. 71-72; also printed in 3 *O. R.,* III, 254-255.

munication by weighted written messages dropped by hand were readily apparent.[61]

Although the practical application of this aerial telegraph was first demonstrated by Lowe a few days later, and was subsequently made effective in the field during combat for the first time in history, the idea was not original with him. In August, 1860, Captain E. B. Hunt, of the regular engineers, had advocated an identical system in an address before the American Association for the Advancement of Science,[62] and the substance of his plan was later published in one of the popular periodicals.[63] Also, J. H. Morrison, of New York, inventor of an apparatus for laying field telegraph lines underground during military operations,[64] was reported to have suggested this form of communication from balloons sometime before Lowe engaged in his practical demonstrations.[65] Likewise, John Johnson, a prominent member of the American Photographical Society, had recommended the use of telegraph with military balloons at a meeting of the Society on June 10, at which the value of aerial photography for military use was under formal consideration.[65a] Though the idea cannot be attributed to Lowe, nevertheless the credit for its first application and practical use belongs solely to him.

The next few days Lowe spent in making arrangements for his projected demonstration. During this time Secretary Chase continued to work in his behalf, and succeeded in arousing the interest of President Lincoln. On June 11 one of the Secretary's assistants sent Lowe encouraging news. " Secretary Chase

[61] The sources for Lowe's plan are the same as those for his actual demonstration. *Infra*, notes 73-80.

[62] New York *World*, August 21, 1860.

[63] E. B. Hunt [Captain, C of E, U. S. A.], " Modern Warfare; Its Science and Art," *New Englander*, XVIII (November, 1860), 927. See also Hunt to the Editors of the *World*, December 6, 1861, New York *World*, December 25, 1861.

[64] Morrison's invention and proposal for military use had also received the support of Professor Henry.

[65] New York *World*, October 24, 1861.

[65a] Minutes of Meeting of June 10, 1861, American Photographical Society, *American Journal of Photography*, New Series, IV (New York, July, 1861), 67; *Photographic Notes*, IV (London, August, 1861), 244.

wishes you to go up to the President's House this evening as early as you can after getting this note," was his welcome message.[66] Lowe eagerly responded and, accompanied by Professor Henry, who served as sponsor for the aeronaut's professional standing and technical ability, was graciously received by the President the same evening. Lincoln listened attentively to the proposals of the two men, and expressed a decided interest in the possibilities of balloons for war service. The interview was concluded with the President's promise of serious consideration of the plans laid before him.[67] Shortly thereafter, the War Department appropriated a sum " not to exceed $200 or $250" for Lowe's use in carrying out the tests and demonstrations.[68] Arrangements for the trials were gradually completed, and by June 13 most of the details had been settled.[69] The grounds of the Columbian Armory were selected as the place of demonstration.[70] Lowe had secured the services of George McDowell Burns, Washington Superintendent of the American Telegraph Company,[71] and one of the company's telegraphers, Herbert C. Robinson, to operate the telegraphic apparatus.[72] Only a few final arrangements remained to be effected.

The preparations for the affair aroused considerable interest and attracted much attention in various sections of the country. For several days prior to the event, the press published numerous notices containing facts, rumors, and comments concerning the coming experiments. It was rumored by some that Lowe had been engaged by the Government to make a reconnaissance and take observations of the Confederate forces at Fairfax Court House and Manassas Junction, for the particular information of General Scott. Lowe was quoted as being confident of ascertaining the exact position and strength of the enemy, their field works and artillery. Some reporters predicted that the ascensions would take place from a Federal camp west

[66] W. D. Garragher to Lowe, June 11, 1861, MS, Lowe Papers; also copy in Lowe's MS Memoirs, p. 70.

[67] Lowe's MS Memoirs, p. 71.

[68] Henry to Cameron, June 21, 1861, 3 O. R., III, 255.

[69] New York *Commercial Advertiser*, June 13, 1861.

[70] Washington *Evening Star*, June 19, 1861.

[71] New York *Tribune*, June 13, 1861.   [72] New York *World*, June 19, 1861.

of Alexandria; others maintained that the balloon would go aloft from the White House lawn. Various estimates of the altitudes to be reached were announced. The proposed use of telegraph drew much favorable comment, and the value of the balloon to the army was predicted in enthusiastic terms. Details of the apparatus to be used were frequently described, even to the weight of the wire for communication. In general the substance of the press notices and comment was favorable and expressed confidence in the success of Lowe's proposed military operations.[73]

Some official military opinion was also quoted. Brigadier General J. K. F. Mansfield, then commanding the Department of Washington, was reported to be much interested in the matter.[74] The seventy-five year old commander-in-chief, General Scott, however, thought that balloons would " not amount to much," the same correspondent declared.[75] The statement as to Scott's opinion proved to be correct, as disclosed by Lowe's later experience with the " Grand Old Man " of the army, and on June 14, Captain A. W. Whipple of the Topographical Engineers informed his chief, Major Hartman Bache, that " the General thinks a balloon of little use in this section of the country." [76]

A few other comments may be interesting. One correspondent, aroused to derision by Beauregard's recent " Beauty and Booty " proclamation,[77] announced that Lowe's demonstra-

---

[73] Boston *Journal*, June 12, 1861; Boston *Transcript*, June 12, 1861; Providence *Journal*, June 13, 1861; Providence *Evening Press*, June 12, 1861; New York *Tribune*, June 13, 1861; New York *Times*, June 12, 1861; Philadelphia *Inquirer*, June 13, 1861; Cincinnati *Commercial*, June 13, 1861; Baltimore *American*, June 13, 1861; Baltimore *Exchange*, June 13, 1861.

[74] Cincinnati *Commercial*, June 17, 1861.   [75] *Ibid.*

[76] Whipple to Bache, June 14, 1861, MS W506, Letters Received, Bureau of Topographical Engineers, War Department Division, National Archives [hereinafter cited as LRTE].

[77] Beauregard had issued on June 5 a proclamation to the people of the Virginia counties adjacent to Washington, excoriating the Federal authorities and the troops which had occupied parts of this region. The Union forces, he declared, had violated all rules of civilized warfare, and their war-cry was " Beauty and Booty."

Proclamation, " To the Good People of the Counties of Loudoun, Fairfax, and Prince William," June 5, 1861, 1 *O. R.*, II, 907.

tion would present "a novel spectacle which will puzzle General Beauregard, and perhaps call forth another proclamation." [78] And still another, a reporter on Halstead's *Commercial* confused Lowe's activities with those of La Mountain, who had just been called for service by General Butler, and stated that Lowe had been ordered to Fortress Monroe for duty there.[79] Interest was also aroused in the Confederate capital, and a notice of Lowe's proposed experiment was published in the *Dispatch* of that city.[80]

The completion of final details delayed the demonstration several more days.[81] At length, on June 18 the much discussed experiment took place at the grounds of the Columbian Armory. The *Enterprise,* fully inflated to its 20,000 cubic foot capacity, decorated with two large American flags and a British union jack, with its car equipped with reconnoitering and signalling apparatus, was prepared for the ascent. Several powerful telescopes and sets of army signal flags were included in the equipment. At the ground station was arranged a reel carrying about half a mile of fine helix wire, insulated with green silk. One end of this wire was attached to a small pocket Morse telegraph transmitting and receiving unit. The other terminal issuing from the reel was connected to a line between the War Department and the Alexandria office of the telegraph company, with an extension to the President's House attached. A battery in the War Department station furnished current for the transmission and reception of messages. When all was ready, Lowe, accompanied by Superintendent Burns and the operator, Robinson, entered the car, taking with them the pocket apparatus. The balloon was then allowed to rise several hundred feet, and the reel paid out the helices as the ground crew likewise let out the mooring ropes. At an elevation of approximately 500 feet,[82] Lowe dictated to Robinson the first

[78] New York *World*, June 12, 1861.
[79] Cincinnati *Commercial*, June 13, 1861.
[80] Richmond *Dispatch*, June 19, 1861.
[81] New York *World*, June 13, 1861.
[82] The newspaper dispatches vary in their reports of the altitude at which the experiment took place. Some gave 450 feet; others 500; and one or two announced 600 feet as the elevation. The figure adopted in the text is an

message ever sent by electric telegraph from the air, addressed
to President Lincoln:

Balloon *Enterprise,*
June 1, 1861.

To the President of the United States:

Sir:

This point of observation commands an area nearly 50 miles in diameter. The city, with its girdle of encampments, presents a superb scene.
I have pleasure in sending you this first dispatch ever telegraphed from
an aerial station, and in acknowledging indebtedness for your encouragement for the opportunity of demonstrating the availability of the
science of aeronautics in the military service of the country.

T. S. C. Lowe [83]

Several additional messages were sent to the officials in the War
Department and to General Scott.[84] In order to test the sending
range from the balloon over longer distances, more messages
were sent to Alexandria and Philadelphia over the regular telegraph lines.[85] President Lincoln was reported to have been
highly pleased with the results of the experiment, indorsing
them with the prediction that the balloon would ultimately
prove of great value in military movements.[86] He was also said
to have telegraphed a reply to Lowe's message.[87] Among the

average of the figures reported. Lowe in his report to the Secretary of War in
1863, gives the elevation as 1000 feet. Lowe to Cameron, May 26, 1863, 3
O. R., III, 254.

[83] The account of Lowe's experiment is based on detailed accounts in: Boston
Transcript, June 20, 1861; New York *Herald,* June 19, 1861; Philadelphia
Press, June 21, 1861; Cincinnati *Commercial,* June 21, 1861; Washington
Evening Star, June 19, 1861; Detroit *Free Press,* June 21, 1861. Full details
of the experiment were also published in the English scientific journal, *The
Engineer,* July 5, 1861.

Lowe's message to Lincoln is published in all the above listed papers, and
also appears in: New York *World,* June, 19, 1861; Boston *Journal,* June 20,
1861; Washington *Daily National Intelligencer,* June 20, 1861; Lowe's MS
Memoirs, p. 73; Frank Moore (ed.), *Rebellion Record,* I [Diary], 108; *Scientific
American,* June 29, 1861.

[84] Washington *National Republican,* June 21, 1861.

[85] New York *Tribune* [weekly], June 22, 1861; Boston *Journal,* June 20,
1861.

[86] Washington *National Republican,* June 20, 1861; New York *World,* June
19, 1861.

[87] New York *Morning Courier and Enquirer,* June 19, 1861; Philadelphia
Ledger, June 19, 1861; Baltimore *American,* June 19, 1861.

176

spectators who witnessed the test was Lowe's adviser, Professor Henry, who was accompanied by several colleagues from the Smithsonian. These gentlemen took a keen interest in the proceedings, and were quoted as having been entirely satisfied with the results.[88]

After the completion of the telegraphic experiment, the balloon, with Lowe and the operators still in the car, was towed through the streets to the President's grounds, where Lincoln inspected it from one of the upper windows of the White House.[89] It was then moored for the night from the White House lawn, to await renewed demonstrations the following day before the President and the cabinet.[90]

The experiments were repeated from the south lawn of the White House grounds on June 19. Several ascensions were made, and the practicability of telegraph was again demonstrated. On this occasion actual observations of Vienna and Fairfax Court House were taken, though nothing of importance was discovered.[91] Lincoln and members of his cabinet witnessed the various ascensions and experiments, and several reporters announced that the President actually made an ascension in person with Lowe, attaining an altitude of 500 feet.[92] Other correspondents reported that Lowe was also taking up passengers for fees in order to cover the expenses of the demon-

[88] New York *Times*, June 19, 1861.

[89] New York *Tribune*, June 19, 1861; Boston *Transcript*, June 20, 1861.

[90] New York *Herald*, June 20, 1861; Cincinnati *Commercial*, June 21, 1861.

[91] Boston *Journal*, June 19, 1861; Boston *Transcript*, June 19, 1861; Providence *Journal*, June 20, 1861; New York *Evening Post*, June 19, 1861; New York *Commercial Advertiser*, June 18, 1861; Philadelphia *Daily News*, June 20, 1861; Philadelphia *Inquirer*, June 20, 1861.

[92] New York *World*, June 20, 1861; Boston *Journal*, June 21, 1861.

"Lofty Position of Mr. Lincoln," was the *World's* headline announcing the President's ascent with Lowe. The same paper on June 27 published another account of Lincoln's ascent, under the heading of "What the President saw from Cloudland."

The story that Lincoln actually ascended seems scarcely credible. Had so distinguished a passenger accompanied Lowe in his ascensions, some mention of the fact would surely have appeared in the aeronaut's memoirs or in his report to Secretary Stanton. Both documents make special mention of ascents by prominent general officers and foreign dignitaries, but are silent as to the alleged ascension of President Lincoln. This fact seems to indicate that the supposed incident was either invented by the war correspondents or was simply a mistaken report.

stration.[93] Among the witnesses to the spectacle was Captain A. W. Whipple, of the Topographical Engineers, with whom Lowe was shortly to have closer contact.[94]

If the preparation for the demonstration drew notice and aroused interest from the press in various localities, the actual results of the experiments elicited even more interest. Favorable comment appeared in the columns of the Northern press all over the country.[95] Editorials praising Lowe's work and expressing certain success in military operations were published by a number of the large dailies. "A Successful and Important Experiment," ran a headline in the Washington Star; [96] "War as a Science—Important Combination of Balloon and Telegraph," was the New York Herald's announcement;[97] "Balloons are proving to be a very useful aid to the army," was the somewhat premature editorial comment of the Philadelphia Ledger.[98] Other press notices referred to the demonstration as "an entire success," [99] "a decided success," [100] and as "eminently successful." [101] One Philadelphia editor, evidently with Beauregard's recent verbal blast in mind, rejoiced that the Confederate army lacked the advantages of balloon observation,

[93] New York Tribune, June 20, 1861.

[94] Lowe to Secretary Stanton, May 26, 1863 [an official report of the operations of the Balloon Corps of the Army of the Potomac prepared at the order of the Secretary of War], 3 O. R., III, 256.

The original MS of this report, in the hand of William Jones Rhees, Chief Clerk of the Smithsonian Institution, is preserved in the War Department Division, National Archives. A copy is in the Lowe Papers. This source will be cited hereinafter as Lowe's Report.

[95] Boston Journal, June 19, 1861; Providence Evening Press, June 19, 1861; Providence Journal, June 19, 1861; New York Evening Post, June 20, 1861; New York World, June 19, 1861; New York Commercial Advertiser, June 19, 1861; New York Morning Courier and Enquirer, June 19, 1861; Philadelphia Daily News, June 20, 1861; Philadelphia Inquirer, June 19, 1861; Philadelphia Public Ledger, June 19, 1861; Lancaster Daily Evening Express, June 19, 1861; Cincinnati Commercial, June 20, 1861; Detroit Free Press, June 19, 1861; Baltimore American, June 19, 1861; Washington National Republican, June 21, 1861; Richmond Examiner, June 28, 1861.

[96] Washington Evening Star, June 19, 1861.

[97] New York Herald, June 20, 1861 [Editorial].

[98] Philadelphia Ledger, June 20, 1861 [Editorial].

[99] Philadelphia Press, June 19, 1861.

[100] New York Herald, June 19, 1861; Detroit Free Press, June 21, 1861; unidentified clipping in James Allen's Scrapbook.

[101] Boston Journal, June 20, 1861.

and with questionable wit announced that the Southern general " has at least an incalculable quantity of gas." [102] Beauregard's discomfiture at first beholding the Yankee balloon on the horizon was also surmised by a fellow editor in Washington, who asserted that

No doubt General Beauregard has looked up on the high position of Professor Lowe with considerable amazement. All his far-reaching guns will fail to reach the messenger, who, from his cloudy seat, spies out the weak points of the traitor's nest.[103]

One factor in the demonstration, the use of telegraph from comparatively high altitudes, was hailed particularly as being of great importance. Before the actual tests, there had been speculation over the feasibility of such procedure. It was questioned whether unknown atmospheric influences might interfere with the operation of the electric current, when the wires were so highly elevated. The highest elevation from the earth of any of the regular commercial lines had been about fifteen feet. The success of the trials, however, dispelled such doubts, and was greeted as an accomplishment of major importance.[104] The tactical value of the balloon and telegraph was also recognized with enthusiasm, and its advantages to an army in the field were pointed out in somewhat exaggerated terms:

What is new and valuable in Professor Lowe's experiments is the combination of the electric telegraph with the balloon, with a view to instant transmission of communication of observations made at distant points. The result of the trials made is sufficiently positive to demonstrate the full practicability of the plan. The unfortunate mischances that have occurred to our arms within the last fortnight will at once show the value of this auxiliary. Had it been in operation our troops would certainly not have been led into such ambushcades as those into which they stumbled at Big Bethel [105] and Vienna.[106]

---

[102] Philadelphia *Press*, June 24, 1861.

[103] Washington *Sunday Morning Chronicle*, June 23, 1861.

[104] Philadelphia *Press*, June 20 and 21, 1861; Washington *Evening Star*, June 19, 1861.

[105] The correspondent refers to the somewhat disastrous engagement at Big Bethel, Virginia, near Newport News and Fortress Monroe, on June 10, in which the Federal losses were 76 killed and wounded, including Major Theodore Winthrop and Lieutenant John T. Greble. Cf. official reports of both Union and Confederate commanders, 1 *O. R.*, II, 77-104.

[106] The correspondent refers to an engagement at Vienna, Virginia, on June 17,

Its present advantage, however, will be to give the head who has devised the plan of campaign the power of directing and controlling its minutest details. By means of telegraph lines established between General Scott's headquarters and the different camps, observations made at the scene of operations can instantaneously be transmitted to him, and the orders based upon them received back with equal rapidity. Thus the blunders of inexperienced subordinates will be avoided, and a celerity and security imparted to each movement of our troops that will protect them against unnecessary exposure.[107]

Similarly other editorials predicted that

With this telegraph apparatus and the means of making an aerial reconnaissance, a general may be accurately informed of everything that may be going on within a long day's march of his position in any direction.[108]

And again, it was declared that

Doubtless with the aid of powerful telescopes, the use of this balloon will enable the commander of a force to inspect thoroughly the interior of works opposed to him, and to know instantly any movements of the enemy; and that, too, without the slightest danger to those making the observations.[109]

A new interpretation of the tactical advantage of the balloon was also set forth by another correspondent, who was strangely careless with his references to his own profession:

If used with skill, this apparatus may obviously render the greatest service and help to equalize the advantages gained to the Confederacy and lost to our side by the perpetual reports of their spies and the blabbing of super-serviceable newspaper reporters.[110]

---

the day before Lowe's demonstration, in which the 1st Ohio Infantry, while making a reconnaissance in force, was ambushed by a masked battery and the 1st South Carolina Infantry, and repulsed with severe losses. Cf. official reports of the Union and Confederate commanders, 1 *O. R.*, II, 124-130.

[107] New York *Herald*, June 20, 1861 [Editorial].

[108] Boston *Transcript*, June 20, 1861; New York *Herald*, June 19, 1861. Note the similarity between this statement and that expressed by William Cooke in his pamphlet in 1783, *supra*, Chapter i, note 8.

[109] Philadelphia *Press*, June 21, 1861; Washington *Evening Star*, June 19, 1861.

[110] Unidentified clipping, dated June 20, 1861, in James Allen's Scrapbook, II, 101.

The favorable reception of Lowe's experiments and the unqualified conviction that the balloon and its auxiliary telegraph would be of valuable service to the army were not confined to the opinions of newspaper editors and their reporters. When the aeronaut's friend Halstead was endeavoring to bring influence to bear in Lowe's favor, Secretary Cameron had requested Professor Henry to witness the experiments and submit an opinion. Henry had examined Lowe's apparatus carefully, and had watched the proceedings at the armory and the White House grounds with critical attention. On June 21 he sent to Cameron a detailed report setting forth his conclusions, summarized in seven major points, all of which were highly favorable to Lowe and the plans he desired to carry out.

In the first place, Henry wrote, the balloon was well built, and once inflated with ordinary illuminating gas, would retain its charge and lifting power for several days. The second point dealt with mobility in the field. Henry declared that in ordinary weather, the inflated balloon could be towed by a detail of men along ordinary roads, over fields, to and from sources of gas, within a radius of twenty miles. Practicability and range of observation occupied the third point, in which it was pointed out that the balloon attached to captive mooring ropes could be allowed to ascend sufficiently high to permit observation within a twenty-mile radius, depending on the state of atmosphere and visibility. Night operations would also be practical, to observe the campfires of a hostile force. Henry's fourth point was an affirmation of what the Secretary already knew, that " it is conclusively proved that telegrams can be sent with ease and certainty between the balloon and the quarters of the commanding officer." The fifth concerned the possibility of free flights over the enemy lines, similar to those carried on by La Mountain later, involving the use of the prevailing upper easterly current for returning to friendly positions. Regarding these tactics, Henry stated that he felt sure, although he had not witnessed actual tests, that this system of reconnaissance could be made to work successfully. The sixth point was a thorough indorsement of balloons for military operations, and of Lowe as a competent aeronaut for this work:

From all the facts I have observed and from information I have gathered, I am sure that important information may be obtained in regard to the topography of the country and to the position and movements of an enemy, by means of the balloon now, and that Mr. Lowe is well qualified to render service in this way by the balloon now in his possession.

The final point of conclusion recommended the manufacture of a portable gas generator to accompany the balloon in the field. " The necessity of generating gas," Henry wrote, " renders the use of the balloon more expensive, but this where important results are required, is of comparatively small importance." [111] Such a complete approval of the experiments and plans must have exceeded Lowe's most sanguine hopes for official recognition.

Even before Professor Henry had written his report of conclusions to Secretary Cameron, the results of Lowe's experiments had come to the attention of Brigadier General Irvin McDowell, recently assigned to command the newly created Department of Northeastern Virginia.[112] McDowell was evidently more than casually interested in the possibilities of the new means of observation. On the day after Lowe's demonstration one of his brigade commanders who had recently reported from " most reliable sources " the presence of 20,000 Confederate troops in the vicinity of Manassas and Fairfax Court House,[113] telegraphed the assistant adjutant general that he was " entirely ignorant of the position of the rebel force." [114] McDowell saw an opportunity to employ the new instrument of war which had just gained such favorable attention, and accordingly instructed Captain Whipple, then on topographical duty with his command, to arrange for Lowe to make an ascension

[111] Henry to Cameron, June 21, 1861, MS H106 [holograph], Letters Received [Irregular], Secretary of War, War Department Division, National Archives [hereinafter cited as LRSW]; MS [copy], Lowe Papers; Lowe's MS Memoirs, pp. 71-73; also printed in 3 O. R., III, 254-255.

[112] General Orders, No. 26, AGO, May 27, 1861, 1 O. R., II, 653.

[113] Colonel Samuel P. Heintzelman to Captain James E. Fry [AAG], June 5, 1861, 1 O. R., II, 666.

[114] Heintzelman to Fry, June 19, 1861 [number omitted, packet No. 1], Telegrams Received, Department of Northeastern Virginia, War Department Division, National Archives [hereinafter cited as TRNV].

from Falls Church to take observations of the Confederate positions and activity.[115] The following day, June 20, Whipple at McDowell's headquarters at Arlington, telegraphed his chief, Major Bache at the Topographical Bureau, asking if the services of Lowe and his balloon could be obtained without delay.[116] Bache immediately replied that he would send Lowe to Arlington, and instructed Whipple to make his arrangements with the balloonist there.[117] He then instructed Lowe to report at once to McDowell, under the orders of the Chief of Topographical Engineers.[118]

Lowe thereupon proceeded to make arrangements for his first military ascension. The *Enterprise* needed reinflation, and could not be taken into the field without replenishing. This delayed matters, and on the following day, when Lowe had not arrived, Whipple telegraphed Lowe directly, ordering him to report with the balloon and telegraph instruments to Arlington.[119] Further delay was occasioned by the inability of the Washington Gas Company to furnish the necessary gas until the afternoon of the following day.[120] Lowe then notified Whipple of the difficulty, and also asked for the sum of $150 for gas and a detail of fifteen men to assist in inflation and to tow the envelope to McDowell's headquarters.[121] The request was referred to Bache,[122] who approved it and designated the 8th New York Infantry for assignment of the detail.[123] Whipple in the meanwhile telegraphed the Bureau demanding to know when Lowe should need the men and where they should report,[124] and the

[115] McDowell to Whipple, June 20, 1861, MS [number omitted], LRTE.

[116] Whipple to Bache, June 20, 1861, MS W538, LRTE.

[117] Bache to Whipple, June 20, 1861, MS Letterbooks of the Bureau of Topographical Engineers, XXII, 460, War Department Division, National Archives [hereinafter cited as LBTE].

[118] E. S. Sanford [clerk BTE] to Lowe, June 20, 1861, MS, Lowe Papers; Pass, issued by the Bureau of Topographical Engineers, for Lowe's passage to Arlington, MS, Lowe Papers.

[119] Whipple to Lowe, June 21, 1861, MS, Lowe Papers; *Lowe's Report,* 3 O. R., III, 255.

[120] *Lowe's Report,* 3 O. R., III, 255; Lowe's MS Memoirs, p. 76.

[121] Whipple to Bache, June 21, 1861, MS W542, LRTE.

[122] Whipple to Bache, June 21, 1861 [second dispatch], MS W540, LRTE.

[123] Whipple to Bache, June 21, 1861 [third dispatch], MS W543, LRTE.

[124] Whipple to Bache, June 21, 1861, 3:30 p. m. [fourth dispatch], quoted in Captain J. C. Woodruff to Lowe, June 21, 1861, 4:30 p. m., MS, Lowe Papers.

acting chief ordered the aeronaut to wire his reply at once, adding that " the sooner the balloon is on the other side of the river, the better." [125] Whipple remained at Arlington, fuming at the delay, and sent two more telegrams asking where was Lowe and why didn't he arrive with the balloon.[126]

At first glance, the assumption of authority by the military officers over Lowe at this time seems irregular and peculiar. The aeronaut had no official status, either as a civil employe of the War Department, or as a member of the military establishment. In reality he was not subject to the orders or instructions of the Topographical Bureau, or in any degree responsible to any governmental bureau or military staff. Yet he received orders and instructions from Bache and Whipple as though he were actually in the army under their jurisdiction, or were a civil employee of their bureau. But the irregularity of the proceedings becomes more understandable in view of the complete novelty of the service Lowe wished to perform, and especially from the fact he was at the time offering his services and seeking an appointment on the basis of what he could accomplish. He had to prove the utility of his apparatus before he could hope for a permanent appointment. His demonstrations in Washington having been successful, it now became necessary to show what he could do in the field; and in order to do so, it was requisite that he place himself at the disposal and subject to the instructions of the military officers.

Finally after the wires between Arlington and the Topographical Bureau's office in Washington had been kept crackling with telegrams, Lowe arrived with the *Enterprise* at McDowell's headquarters on the afternoon of June 22.[127] The movement of the balloon and its ground crew of troops excited interest in the city, and the various war correspondents quickly informed their papers of the unusual event.[128] Shortly after his

[125] Woodruff to Lowe, June 21, 1861, 4:30 p. m., MS, Lowe Papers.

[126] Whipple to Woodruff, June 21, 1861, MS W539, LRTE; Woodruff to Whipple, June 21, 1861, MS LBTE, XXII, 463; Whipple to Woodruff, June 21, 1861, MS [number omitted], LRTE; Woodruff to Whipple, June 21, 1861, MS LBTE, XXII, 462.

[127] *Lowe's Report,* 3 *O. R.,* III, 255.

[128] Boston *Journal,* June 24, 1861; Boston *Transcript,* June 24, 1861; Provi-

arrival, Lowe made an ascension from Arlington to take observations of the Confederate forces from this point. The results were reported in the press as being of no particular interest.[129]

At four o'clock the next morning, Whipple ordered the aeronautic party to Falls Church. When the balloon detail reached the line of the Alexandria and Loudoun Railroad, guards informed Lowe that the country in front into which the party was marching was not picketed by Federal troops. Since woodland prevented a change in the route, Lowe decided to proceed, and had the aerostat raised in order to reconnoitre the ground as the party advanced, keeping a sharp lookout from the car. Arriving at Bailey's Cross Roads, the party learned from local residents that a Confederate patrol had just left the vicinity, and having seen the approaching balloon, believed it to be accompanied by a large force.[130]

Whipple meanwhile appears not to have known of the balloon's whereabouts, and again telegraphed the Topographical Bureau, stating that he was sorry to lose the advantage of the day's exceptional visibility. " Do you know what prevents the ascension? " he asked his chief.[131] A short while later, about ten o'clock, Lowe's party arrived at Falls Church.[132]

Whipple had awaited the detail at Camp Tyler, near Falls Church. Despite his previous impatience at the delays, he allowed the party a rest period after their march, and deferred the ascensions until Lowe and his crew had taken refreshment.[133] After this respite, the balloon was prepared for ascen-

dence Daily Post, June 24, 1861; New York World, June 24, 1861 [two dispatches]; Philadelphia Ledger, June 24, 1861; Baltimore American, June 24, 1861.

[129] Boston Transcript, June 24, 1861 [second dispatch]; Providence Journal, June 24, 1861; New York Evening Post, June 24, 1861 [two dispatches]; New York World, June 24, 1861 [second dispatch]; New York Commercial Advertiser, June 24, 1861; Philadelphia Daily News, June 24, 1861 [two dispatches]; Philadelphia Press, June 24, 1861 [special dispatch].

[130] Lowe's Report, 3 O. R., III, 255; Lowe's MS Memoirs, p. 76; Providence Press, June 24, 1861.

[131] Whipple to Bache, June 23, 1861, MS W536, LRTE.

[132] Lowe's Report, 3 O. R., III, 255. Philadelphia Press, June 25, 1861.

[133] " Mack " [Special correspondent] to Editors of the New York Evening Post, June 24, 1861, published June 26.

sion, and Lowe's trial observations in the military service were again attempted. The fine weather of the previous day had now changed, and stiff winds prevented the attainment of sufficient elevation to take observations of any importance.[134] Towards sundown several additional attempts were made.[135] At six o'clock Lowe again went aloft, rising to an altitude of several hundred feet, but the unsteady condition of the atmosphere was still a disturbing factor. One correspondent who witnessed the ascent declared that Lowe was unable to see anything at all of importance, and asserted that he had not brought enough cable to permit higher ascension. According to this critic, the observers could not distinguish even the village of Fairfax Court House, less than eight miles distant.[136] Large clouds of dust, however, were discerned at some distance in this general direction, and were believed to indicate the presence of large bodies of cavalry.[137] Night ascensions were also tried at about 10 p. m., but neither campfires nor any unusual lights were seen in the direction of the hostile territory.[138] Another correspondent, present during the operations, attributed the failure of observation to the thickly wooded nature of the surrounding country.[139] The telegraph was again tried with complete success,[140] and communication was established from the balloon with the Department headquarters at Arlington.[141]

Whipple had accompanied Lowe in the ascensions, taking with him a subordinate topographical officer.[142] On the following day he reported to Major Bache that he had gone aloft several times, but had not been able to observe the Confederate positions because of low altitude, occasioned by wind inter-

---

[134] Boston *Transcript*, June 24, 1861; New York *Morning Courier and Enquirer*, June 24, 1861; Philadelphia *Inquirer*, June 24, 1861; Philadelphia *Ledger*, June 24, 1861 [second dispatch]; Philadelphia *Press*, June 24, 1861.

[135] Special Correspondent to New York *Evening Post*, June 24, 1861, published June 26.

[136] Washington *Evening Star*, June 24, 1861.

[137] New York *World*, June 25, 1861; Detroit *Free Press*, June 27, 1861.

[138] Washington *Star*, June 24, 1861; Philadelphia *Press*, June 25, 1861.

[139] New York *Evening Post*, June 26, 1861.

[140] New York *Times*, June 24, 1861.

[141] Captain James B. Fry [AAG] to Lowe, June 23, 1861, MS, Lowe Papers.

[142] Cincinnati *Commercial*, June 24, 1861; New York *Times*, June 24, 1861.

ference. He did not mention the factor of defilade caused by woodland.[143]

Operations were continued on the 24th, and at this time Brigadier General Daniel Tyler, then commanding a brigade at Falls Church, sent up Major Leyard Colburn of the 2nd Connecticut Infantry to make a map of the surrounding country.[144] The map he produced was reported to be remarkably correct, so much so that local residents familiar with the adjacent territory vouched for its accuracy and identified roads, streams, and even houses on it.[145] The map drew praise from General Tyler also, and caused him to alter his views considerably concerning the military value of balloons. On the same day that the sketch was made, he telegraphed General McDowell, confirming the accuracy of the map, and confessing his belief in the balloon:

I have not been much of a convert to ballooning in military operations, but the last ascent made by Major Colburn of the 2nd Connecticut Volunteers this p. m. and the map of the country, rough as it is, which he made during the ascent, convince me that a balloon may at times greatly assist military movements. By Major Colburn's map we got the location of roads not heretofore noted, which Perkins and others [i. e., other local residents] agree to be correct.[146]

Whipple was also sufficiently impressed to write to Bache, asking for an escort wagon and two horses, to facilitate the work of the balloon party.[147] The request was approved, and arrangements for the procurement of the equipment, as well as a driver, were made the same day.[148]

The results of the ascensions of June 24, other than the sketching of Major Colburn's map, do not appear in the available official reports. The various war correspondents announced that information of the enemy obtained from the air was withheld from the public, and remained confidential with the mili-

[143] Whipple to Bache, June 24, 1861, MS W543, LRTE.
[144] Lowe's Report, 3 O. R., III, 255.   [145] Boston Journal, June 27, 1861.
[146] Tyler to McDowell, June 24, 1861, MS 132, LRNV.
Mr. Perkins was a resident of the vicinity, whose family owned land known as Perkins' Hill, on the Alexandria Turnpike near Taylor's Tavern, a mile and a half from Falls Church.
[147] Whipple to Bache, June 24, 1861, MS W543, LRTE.
[148] Bache to Whipple, June 24, 1861, MS LBTE, XII, 465.

tary officers.[149] Later dispatches, however, declared that Lowe had discovered scattered camps of the enemy, the largest containing about twenty tents, in the vicinity of Fairfax Court House. This information, it was asserted, caused the commanding officer to send a troop of the 2nd U. S. Cavalry, with a battalion of Connecticut infantry, under Lieutenant Charles H. Tompkins, to make a reconnaissance in force in this direction.[150] Official dispatches concerning this movement, however, are not available.

During these aerial operations at Falls Church, the first mention of Confederate balloon activities seems to have been made. Although a Confederate signal balloon had been reported visible beyond Chain Bridge in the direction of the Leesburg Pike as early as the night of July 14,[151] the apparatus observed on this occasion might well have been some form of pyrotechnics mistaken for a balloon, and could hardly be classified as real aeronautic activity. But during Lowe's activities of the 23rd and 24th, several correspondents reported that the Confederates had an observation balloon up in the direction of Fairfax.[152] Lowe makes no mention of this occurrence in his full report of his operations during the war, and one paper declared a few days later that the supposed secessionist balloon was Lowe's *Enterprise*, mistaken by reporters for a Confederate aerostat.[153] Another journal, how-

[149] Boston *Transcript*, June 25, 1861; Boston *Journal*, June 25, 1861; Providence *Journal*, June 25, 1861; Providence *Daily Post*, June 25, 1861; New York *Morning Courier and Enquirer*, June 25, 1861; Philadelphia *Ledger*, June 25, 1861; Philadelphia *Press*, June 25, 1861; Cincinnati *Commercial*, June 26, 1861.

[150] Boston *Transcript*, June 25, 1861 [second dispatch]; Philadelphia *Daily News*, June 26, 1861; New York *Evening Post*, June 26, 1861; New York *Commercial Advertiser*, June 25 and 26, 1861; New York *Tribune*, June 26, 1861; New York *World*, June 26, 1861; Cincinnati *Commercial*, June 26, 1861 [second dispatch]; Baltimore *American*, June 26, 1861; Providence *Press*, June 26, 1861.

[151] Boston *Journal*, June 17, 1861.

[152] Boston *Transcript*, June 25, 1861 [first dispatch]; Boston *Journal*, June 25, 1861; Providence *Journal*, June 25, 1861; Providence *Daily Post*, June 25, 1861; New York *Morning Courier and Enquirer*, June 25, 1861; Philadelphia *Ledger*, June 25, 1861; Philadelphia *Press*, June 25, 1861; Cincinnati *Commercial*, June 26, 1861 [first dispatch].

[153] Boston *Journal*, June 27, 1861.

ever, asserted that it was quite probable that the Confederates were using a balloon at this time, " since several aeronauts had offered their services to the Southern Government as early as May, at a considerale time previous to the offers made to the Federal Government by Professor Lowe and others." [154] The Richmond papers, however, were entirely silent on the subject of a balloon in Confederate service, but gave considerable attention to the activities of Lowe.[155] The *Examiner* also published a notice on the potential value of balloons to both armies, adding after a brief review of earlier European experience, that " the experiments already made with them in the United States in the present war, justify the belief that they will be found eminently useful." [156] A few days later several prominent Northern papers again called attention to the presence of a balloon reconnoitering from the Confederate lines.[157] Whether or not the Southern forces actually had a balloon in operation at this time cannot be definitely determined. It can be established that some months later they did, but the exact beginning of their aeronautic operations is uncertain. On August 22 General Johnston wrote to Beauregard that " it seems to me that the balloon may be useful. . . . Let us send for it; we can surely use it advantageously." [158] From the wording of the message the deduction might be drawn that the balloon referred to was one already in service elsewhere. Or the conclusion might also be reached that the procurement of the aerostat was an entirely new project.

Although Lowe's experiments in Washington and his subsequent trial operations in the field had generally received enthusiastic praise and approval from the public at large, the aeronaut had his share of ridicule and jesting at the hands of the public press. One of the Washington papers which had hailed

[154] Providence *Daily Post*, June 28, 1861.

[155] Richmond *Enquirer*, June 27, 1861 [three dispatches] ; *ibid.*, July 11, 1861; Richmond *Examiner*, June 27 and 28, 1861; Richmond *Dispatch*, June 27, 1861.

[156] Richmond *Examiner*, June 26, 1861.

[157] Boston *Transcript*, July 5, 1861; Boston *Journal*, July 5, 1861; New York *Evening Post*, July 5, 1861; New York *Commercial Advertiser*, July 5, 1861; Philadelphia *Daily News*, July 5, 1861; Philadelphia *Press*, July 5, 1861.

[158] Johnston to Beauregard, August 22, 1861, Catalogue of MSS offered for sale by Stan V. Henkels, October, 1915, Beauregard Papers, Division of MSS, Library of Congress.

his experiments at the Armory grounds with generous praise, humorously reported that

a late rumor seriously believed by some individuals is that General Scott went up in Professor Lowe's balloon and discovered Jeff Davis, Lee, and Beauregard at breakfast together at Manassas Junction, with a mere sprinkling of 80,000 men in the tented fields surrounding that locality.[159]

Another journal jocularly suggested that Lowe be given the imposing title of " Brigadier General of the Skies," [160] and even Halstead's *Commercial,* despite its editor's efforts on Lowe's behalf, chaffed his protegé with a story that an army in balloons, designated as the " Flying Aerials," would set out from " Camp Ohwhatawhopper " to relieve Fort Pickens in Florida.[161] Another wag's handiwork in the form of a comic poem ridiculing Lowe with a pun directed at the balloonist's name was also displayed in the *Commercial's* columns.[162] This latter jest, however, was soon parried in verse form by one of Lowe's admirers.[163] Another writer on Halstead's paper also submitted a letter containing grotesquely comic " war news," in which the Union generals were freely chaffed, and Lowe's balloon likewise received a full portion of ridicule. A mocking imitation of Lowe's first aerial message to President Lincoln was included.[164] The reported balloon activities of the Confederates in turn produced more humorous comments, to the effect that Lowe and his enemy aeronaut were

[159] Washington *Star,* June 26, 1861.
[160] Boston *Transcript,* June 25, 1861.
[161] Cincinnati *Commercial,* June 27, 1861.
[162] Cincinnati *Commercial,* June 28, 1861. The verse, written anonymously under the title of " On War Balloons," was not complimentary to Lowe:

> Professor Lowe would fain get high
> At Government expense;
> With big balloon he'd scale the moon
> To spy Virginia fence;
> To spot the camps of rebel scamps
> With telegraph and glass—
> You ask me, friend, how will this end?
> And I reply—in gas!

[163] Cf. verse, " Balloons," reprinted in E. G. Squire (ed.), *Leslie's Pictorial History of the American Civil War* (New York, 1862), p. 80.
[164] Cincinnati *Commercial,* July 5, 1861.

190

preparing to fight a duel from their respective balloons,[165] and
a Philadelphia correspondent remarked that a bullet piercing
either of the balloons would " let out as much gas as if it went
through the head of Governor [Henry A.] Wise.[166]

Having completed the ascensions desired by General
McDowell at Falls Church, Lowe returned with his balloon to
Washington on June 25.[167] Here, he hoped with the backing
previously enjoyed, and with the results of the practical demonstrations he had made both at the Capital and in the field, to
secure an appointment with the army and to carry out his plans
for an aeronautic branch of the military service.

At first, it appeared that his hopes would be realized. On
June 26, Whipple notified him that the Topographical Bureau
had decided to adopt the balloon for army service, and
requested that he submit a full report on his proposed methods
of operation, and an estimate of construction costs.[168] Lowe
complied at once, not knowing that the Topographical Bureau
had negotiated some time before with the Lancaster aeronaut
John Wise on the same subject.[169] Much to his disappointment,
Lowe learned when he later called on Whipple to ask for a
decision, that the Bureau officials had decided to give the order
for a government balloon to Wise. The latter aeronaut,
Whipple informed Lowe, had submitted an estimate lower by
almost $200, but suggested that Lowe might well be employed
to operate the new balloon, even though he had not been given
the order to build it. Although Wise was senior to Lowe in the
profession by more than twenty years, the latter refused to
recognize the value of his long experience, and firmly declined
to consider the suggestion. " I would not be willing to expose
my life and reputation by using so delicate a machine . . . made

[165] Boston *Transcript*, July 1, 1861, quoting a dispatch from the Louisville
Journal.
[166] Philadelphia *Ledger*, July 1, 1861.
[167] Boston *Journal*, June 26, 1861; Boston *Transcript*, June 26, 1861; Providence *Daily Post*, June 26, 1861; Providence *Journal*, June 26, 1861; New York
Morning Courier and Enquirer, June 26, 1861; New York *World*, June 26, 1861
[second dispatch]; Philadelphia *Ledger*, June 26, 1861; Philadelphia *Press*,
June 26, 1861; Baltimore *American*, June 26, 1861 [second dispatch].
[168] *Lowe's Report*, 3 *O. R.*, III, 255.   [169] See Chapter iii, *supra*, pp. 60-61.

by a person in whom I had no confidence," was his blunt answer.[170] Whipple accordingly considered the matter closed, and employed Wise for service with the army in the coming campaign in Virginia.

Lowe, however, was still convinced that he could eventually gain recognition and an appointment. He revarnished the envelope to give it greater retentive qualities, and then at his own expense began a new series of demonstrations at the grounds of the Smithsonian.[171] A number of scientific men from this organization and army officers, among whom was Whipple, witnessed the new trials. Lowe records that the majority of these individuals strongly recommended that his system be adopted into the service.[172]

In the meanwhile preparations for General McDowell's grand movement into Virginia were nearing completion, and Wise was expected daily to report for duty with the army. But the construction of the balloon had taken longer than had been expected, and when Wise had failed to report by July 17, Whipple communicated with Lowe, directing him to join the advancing Union column.[173]   Again in this instance, it appears curious and irregular that the military officers ordered Lowe on active service when he had no status or official connection with the army. At the time, Lowe was unprepared to take the field on such short notice, and with Whipple away from Washington with the advance elements of the army, the aeronaut had difficulty in obtaining a detail to assist with inflation and transportation. He employed assistants independently to prepare the balloon for the field, and was in process of inflating it when Wise arrived in Washington. The director of the gas company then informed Lowe that Wise's balloon would be used in the campaign, and that his would not be required. Keenly disappointed, he ceased all preparations for field service.[174]

[170] *Lowe's Report,* 3 *O. R.,* III, 256.

[171] *Lowe's Report,* 3 *O. R.,* III, 256; Philadelphia *Press,* July 11, 1861; New York *Herald,* July 14, 1861. Lowe to Bache, July 29, 1861, MS [number omitted], LRTE; also MS [copy], Lowe Papers; also printed in 3 *O. R.,* III, 256.

[172] *Lowe's Report,* 3 *O. R.,* III, 256.   [173] *Lowe's Report,* 3 *O. R.,* III, 256.

[174] *Ibid.;* Lowe to Bache, July 29, 1861. See references in note 171.

When the news of Wise's failure to reach McDowell's army became known, Lowe resolved to take his balloon to the front on his own account. This resolution was thoroughly justified by Whipple's previous instructions issued before Wise reached Washington.[175] Colonel William F. Small, of the 26th Pennsylvania Infantry, assisted by assigning a detail of twenty men to aid with inflation and to tow the balloon. Lowe reached Falls Church on the evening of July 21, only to meet the avalanche of McDowell's rout back to Washington. It was useless to go on. Lowe decided, however, to remain here, even after most of the shattered army had streamed past, thinking that he might render service by obtaining information of the approach of any hostile pursuing columns that might be in motion. But when the pickets supporting the remnants of the Union rearguard were withdrawn, he left Falls Church for Arlington, and arrived at Fort Corcoran with his inflated balloon at eight o'clock the next evening.[176]

He remained at Fort Corcoran until the morning of July 24, and learning of the excitement and panic in Washington because of the supposed advance of the victorious Confederate army, made a free ascension from the fort shortly after daylight, to observe the reported movement. Reaching an altitude he asserts to have been over three miles, Lowe observed massed Confederate bivouacs at Manassas, and some movement of troops between the junction and Fairfax.[177] A considerable force was also noted encamped at Fairfax Court House, one field containing about 200 tents. Cavalry videttes were seen on the road a mile from the Court House towards Washington, and two regiments of infantry with a supporting body of horse

[175] It seems strange that Lowe did not go on to the army on the basis of Whipple's original instructions, despite the arrival of Wise and the instructions of the Gas Company official. The original of Whipple's order to Lowe, however, is missing, and it may well have contained a provision rescinding the orders if Wise should arrive in time. Such instructions may also have been given to the director of the Gas Company.

[176] Lowe to Bache, July 29, 1861; see references in note 171. Baltimore *American*, July 27, 1861; Cincinnati *Commercial*, July 29, 1861.

[177] Lowe to Bache, July 29, 1861. See references in note 171. Boston *Transcript*, July 26, 1861; Providence *Post*, July 27, 1861; Providence *Press*, July 25, 1861.

were between Centreville and the Court House.[178] There seemed to be no concentrated movements of troops, however, and the rumor of the hostile army being in full march on the Capital proved to be without foundation. This information, reported to the authorities next day, Lowe declares, restored confidence and aided in dispelling the prevailing panic.[179]

The reason for cutting loose from his cables is not clear. Possibly he felt that the importance of the occasion required a higher elevation for reconnaissance than he could reach by captive ascension. Regardless of reasons, Lowe very nearly came to an untimely end. The easterly current which he entered at his high elevation carried him in front of the Federal picket lines near Alexandria, and descending, he was mistaken by the outposts for the supposed Confederate balloonist whose activities had again been reported early in July. The troops opened fire on the balloon as it neared the earth, and Lowe heard faint shouts of " Show your colors," followed by the vicious whistling of minié balls. Having no national flag with him, he released ballast and continued on, landing on the Mason plantation over two miles outside the Federal lines. With the assistance of a party from the 31st New York Infantry, Lowe brought his damaged balloon back to Washington and reported his observations.[180] As a result of the attack on his balloon by friendly troops, he sent to the War Department a statement emphatically denying that the Confederates possessed any balloons or aeronauts, and assured the officials that all " air vessels seen thus far, and probably all that will be seen hereafter, are for Union purposes." [181] A public statement to this effect was also prepared for insertion in the Washington press.[182]

On the afternoon following the free ascent, Lowe's hopes for

---

[178] Boston *Journal*, July 25, 1861; Boston *Transcript*, July 25, 1861; Providence *Post*, July 26, 1861; Cincinnati *Commercial*, July 29, 1861; Baltimore *American*, July 27, 1861; New York *Herald*, July 25, 1861.

[179] *Lowe's Report*, 3 *O. R.*, III, 258.

[180] *Ibid.*; New York *World*, July 25, 1861.

[181] Lowe's statement to the War Department, July 25, 1861, MS, Lowe Papers.

[182] Copy of Lowe's statement to the War Department, with title, " The Balloon of Wednesday," evidently prepared as a press release, MS, Lowe Papers. The writer has been unable to find this statement published in any of the available Washington papers of the period.

an army appointment were raised by the receipt of a message from President Lincoln, directing him to come to the executive mansion that evening.[183] Lincoln had interviewed him previously, immediately following the experiments on the White House grounds, and had expressed interest in Lowe's plans for an aeronautic branch of the military service. On this former occasion he had caused his secretary to give Lowe a letter of introduction to General Scott, but the septuagenarian general-in-chief had paid no attention to the aeronaut's request for a hearing. The "Grand Old Man" of the army, Lowe later wrote concerning the incident, "was now very infirm, pompous, and with many of the affectations that sometimes go with extreme age, and it was extremely difficult to engage his attention. His mind was centered on the make-up of an army as he had always known it, and he did not care for innovations. It was evident that the General of the Army had no interest in a balloon corps." [184] During this second interview at the White House, the President, whom Lowe found deeply distressed over the recent disaster at Manassas, suggested that perhaps the outcome of the battle might have been different had McDowell had accurate information of the Confederate positions and movements. A balloon observer, he mused, might well have given him such information. "Professor," Lincoln concluded, "I wish you would go and confer with General Scott again." [185] Lowe must have related his past unsuccessful experience with that worthy officer, whereupon the President took a card and wrote in his customary irregular hand:

Will Lieut. Genl. Scott please see Professor Lowe once more about his balloon?

A. Lincoln [186]

July 25, 1861.

The following morning Lowe called at Scott's headquarters

---

[183] Lowe's MS Memoirs, p. 78.

[184] *Ibid.*, p. 74. Lincoln was particularly interested in Lowe's plan for directing artillery fire by aerial observation.

[185] Lowe's MS Memoirs, p. 79.

[186] Lincoln to Scott, July 25, 1861, MS [holograph], Lowe Collection, Aircraft Building, United States National Museum [Smithsonian Institution].

and sent in the President's card. An orderly returned with it and stated that the General was engaged and could not see Mr. Lowe. A few hours later Lowe again called, only to receive the same answer. Returning several hours later, he was informed that General Scott was eating his lunch. A fourth attempt to gain the interview requested by the President of the United States drew forth the answer that the General of the Army was asleep and could not be disturbed. This was the last straw.

Fuming with indignation and annoyance, Lowe returned to the White House and reported the result of his repeated visits. Lincoln, so Lowe reports in his memoirs, first looked at him sharply; then he laughed, rose to his angular height of more than six feet, clapped on his fuzzy old stovepipe, and said, "Come on." [187] The two marched straight to General Scott's headquarters, where the sentry who had just refused Lowe admission four times in twice as many hours, hastily turned out the guard for the President. In the interview that followed, Scott was courteous to a fault, and promised to take official action on the question of balloons at once.[188]

The General kept his promise. The next day, July 26, Major Bache telegraphed Whipple at Arlington: "Report at the Bureau at once—subject: balloon matters." [189] Whipple replied that he would be unavoidably detained on special duty,[190] whereupon Bache instructed him to report early next morning.[191] The conference in Bache's headquarters next day resulted in the long-desired offer of government employment for which Lowe had so earnestly hoped. The numerous setbacks and delays, Lowe recorded, "were maddening. History was being made every hour, and I was chafing to get into action." [192] But again, the terms which accompanied the offer, though ostensibly generous, were not such as fitted Lowe's plans. Whipple telegraphed on the 29th that if Lowe would repair and inflate his balloon and bring it to Arlington, the Government would

[187] Lowe's MS Memoirs, p. 80.   [188] Ibid., pp. 80-81.
[189] Bache to Whipple, July 26, 1861, MS LBTE, XXII, 512.
[190] Whipple to Bache, July 27, 1861, MS [number omitted], LRTE.
[191] Bache to Whipple, July 27, 1861, MS LBTE, XXII, 513.
[192] Lowe's MS Memoirs, p. 78.

engage him at the rate of $30 per day for every day that the balloon should be used, and also pay for the gas for inflation. A detail of twenty men would be provided as a ground crew, and he was expected to report within twenty-four hours.[193]

Such an arrangement did not appeal to Lowe for several reasons. The amount of pay was definitely tempting, but he looked to regular service that would lead to the formation of an organized balloon corps, and not employment involving per diem status and ascensions at irregular intervals. It was obvious also, from the rate of pay, more than that of a major general of the line, [194] that Whipple did not expect him to perform duty every day. Common sense told him that not even in wartime would the military bureaucracy employ a civilian balloonist permanently at a figure greater than that of the commanding officer of an army. Accordingly, Lowe answered that he could not accept the offer on the terms proposed, but countered with the suggestion that the Government authorize him to build a balloon specially designed for field service, and pay him at the rate of $10 per day in regular service. On these terms he guaranteed successful results, and submitted a statement of costs and time of construction.[195]

Without waiting for a reply, however, Lowe complied with Whipple's instructions, repaired and inflated his balloon, and set out for Arlington with a detail of inexperienced volunteers. The way was rough and the party was unfortunately overtaken by a thunderstorm, which, because of the inexperience of the ground crew, caused Lowe to discharge gas to prevent serious damage.[196] Whipple's version of the affair was different. Impatient with the recent misfortunes that had marked Wise's operations, he telegraphed Captain Woodruff at the Topo-

---

[193] Whipple to Lowe, July 29, 1861, MS, Lowe Papers; also printed in 3 *O. R.*, III, 258.

[194] The base pay of a major general at this time was $7.33 per day, with an allowance of $14.17 for subsistence, servants, etc., making a total of $21.50. Cf. *Pay Table, United States Army, 1863-1865* [compiled by J. Lowenthal under direction of the Chief Paymaster, U. S. A.], Washington, 1865.

[195] *Lowe's Report,* 3 *O. R.*, III, 258.

[196] *Ibid.*; Henry to Whipple, August 2, 1861, MS W918, LRTE; also printed in 3 *O. R.*, III, 258-259.

graphical Bureau that " Lowe's balloon filled at our expense has burst. I wish to have nothing more to do with it." [197]   With this unfortunate turn of affairs, it appeared that Lowe's long-cherished appointment had been lost at the very point of realization.

Lowe's disappointment was bitter. He returned to Washington and related his misfortune to his firm friend and adviser, Professor Henry, who promptly wrote to Whipple in an effort to persuade him to change his hasty decision.

I regret very much [Henry wrote] to learn from Mr. Lowe that you think of giving up the balloon operations, and I write to express the hope that you will make further attempts.   A single successful observation will fully repay all that you have yet expended.

The experiment of Wednesday was rendered abortive by the accidental occurrence of a thunderstorm which could not be foreseen.   At this season of the year thunderstorms generally occur in the after part of the day or night, and I would therefore advise that the balloon be filled immediately after the clearing off of the sky and then used as soon as possible after daylight the next day.

Mr. Lowe came to this city with the implied understanding that, if the experiments he exhibited to me were successful, he would be employed.   He has labored under great disadvantages, and has been obliged to do all that he has done, without money. From the first he has said that the balloon he now has was not sufficiently strong to bear the pressure of a hard wind, although it might be used with success in favorable situations and in perfectly calm weather.   I hope that you will not give up the experiments and that you will be enabled, even with this balloon, to do enough to prove the importance of this method of observation, and to warrant the construction of a balloon better adapted to the purpose.[198]

Whipple, with due regard for Henry's position, as well as the ingrained procedure of " through channels," referred the letter to his immediate chief.[199]   Major Bache had a profound respect for the Smithsonian Institution, and for Professor Henry, who knew and collaborated with the Major's nephew, Dallas Bache, of the Coast Survey.   Perhaps Major Bache was also influenced

[197] Whipple to Woodruff, August 1, 1861, MS W912, LRTE.

[198] Henry to Whipple, August 2, 1861, MS W918, enc., LRTE; also printed in 3 O. R., III, 258-259.

[199] Whipple to Bache, August 2, 1861, MS W918, LRTE.

by the letter of recommendation on Lowe's behalf he had recently received from Dr. Cresson, of the Franklin Institute, who had warmly declared that the aeronaut was " unsurpassed " in the requisite qualifications for military aeronautics.[200] Accordingly, when Lowe called on Bache on August 2, to make a final appeal for a reconsideration of the matter, the major received him courteously and favorably reconsidered his proposals.[201] He ordered Whipple to employ Lowe to build a new balloon for the Government, designed especially for the rigors of field service, constructed of the best India silk according to an accepted sample. Lowe was to be paid $5 per day during the period of construction, and as soon as he should have the balloon in operation, this rate was to be increased to $10.[202] Whipple prepared the formal papers of authorization the same day,[203] and Lowe's long struggle for a post in the military service was ended. He had finally gained his first objective, and was now a military aeronaut employed with official status. It remained for him to justify the confidence that his friends and backers had placed in him. Also, there lay before him the task of which he had dreamed during his journey back from war-fevered South Carolina more than three months before: the organization of aeronautics as a branch of the military service.

[200] Cresson to Bache, July 1, 1861, MS C556, LRTE.

[201] *Lowe's Report,* 3 *O. R.,* III, 259.

[202] Bache to Whipple, August 2, 1861, MS LBTE, XXIII, 1; also MS [copy] L100, LRSW (1864).

[203] Whipple to Lowe, August 2, 1861, MS, Lowe Papers; also printed in 3 *O. R.,* III, 259.

PLATE I

Louis Bernard Guyton de Morveau
Early advocate of military balloon reconnaissance,
French Revolutionary Wars, 1794-1798

From a print in the Tissandier Collection, Division of Aeronautics, Library of Congress

PLATE II

COLONEL JEAN MARIE-JOSEPH COUTELLE

Commander of the French Balloon Companies in the
Revolutionary Wars, 1794-1798

From a print in the Tissandier Collection, Division of
Aeronautics, Library of Congress

PLATE III

BALLOON *Entreprenant* DURING THE BATTLE OF FLEURUS, JUNE 26, 1794

From a copy of an original watercolor by Nicolas Conté, Tissandier Collection,
Division of Aeronautics, Library of Congress

PLATE IV

NICOLAS CONTÉ
Director of École Nationale Aérostatique
From a print in the Tissandier Collection, Division of Aeronautics, Library of Congress

PLATE V

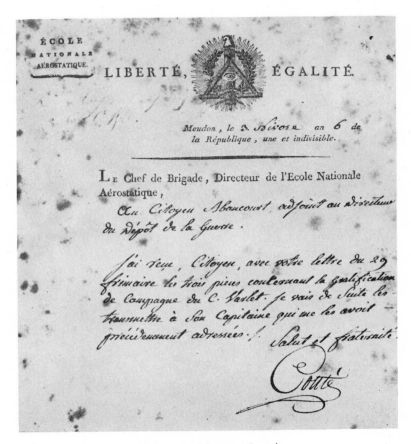

LETTER OF NICOLAS CONTÉ
Note the formal letterhead of the *École Nationale Aérostatique*
From the original MS in the Tissandier Collection, Division of Aeronautics, Library of Congress

PLATE VI

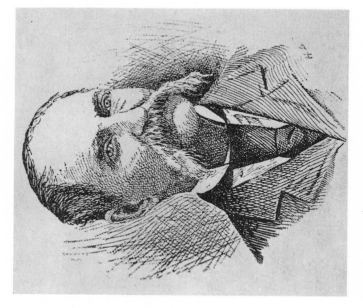

JAMES ALLEN
Aeronaut, Army of the Potomac

From a drawing in the Division of Aeronautics, Library of Congress

LIEUTENANT (later Maj. Gen.) HENRY L. ABBOT
In charge of Allen's balloon experiments, June-July, 1861

Collection of U. S. Army Signal Corps, No. BA88

PLATE VII

JOHN WISE

Aeronaut with the Federal Army, July-August, 1861

From an engraving in Wise, *System of Aeronautics*

PLATE VIII

MAJOR ALBERT J. MYER

In charge of Wise's balloon detachment in the first Manassas Campaign

Collection of the U. S. Army Signal Corps, No. BA35

CAPTAIN AMIEL W. WHIPPLE

In charge of the experimental balloon operations, June-September, 1861

From a portrait in Miller's *Photographic History of the Civil War*

PLATE IX

JOHN WISE

From a copy of a contemporary woodcut in the Division of Aeronautics, Library of Congress

PLATE X

JOHN LA MOUNTAIN

Aeronaut, Department of Virginia, and Army of the Potomac,
July, 1861–February, 1862

From a copy of a contemporary woodcut in the Division of Aeronautics,
Library of Congress

MAJOR GENERAL BENJAMIN F. BUTLER

Early advocate of balloon reconnaissance and in charge
of La Mountain's operations at Fortress Monroe

Collection of U. S. Army Signal Corps, No. BA1143

PLATE XI

U. S. S. *Fanny* (foreground)
Vessel used in La Mountain's ascensions from Hampton Roads, 1861
Collection of Naval Records and Library, No. OS11607

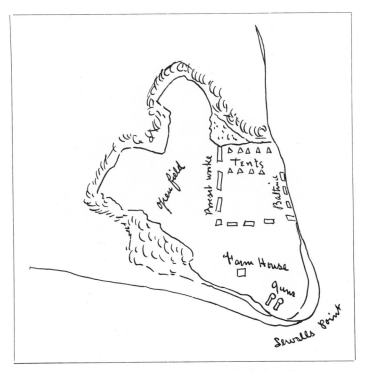

AERIAL SKETCH MAP OF SEWALL'S POINT, VA.
Drawn by John La Mountain from an elevation of 3,500 feet,
August 10, 1861
Misc. File, " Balloon Corps," War Department Division, National Archives

PLATE XII

LA MOUNTAIN'S BALLOON *Atlantic* IN ASCENSION NEAR FORTRESS MONROE

From a war correspondent's drawing in *Leslie's Illustrated Newspaper*, August 31, 1861

PLATE XIII

T. S. C. Lowe
Period of the trans-Atlantic project
From an original photograph in the Lowe Papers

PLATE XIV

INFLATION OF THE TRANS-ATLANTIC BALLOON *Great Western*

September, 1860

Lowe Collection, U. S. National Museum, No. 30917-A

PLATE XV

INFLATION OF THE BALLOON *Enterprise*
Cincinnati, April 20, 1861
Lowe Collection, U. S. National Museum, No. 30917-B

PLATE XVI

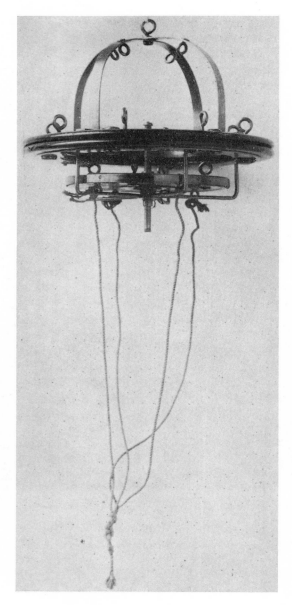

VALVE OF THE BALLOON *Enterprise*

Lowe Collection, U. S. National Museum, No. 30915-A

PLATE XVII

PROFESSOR JOSEPH HENRY

Secretary of the Smithsonian Institution
Lowe's friend and advisor throughout the war years

From a portrait in U. S. National Museum

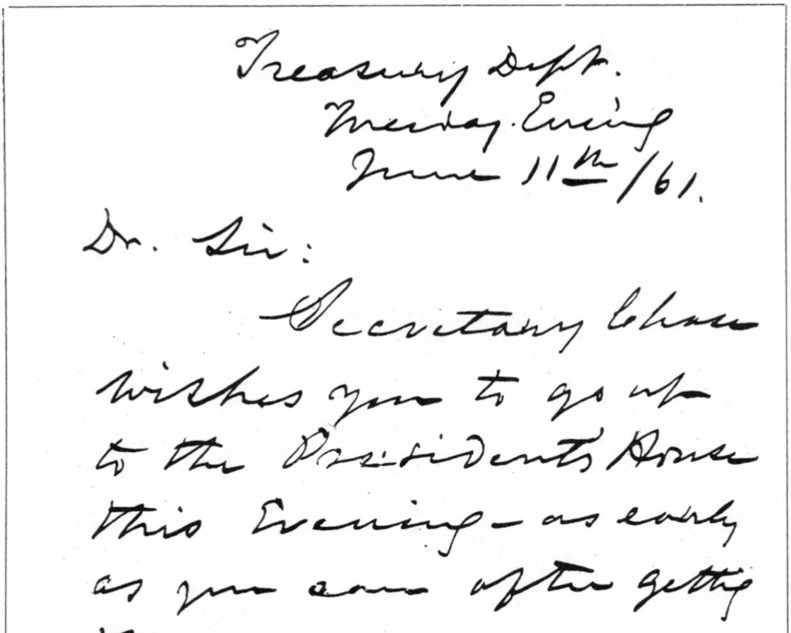

PLATE XVIII

Treasury Dept.
Monday Evening
June 11th /61.

Dr. Sir:

Secretary Chase wishes you to go up to the Presidents House this Evening — as early as you can after getting this note.

Respectfully,
W. D. Gallagher.

Prof. Lowe,
National Hotel.

SECRETARY CHASE'S MESSAGE TO LOWE

Lowe Collection, U. S. National Museum, No. 30917-G

PLATE XIX

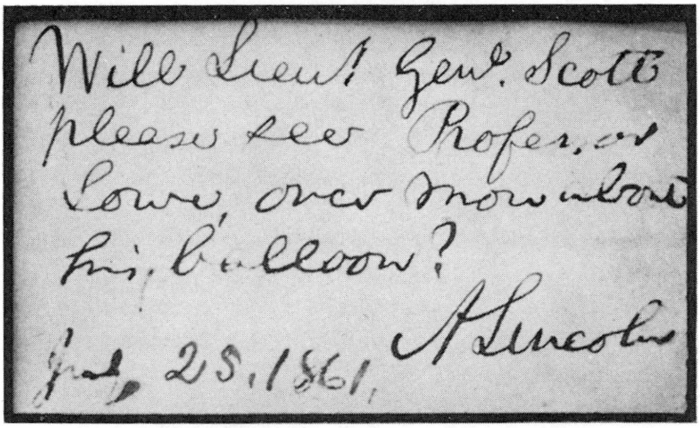
Will Lieut. Genl. Scott please see Professor Lowe, once more about his balloon?
A. Lincoln
July 25, 1861.

President Lincoln's Card to General Scott

Lowe Collection, U. S. National Museum, No. 30915-L

PLATE XX

MAJOR GENERAL FITZ JOHN PORTER

Collection of U. S. Army Signal Corps, No. B5608

MAJOR GENERAL GEORGE B. MCCLELLAN

Collection of U. S. Army Signal Corps, No. B4264

PLATE XXI

BALLOON *Union* AT GENERAL MCDOWELL'S HEADQUARTERS, OCTOBER, 1861

From a sketch by Ed. Pietsch, War Correspondent of *Harper's Weekly*, published October 21, 1861

PLATE XXII

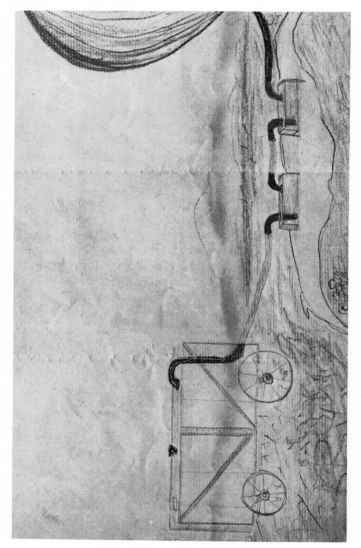

LOWE'S ORIGINAL ROUGH SKETCH OF GENERATING APPARATUS

Showing mobile generator, purifier, cooler, and couplings

Lowe Collection, U. S. National Museum, No. 30914-A

PLATE XXIII

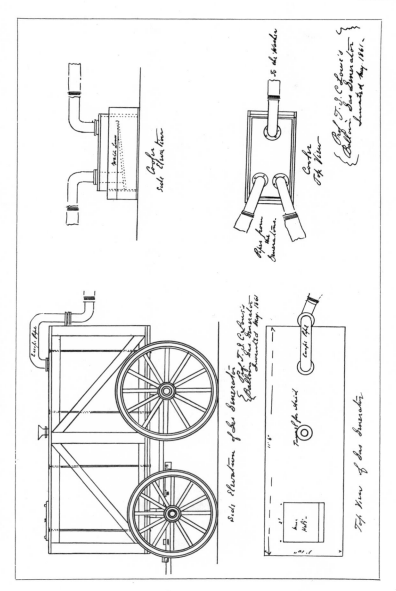

SCALE DRAWINGS OF GENERATOR AND COOLER

Drawn by J. M. Bancroft, January, 1862

Lowe Collection, U. S. National Museum, No. 30914

PLATE XXIV

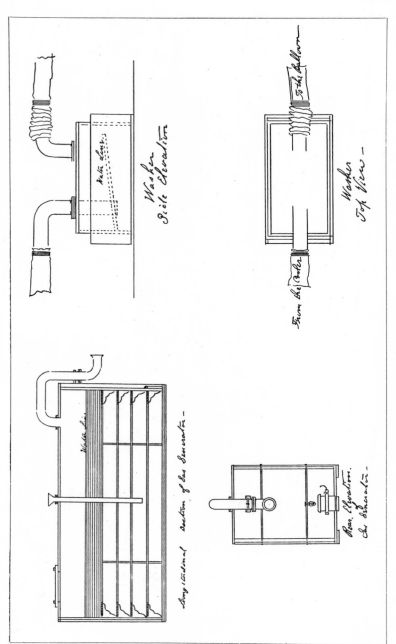

SCALE DRAWINGS OF GENERATOR AND PURIFIER

Drawn by J. M. Bancroft, January, 1862

Lowe Collection, U. S. National Museum, No. 30914-C (left), and 30913-G (right)

PLATE XXV

Brady Photograph of Complete Generating Apparatus

Showing mobile generator, cooler, force pump, purifier, and couplings

Lowe Collection, U. S. National Museum, No. 30914-B

PLATE XXVI

VALVE OF THE TRANS-ATLANTIC BALLOON *Great Western*

Lowe Collection, U. S. National Museum, No. 30915

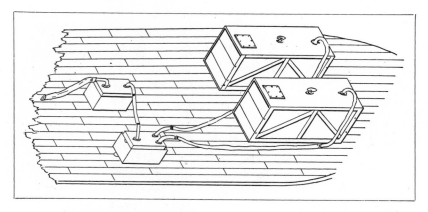

GENERATORS MOUNTED ON DECK OF BALLOON-BOAT *G. W. Parke Custis*

Drawn by J. M. Bancroft, January, 1862

Lowe Collection, U. S. National Museum, No. 30913-E

PLATE XXVII

BALLOON-BOAT *G. W. Parke Custis*, WITH BALLOON *Washington*
IN ASCENSION NEAR BUDD'S FERRY, MARYLAND

Lowe Collection, U. S. National Museum, No. 30913-A

PLATE XXVIII

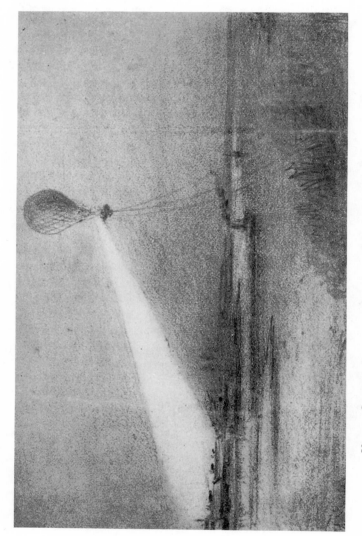

USE OF CALCIUM LIGHT DURING NIGHT ASCENSIONS, LOWER POTOMAC RIVER

From an original pencil sketch by Arthur Lumley, Lowe Papers

PLATE XXIX

T. S. C. Lowe

Collection of the U. S. Army Signal Corps, No. B2514

PLATE XXX

MAJOR GENERAL ANDREW ATKINSON HUMPHREYS

In charge of the Balloon Corps during the Peninsular Campaign, April-August, 1862

Collection of the U. S. Army Signal Corps, No. BA353

MAJOR GENERAL GOUVERNEUR K. WARREN

In charge of the Balloon Corps, June, 1863

Collection of the U. S. Army Signal Corps, No. B4140

PLATE XXXI

CAPTAIN (LATER BRIGADIER GENERAL) CYRUS B. COMSTOCK

In charge of the Balloon Corps, April-June, 1863

Collection of the U. S. Army Signal Corps, No. 94290

PLATE XXXII

COLONEL (later Maj. Gen.) RUFUS INGALLS
In administrative charge of the Balloon Corps,
March, 1862–April, 1863
Collection of U. S. Army Signal Corps, No. BA262

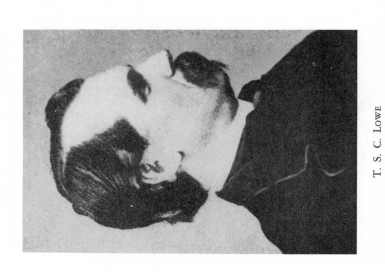

T. S. C. LOWE
A post-war portrait
From a photograph in the Lowe Papers

PLATE XXXIII

MAJOR GENERAL ANDREW A. HUMPHREYS

MAJOR GENERAL FITZ JOHN PORTER

From original Brady photographs in the Lowe Papers

JOHN B. STARKWEATHER

PLATE XXXIV

T. S. C. LOWE

Period of the Peninsular Campaign

From an original photograph in the Lowe Papers

PLATE XXXV

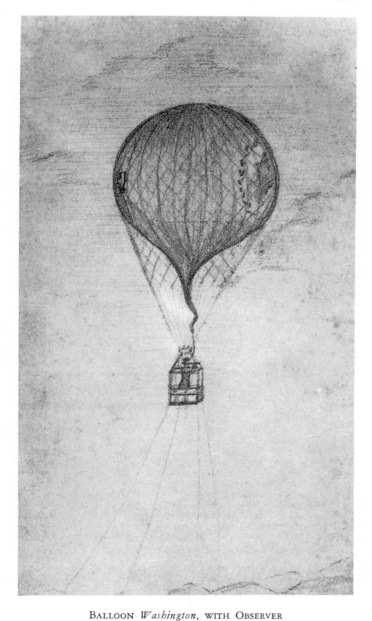

BALLOON *Washington*, WITH OBSERVER

From an original war correspondent's pencil sketch in the Lowe Papers

PLATE XXXVI

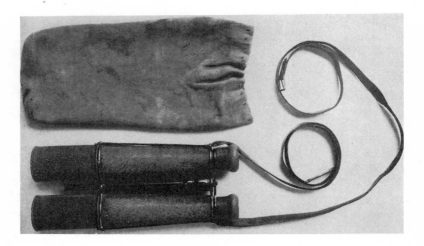

LOWE'S FIELD GLASSES

In the Lowe Collection, U. S. National Museum, No. 30915-B

In the case with these glasses, before they were presented to the
National Museum, was kept the following notation
in Lowe's hand:

This double telescope was the companion of Prof. Lowe during his
three years' service as Chief Aeronaut in the Army of the Potomac—and
through which he viewed every battle of that famous army from the
commencement of the Rebellion in 1861 to the battle of Gettysburg in
1863. Nearly all the prominent generals of the army have taken
observations through this glass, among whom were McClellan and Fitz
John Porter, Generals Smith, Sedgwick, Griffin, Berry, Reynolds, Heintzelman, Stoneman, Upton, Whipple, Humphreys, Ingalls, Hooker, and
scores of others. Also prominent foreign visitors have looked through
this glass from the balloons, among whom, the Count of Paris and
Duc de Chartres made daily observations while on the Peninsula,
Prince de Joinville, Captain Beaumont of the Royal Engineers, etc. etc.

PLATE XXXVII

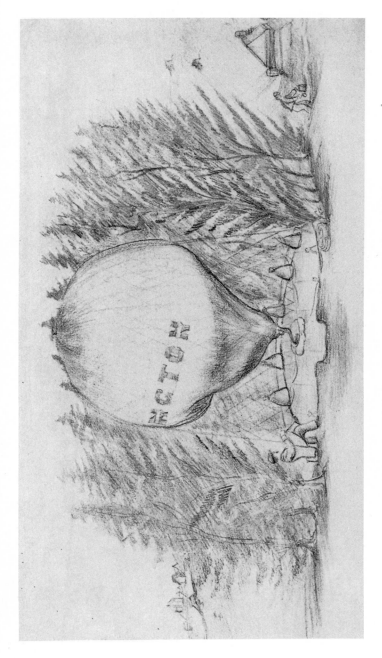

BALLOON *Washington* ENCLOSED IN ARTIFICIAL WIND SCREEN

These screens were made by cutting stout evergreens which were planted in trenches about the point of inflation or anchoring area, thus affording protection from wind

From an original war correspondent's pencil drawing in the Lowe Papers

PLATE XXXVIII

HEADQUARTERS OF THE CHIEF AERONAUT AND BALLOON CORPS, BEFORE YORKTOWN, 1862

Lowe is shown seated under the fly of the wall-tent

From an original pencil drawing by Arthur Lumley in the Lowe Papers

PLATE XXXIX

CAMP LOWE

Headquarters of the Balloon Corps at White House, Virginia, May 18, 1862, Peninsular Campaign

From an original pencil drawing by Arthur Lumley in the Lowe Papers

PLATE XL

T. S. C. LOWE

Period of the Peninsular Campaign

From an original photograph in the Lowe Papers

PLATE XLI

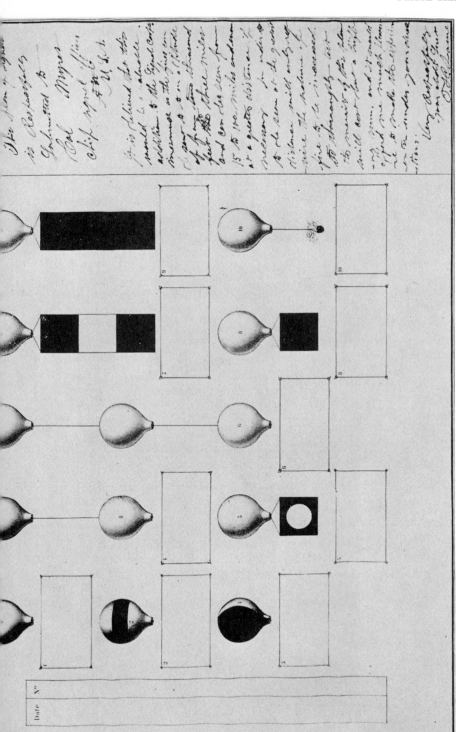

PLAN OF LONG DISTANCE AERIAL SIGNALLING   Proposed by Lowe in 1863   Lowe Collection, U. S. National Museum, No. 30913-D

PLATE XLII

EBENEZER SEAVER AND JOHN B. STARKWEATHER
(left)   (right)

From an original photograph presented to Lowe by Seaver and Starkweather, 1862

Lowe Papers

PLATE XLIII

BALLOON *Intrepid* ABOUT TO ASCEND NEAR VIENNA, VIRGINIA

From a war correspondent's sketch, Lowe Collection, U. S. National Museum, No. 30915-E

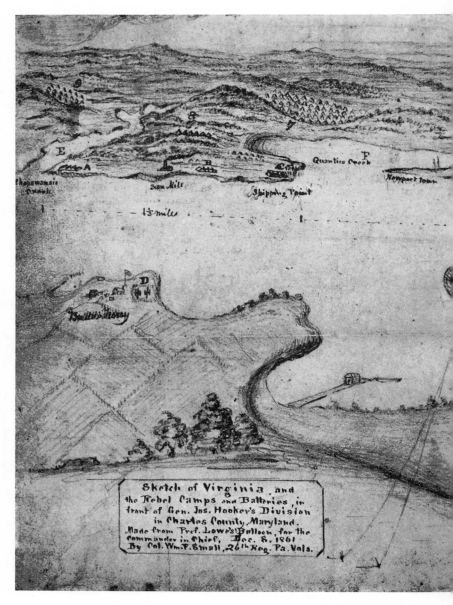

Within the sketch, the following labels appear:

Quantico Creek

E

F

Newport tour

Chapawomsic Creek

A

B

Saw Mill

Shipping Point

— 1½ mile

D

Budd's Ferry

Sketch of Virginia, and
the Rebel Camps and Batteries, in
front of Gen. Jos. Hooker's Division
in Charles County, Maryland.
Made from Prof. Lowe's Balloon, for the
Commander in Chief, Dec. 8, 1861
By Col. Wm. F. Small, 26th Reg. Pa. Vols.

PANORAMIC SKETCH OF CONFEDE
O

Drawn by Colone

MS H 479, enc. 3, Letters Received, Arm

PLATE XLIV

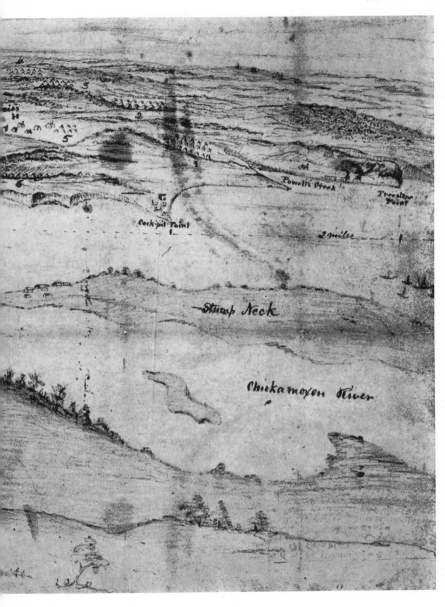

CAMPS ON THE LOWER POTOMAC

.RY

December 8, 1861

r Department Division, National Archives

PLATE XLV

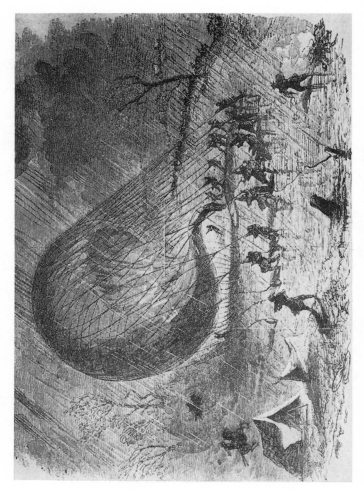

BALLOON *Eagle* DURING A STORM

From a war correspondent's drawing

Lowe Collection, U. S. National Museum, No. 30914-D

CHAPTER VI

TWO MONTHS OF PROGRESS: LOWE'S OPERATIONS,
AUGUST–SEPTEMBER, 1861

Whipple's instructions of August 2 concerning the construction of the new balloon were detailed and explicit. The envelope was to be of not less than 25,000 cubic-foot capacity, built of the best available India silk, in strict conformity to a sample approved by the Topographical Bureau. The rigging and cordage were to be of the best linen, including three manilla guys from 1,200 to 1,500 feet in length. Lowe was to collect his materials in Philadelphia, where the balloon was to be built, notify the Bureau when they should be ready for use, and then proceed with construction. The materials were to be subject to inspection by an officer of the Bureau, with the understanding that construction should be suspended if this officer disapproved their quality and specifications. Lowe was given a free hand in the purchase of all equipment and in the employment of assistants, and the bills thus incurred were to be forwarded for payment to Whipple through Major Bache.[1]

Whipple also provided Lowe with a blanket order authorizing him to purchase the necessary silk, linen, cordage, and other materials, which was binding on the Government to pay " all reasonable bills " incurred in the process of construction.[2]

In compliance with these instructions Lowe left Washington for Philadelphia on August 5 to enter upon his first official

---

[1] Whipple to Lowe, August 2, 1861, MS, Lowe Papers; also MS [copy] L852 (1863), Letters Received, Secretary of War, War Department Division, National Archives [cited hereinafter as LRSW]. Also printed in *Official Records of the Union and Confederate Armies* (Washington [Government Printing Office], 1880 *et seq.*), Series 3, Vol. III, 259 [cited hereinafter as *O. R.*, with series number in arabic preceding abbreviated title].

[2] Authorization order, August 2, 1861, signed by Captain A. W. Whipple, MS, Lowe Papers; also MS [copy] L852, enc. (1863), LRSW; also MS W938, enc., Letters Received, Bureau of Topographical Engineers, War Department Division, National Archives [cited hereinafter as LRTE]. Also printed in 3 *O. R.*, III, 259.

duties in the military service.[3] By the 7th he had obtained the silk for the envelope and had begun construction at the Philadelphia Assembly Buildings.[4] Within a week he had completed the envelope and was assembling the rigging and other attachments. In the meanwhile Whipple secured leave from the Bureau and arrived in the city on the 13th to inspect the materials.[5] Finding them satisfactory, he arranged for the settlement of Lowe's accounts and returned to Washington to report on his inspection.[6]

Both workmanship and materials were entirely satisfactory, he declared. The silk matched the approved sample in quality, and the basket and cordage were " well made and strong." The envelope had received its first coating of varnish, and as soon as the second should be applied and dried, the balloon would be ready for service. Lowe had assured him that in the absence of unforeseen contingencies, he would be prepared to take the field within a week.[7] Five days later the final details were complete, and on August 21 Lowe left Philadelphia for the Capital with the new aerostat, which he named *Union*.[8] The time from Lowe's departure from Washington with Whipple's instructions to his arrival with the new balloon had been only sixteen days. Since several days were required for the drying of the two applications of varnish, the rapidity with which he completed

[3] Washington *National Republican,* August 7, 1861; Washington *Star,* August 7, 1861; Philadelphia *Inquirer,* August 6, 1861; Philadelphia *Press,* August 7, 1861.

[4] Whipple to Bache, August 7, 1861, MS W938, LRTE; Lowe to Bache, August 13, 1861, MS L960, LRTE.

[5] Bache to Whipple, August 8, 1861, MS Letterbooks, Bureau of Topographical Engineers, War Department Division, National Archives, Vol. XXIII, 10 [cited hereinafter as LBTE]. Whipple to Captain James B. Fry [AAG, Dept. NE Va.], August 9, 1861, MS 564, Letters Received, Department of Northeastern Virginia, War Department Division, National Archives [cited hereinafter as LRNV]. Fry to Whipple, August 10, 1861, MS Letterbooks, Department of Northeastern Virginia, Vol. III, 3 [cited hereinafter as LBNV]. Bache to Lowe, August 14, 1861, MS LBTE, XXIII, 23.

[6] Lowe to Bache, August 13, 1861, MS L960, LRTE; Whipple to Major Israel C. Woodruff, August 16, 1861, MS W972, LRTE; Bache to Lowe, August 16, 1861, MS LBTE, XXIII, 28.

[7] Whipple to Bache, August 17, 1861, MS W992, LRTE.

[8] Philadelphia *Press,* August 22, 1861; Lowe to William Paullin [assistant aeronaut], October 11, 1861, MS, Lowe Papers.

the project seems to indicate efficiency and energy in the prosecution of the work.

Lowe's arrival in Washington with the new balloon immediately raised the question of assigning a detail as a ground crew. Whipple had anticipated this need before the balloon was finished, and on August 17 had requested Bache to arrange for the assignment of a detachment for this purpose.[9] He doubtless recalled the difficulties encountered in securing competent men for this work during Wise's brief period of service, and also during the recent trials with Lowe's first balloon. His subordinate, Lieutenant Abbot, had also expressed himself emphatically on the subject during James Allen's early attempts. Whipple had raised the question in a message to Bache as early as August 2,[10] and now, after Lowe's definite appointment, anticipating the same old difficulty, he warned Bache that " without some permanent arrangement of this kind I fear we shall not be able to reap the benefits anticipated from aerial reconnaissance." [11]

Since Lowe's operations were to be under the direction of the Topographical Bureau, the logical source of troops for his ground crew would have been this branch of the service.[12] But at this time there were no enlisted men attached to the Topographical establishment, and though one company was soon recruited for this branch, their routine work with the engineer service required their undivided duty.[13] Consequently when Whipple asked that the Bureau arrange for a detachment for balloon duty, Bache requested General McDowell to provide the men, and suggested that they be drawn from one of the vol-

[9] Whipple to Bache, August 17, 1861, MS W992, LRTE.

[10] Whipple to Bache, August 2, 1861 [telegram], MS W4, Letters Received, Army of the Potomac, War Department Division, National Archives [cited hereinafter as LRAP].

[11] Whipple to Bache, August 17, 1861, MS W992, LRTE.

[12] Although the balloon service was initially placed under the loose direction of the Topographical Engineers, it did not become an integral part of this branch. Lowe's status and that of the balloon corps subsequently created are treated in a succeeding chapter on administration.

[13] *Report of the Secretary of War, 1860* [Sen. Doc. 2, 36 Cong. 2 Sess.], pp. 592-593. *Report of the Secretary of War, 1861* [Sen. Doc. 1, 37 Cong. 2 Sess.], p. 128. See also footnote 43, Chapter ii, *supra.*

unteer regiments of engineers. Unless these men were made available, he declared, " it is apprehended that substantial benefit from aerial ascensions will not be secured." [14]

The proverbial red tape of military administrative machinery then performed its traditional function. McDowell's adjutant general disapproved the request and refused to assign the men.[15] For eleven days letters and telegrams, with accumulated indorsements, passed to and from Whipple, Bache, McDowell's headquarters, and finally the headquarters of the General-in-Chief, before the assignment of one lieutenant and thirty-three men could be secured.[16] Not until August 22, after the matter had been carried to the highest link in the chain of command,

[14] Bache to McDowell, August 17, 1861, and enc., MS LBTE, XXIII, 30; Bache to McDowell, August 17, 1861, and enc., MS E11, LRNV.

[15] Fry to Bache, August 18, 1861, MS [number omitted], LRTE.

[16] The workings of military administration through channels at this time, and its effect on this one phase of the balloon operations may be illustrated by a summary of what took place before the ground crew was finally assigned.

When McDowell's headquarters refused to assign the men requested by Bache, Whipple proceeded on his own responsibility to secure their services. He obtained the consent of the commander of the 14th New York State Militia to detach one company under Captain Richard Butt for balloon duty, and also secured the approval of Brigadier General E. D. Keyes, to whose brigade the regiment was attached. Whipple then notified McDowell's headquarters that both regimental and brigade commanders were willing to assign the company, and asked McDowell's approval of the transfer. Receiving no reply, he then wrote to Major Woodruff, Chief of the Bureau, and complained that Lowe was waiting idly in Washington and could perform no service because McDowell would not provide the ground crew. He urged that Bache take the matter up directly with McDowell's superior, General McClellan. Bache then wrote to McClellan's adjutant general, and explained the problem. The letter was referred to Lieutenant Colonel John N. Macomb, chief topographical officer on McClellan's staff, who approved the assignment of a detail and concurred with Bache that the troops be drawn from some volunteer regiment of engineers. Four days after Macomb's favorable indorsement, McClellan's headquarters ordered McDowell's adjutant general to assign an officer and thirty-three men to report to Whipple for balloon duty.

The above summary is based on letters and papers of the Army of the Potomac, Department of Northeastern Virginia, and Bureau of Topographical Engineers, in the War Department Division of the National Archives, especially the following documents: Whipple to Fry, August 22, 1861, MS W28, enc., LRNV; Whipple to Fry, August 22, 1861, MS W28, LRNV; Whipple to Woodruff, August 23, 1861, MS W1023, LRTE; Bache to Major Seth Williams [AAG, AP], August 24, 1861, MS [number omitted], LRAP; Macomb to Williams, August 24, 1861, MS M46, LRAP.

was the ground crew ordered to report to Whipple.[17] That it should have required this much negotiation and correspondence for nearly two weeks serves well to illustrate at least one of the problems that arose in the operation of the balloons.   This early incident is of special importance, since it marked the beginning of Lowe's difficulties over ground crews which were never properly solved even after the balloon service was expanded to a corps.

The problem of ground troops having finally been overcome, Whipple requested the chief of the Bureau to have the balloon sent into the field at once. " A balloon to overlook the enemy now is important. Will you direct Lowe to fill [it] with gas and bring [it] over to Fort Corcoran or here as quickly as he can," he telegraphed Major Woodruff from McDowell's head-quarters at Arlington on the 28th.[18] Woodruff immediately complied and sent Lowe an order for gas.[19] The following day the *Union* was towed through the streets of the Capital to Georgetown, from which point it was taken across the Alexandria Aqueduct to Fort Corcoran and the vicinity of Arling-ton.[20] Here on the afternoon of August 29 its "maiden" ascension was made, during which the new balloon and its aeronaut received their first baptism of fire.[21]

The reconnaissance occupied an hour or more, and disclosed minor movements along the Confederate lines, as well as a

[17] Williams to Captain H. W. Kingsbury [AAAG, Dept. NE Va.], August 28, 1861, Papers of George B. McClellan, Division of MSS, Library of Congress, Series 1, Vol. XXV, MS 5556 [cited hereinafter as McClellan Papers]. Williams to Whipple, August 28, 1861, MS 5550, McClellant Papers, Series 1, Vol. XXV; Lieutenant O. M. Poe to Macomb, August 28, 1861, MS W1023½, LRTE.

The regiment from which the detail was selected is not mentioned, though it appears to have been drawn from the 14th New York. Several weeks later General Keyes requested its return to duty with his brigade, an indication that it was from the regiment Whipple originally approached.

[18] Whipple to Woodruff, August 28, 1861, MS M62, LRAP.

[19] Woodruff to Lowe, August 28, 1861, MS, Lowe Papers; also printed in 3 *O. R.*, III, 260.

[20] Washington *Star*, August 30, 1861; Philadelphia *Press*, August 31, 1861.

[21] New York *Times*, August 30, 1861; New York *Commercial Advertiser*, August 30, 1861; Boston *Journal*, August 30, 1861; Boston *Transcript*, August 30, 1861.

concentrated effort at entrenchment on Munson's, Clark's, and Upton's Hills. On the first of these hills Lowe estimated that a thousand men were engaged in erecting field works, and two guns were noted already in position. During the ascent a Confederate rifled battery was brought to bear on the balloon and opened fire with solid shot and shell.[22] None of the projectiles took effect, though Captain E. P. Alexander, of the Confederate Engineers, later to become an army balloonist of parts himself, wrote to his father that " we sent a rifle shell so near old Lowe and his balloon that he came down as fast as gravity could bring him." [23]

The situation that developed in northern Virginia during the fall and winter of 1861, and the operations that ensued were not conducive to any spectacular discoveries from the air. After the abortive Manassas campaign both armies settled down to the task of reorganization and preparation for serious warfare. They occupied more or less fixed positions on the general line of the Potomac, maintained loose contact, and watched each other from various points of vantage. No major combat operations or troop movements leading to action were effected by either side; and save for several skirmishes and one serious engagement at Ball's Bluff in late October, the period in the Potomac region was practically devoid of actual hostilities. Consequently Lowe's observations, when viewed in the perspective of military history, do not reveal any information of great importance or remarkable significance.

Nevertheless Lowe's operations in the first months of his field service show definite progress in the establishment and recognition of the observation balloon as an important and valuable adjunct to the military service. In the first place he demonstrated beyond all doubt that in reasonable weather he could make continuous daily ascensions and keep the enemy lines under constant observation for indefinite periods of time. A

[22] *Lowe's Report*, 3 *O. R.*, III, 260; New York *Tribune*, August 31, 1861; *ibid.*, September 1, 1861; Hartford *Daily Courant*, September 2, 1861; New York *Herald*, September 3, 1861.
[23] Alexander to A. L. Alexander, September 8, 1861, MS, Mackall Collection, Library of the Johns Hopkins University.

day by day survey of his operations from August 29, when the first ascension with the *Union* was made, to October 1, when Lowe left for Philadelphia to begin construction of additional balloons, reveals that he made ascensions on twenty-three of the thirty-four days. The *Union* was aloft every day, sometimes several times a day, from August 29 to September 9, during which time observations were constantly taken and reported to the Union high command.[24] A heavy gale on the evening of the 10th compelled Lowe to discharge gas in order to prevent damage to the envelope, but by the morning of the 12th he had completed reinflation and the balloon was again ready for field operations.[25] Ascensions were resumed without interruption until the 17th, when replenishing again became necessary, and the ground crew was relieved temporarily to receive pay and fresh clothing.[26] The *Union* was back in the field on the 19th,[27] and was sent aloft almost daily until October 1, whenever the various general officers requested observations.[28] The operations during this period thus established conclusively that the service, when competently managed, could be dependable and continuous in the field.

Lowe also demonstrated that a captive balloon could be safely and efficiently towed to and from various positions, and that observations could be taken from almost any reasonable point selected by the commanding officers. The unfortunate accidents experienced by Allen and Wise in transporting their inflated balloons were not repeated during this period of Lowe's activities, though the facilities at Lowe's disposal were

[24] These statements are based on a detailed study of dispatches in the contemporary newspapers, *Lowe's Report* and its accompanying documents; MS, letters, reports, and other papers in the records of the Army of the Potomac, the Department of Northeastern Virginia, the Bureau of Topographical Engineers, and the Lowe Papers.

[25] Lowe to Porter, September 11, 1861, MS, Letters Received and Miscellaneous Papers of Porter's Division, Army of the Potomac, War Department Division, National Archives [cited hereinafter as LRPD]. Washington *Star*, September 11, 1861; Baltimore *American*, September 12, 1861; Philadelphia *Inquirer*, September 13, 1861; Lowe to Porter, September 11, 1861 [second telegram of this date], MS, LRPD.

[26] Lowe to Porter, September 17, 1861, MS, LRPD.

[27] Washington *Star*, September 20, 1861.

[28] See note 24, *supra*.

virtually the same as those of his two predecessors. Marches which involved towing the inflated *Union* were frequent incidents to routine field service. The aerostat was taken to and from Fort Corcoran and Washington; from Fort Corcoran and Arlington out to Ball's Cross Roads and back; over to Chain Bridge and across to the highlands on the Virginia side; and back into Washington for reinflation on several occasions.[29]

Only once did the balloon ever get out of control, and then not seriously. On the way through Washington to Fort Corcoran after the first reinflation, it escaped from the ground crew while Lowe was in the car; but it was recovered after a short chase of only a few blocks.[30] No damage occurred at this time, or in any of the other numerous marches that became part of Lowe's daily routine. After the failures of his predecessors in this respect, Lowe's demonstrations that his balloon could be maneuvered without accident or damage constituted decided progress towards his ultimate objective: proving to his superiors that balloon operations in the field were reliable and practical.

Although Lowe's observations did not reveal information of startling significance, his reports of enemy activity and position when compared with Confederate documents now available, were reasonably accurate. In one or two minor cases he was

---

[29] Washington *Star*, August 30, 1861; Philadelphia *Press*, August 31, 1861; *Lowe's Report*, 3 *O. R.*, III, 260; New York *Commercial Advertiser*, September 3, 1861; *ibid.*, September 6, 1861; Porter to McClellan, September 9, 1861, MS, LRPD; Porter to Lowe, September 11, 1861, MS, Lowe Papers; Washington *Star*, September 11, 1861; Baltimore *American*, September 12, 1861; Philadelphia *Inquirer*, September 13, 1861; Smith to Porter, September 13, 1861, MS, LRPD; Smith to Porter, September 13, 1861 [second telegram], MS, LRPD; Philadelphia *Daily News*, September 14, 1861; New York *Evening Post*, September 13, 1861; New York *Commercial Advertiser*, September 13, 1861; Lowe to Porter, September 17, 1861, MS, LRPD; Washington *Star*, September 20, 1861; Lowe to Porter, September 24, 1861, 3 *O. R.*, III, 263; New York *Tribune*, October 1, 1861; Philadelphia *Ledger*, October 1, 1861; Detroit *Free Press*, October 2, 1861; Providence *Press*, August 30, 1861; Providence *Post*, September 26, 1861; " Sidney " [war correspondent with 2nd R. I. Infantry] to editors of the Providence *Post*, September 12, 1861, published September 16; New York *Herald*, August 30, 1861; *Lowe's Report*, 3 *O. R.*, III, 264.

[30] New York *Commercial Advertiser*, September 6, 1861; Philadelphia *Inquirer*, September 6, 1861; Washington *Star*, September 5, 1861.

deceived by deliberate attempts of the enemy to frustrate his observations, but in the main his estimates of situations observed were essentially correct. A comparison of some of his more detailed reports with Confederate testimony relating to the same period will serve to illustrate this phase of Lowe's successful operations during the formative weeks of his service.

Inspection of the enemy lines opposite Fort Corcoran during the first ascension on August 29 had revealed troops engaged in building field fortifications on the high ground of Munson's, Mason's, and Upton's Hills, and the presence of two guns mounted in position.[31] Observations taken from Ball's Cross Roads on the following day, two and a half miles nearer the enemy positions, disclosed similar activities progressing under the efforts of a large body of men. Lowe also distinguished what he thought to be a piece of field ordnance being dragged into position by a team of six horses.[32] Continued observation for the next several days showed further occupation and entrenchment of these hills, as well as the occupation of numerous and extensive encampments south of the Alexandria-Leesburg turnpike, in the general direction west and northwest. A column of troops was also observed in motion on the same road, going southeast as though marching from Leesburg to threaten Chain Bridge.[33]

A correspondent of the New York *World,* in evaluating these observations, asserted that Beauregard's plan was to use these entrenched positions on the high ground near Washington as a base of offensive operations, and declared that his theory was confirmed by the nature of Lowe's disclosures from the air.[34] As a matter of fact, the correspondent was a better tactician than would be generally supposed. Beauregard was at this time urging that a decisive action be forced, by threatening the Capital from these positions while a strong column should move

---

[31] *Lowe's Report,* 3 *O. R.,* III, 260; New York *Tribune,* September 1, 1861; *ibid.,* August 31, 1861; Hartford *Daily Courant,* September 2, 1861.

[32] New York *Tribune,* September 1, 1861 [second dispatch].

[33] Philadelphia *Press,* September 3, 1861; Philadelphia *Ledger,* September 4, 1861; Detroit *Free Press,* September 5, 1861; New York *World,* September 3, 1861; Baltimore *American,* September 4, 1861.

[34] New York *World,* September 3, 1861 [special correspondence].

south of Occoquan to roll up the Federal left flank. The plan, however, was disapproved by General Johnston.[35]

Information from Confederate sources confirms the general accuracy of Lowe's observations. Shortly after the battle of Manassas, the brigade of Brigadier General James Longstreet, consisting of four infantry regiments and a battery of artillery, had been gradually advanced toward the Federal capital.[36] On August 9 orders were issued for this command to occupy Fairfax Court House,[37] and by the middle of the month the picket lines had reached Mason's and Munson's Hills, within five miles of Washington.[38] The main body of the brigade soon followed, and established strong outposts of an advanced line in front of the Confederate army which had occupied Fairfax Court House and surrounding country by the end of August.[39]

Field works were thrown up to protect these advanced positions, along which Beauregard recommended the posting of three additional brigades.[40] The day before Lowe made his initial ascent in the *Union,* Beauregard had written to a relative that " my advanced forces are at this moment near Arlington Heights at Munson's Hill," [41] and three days later, even while Lowe was reporting visible Confederate activity in this vicinity, the Southern commander's adjutant announced that " the advanced brigades of this army are now bivouaced in full view

[35] Colonel Alfred Roman [IG, Beauregard's Staff], *The Military Operations of General Beauregard in the War Between the States* (New York, 1884), I, 136.

[36] Lieutenant General James Longstreet, *From Manassas to Appomattox* (Philadelphia, 1896), pp. 57, 59.

[37] Special Orders, No. 216, 1st Corps, Army of the Potomac [CSA], August 9, 1861, MS Orderly Book (G and SO, June-December, 1861), Papers of Pierre Gustave Toutant Beauregard, Division of MSS, Library of Congress [cited hereinafter as Beauregard Papers].

[38] Brigadier General Edward Porter Alexander, *Military Memoirs of a Confederate* (New York, 1907), pp. 54-55.

[39] Major Thomas Jordan [AAG, 1st Corps] to Colonel Nathan G. Evans, September 2, 1861, MS Letterbook, 1st Corps, Army of the Potomac [CSA], Beauregard Papers; General Joseph E. Johnston, *Narrative of Military Operations During the Late War Between the States* (New York, 1874), p. 69.

[40] Beauregard to Johnston, September 6, 1861, Roman, I, 476.

[41] Beauregard to A. A. Toutant, August 28, 1861, Catalogue of MSS offered by Stan V. Henkels [Philadelphia, October 25, 1915], p. 10, No. 45, Beauregard Papers.

of the capital of the United States."⁴² The same officer also notified Colonel Nathan G. Evans at Leesburg the following day that "the enemy is much exercised by the [our] movements in front of Fairfax Court House and the occupation of Mason's and Munson's Hills."⁴³ And Captain Alexander, whose interest in the pot-shooting at Lowe's balloon has already been noted, also recorded in his correspondence that the Confederate army was "moving very close to the enemy and fortifying in plain view of them and the capital."⁴⁴

Working parties from Longstreet's brigade, therefore, were the troops entrenching these positions that Lowe had observed and reported during his several reconnaissances. The guns noted in position during his first ascent can be identified as two of the four pieces of Major J. B. Walton's battalion of the celebrated Washington Artillery, which had been assigned to Longstreet's brigade by special orders on August 21.⁴⁵ The piece Lowe observed being maneuvered into position during his subsequent ascension may well have been another of these four guns. The encampments noted at various points south of the Alexandria-Leesburg pike were doubtless the positions of occupation assumed by various units of the Confederate army at Fairfax Court House, Vienna, Flint Hill, Centreville, Germantown, and Sangster's Cross Roads, during August.⁴⁶ The column of infantry reported on the Leesburg turnpike cannot be satisfactorily identified. Evans' brigade had occupied Leesburg early in August,⁴⁷ and later took part in the action at Ball's Bluff in late October.⁴⁸ But the movement of a unit of unde-

---

⁴² Jordan to the Reverend I. Michelbacher, September 1, 1861, MS Letterbook, 1st Corps, p. 120, Beauregard Papers.

⁴³ Jordan to Evans, September 2, 1861, MS Letterbook, 1st Corps, p. 125, Beauregard Papers.

⁴⁴ Alexander to A. L. Alexander, September 8, 1861, MS, Mackall Collection, Library of the Johns Hopkins University.

⁴⁵ Special Orders, No. 257, 1st Corps, Army of the Potomac [CSA], August 21, 1861, MS Orderly Book, p. 186, Beauregard Papers.

⁴⁶ Beauregard to Johnston, August 11, 1861, Roman, I, 473; see also *Atlas to Accompany Official Records* (Washington [Government Printing Office], 1891-1896), Vol. I, Plate vii, Sheet 1.

⁴⁷ Beauregard to Johnston, August 11, 1861, Roman, I, 473.

⁴⁸ Reports of Brigadier General Nathan G. Evans, October 31, 1861, *et seq.*, 1 *O. R.*, V, 348-353.

fined size apparently from Leesburg on September 3, cannot be definitely placed in the shifts that occurred among the Confederate units along the general line of the Potomac at this time.

Other observations taken during routine ascensions are interesting as examples of what was seen from the air, and also illustrate the general accuracy of Lowe's reports. On September 13 he went up from Chain Bridge and reported the presence of a few scattering videttes and pickets in the vicinity of Vienna. No troop movements were visible between this point and the Potomac.[49] Later the same afternoon he again ascended from the highlands and found the appearance of the previously noted Confederate camps to be unchanged. Then a puff of smoke appeared beyond Hall's Hill, followed by the distinct report of a cannon. A short while later Lowe picked up in his glass a battery of four guns in the vicinity of Falls Church. The greatest number of camp smokes appeared at Manassas, Centreville, and Fairfax Court House. A lesser distribution was noted at Munson's Hill, where Lowe estimated that a regiment occupied the west slope of the position.[50]

Coming down, Lowe changed the position of his balloon and again went up at dusk, remaining aloft until well after dark. The first unusual event was the appearance of a large fire, thought to be a burning building on the Alexandria-Leesburg pike just east of Falls Church. The appearance of the fire was immediately followed by the faint sound of cheering from what seemed to be large bodies of men. A second fire similar to the first soon cut the darkness in the same vicinity, quickly followed by a third on Hall's Hill. More cheering was wafted by the night breezes toward the observer, and three cannon were fired from Hall's Hill toward Washington, near the burning building, accompanied by continued shouting. The darkness, Lowe reported, prevented any satisfactory estimate of the force engaged in these unusual activities.[51]

[49] Philadelphia *Daily News*, September 14, 1861; New York *Evening Post*, September 13, 1861; New York *Commercial Advertiser*, September 13, 1861; Providence *Press*, September 14, 1861; Lowe to Porter, September 13, 1861, MS, LRPD.

[50] Lowe to Porter, September 13, 1861 [second telegram], MS, LRPD.

[51] *Ibid.*

In general, the information secured during these observations matches the available Confederate sources reasonably well. The large number of camp smokes discerned at Manassas, Centreville, and Fairfax Court House can be readily identified with the main force of the Confederate army which occupied these areas at the time.[52] The discovery of four guns in position near Falls Church, when observations several days before revealed only pickets in this vicinity,[53] indicated that a larger force had recently occupied this ground since the previous ascension. Small outposts and pickets are generally not accompanied by field batteries. Such a deduction may be substantiated by the fact that Longstreet's brigade, relieved from the line of Mason's and Munson's Hills, had been ordered to Falls Church on the day before Lowe discovered the battery there.[54]   The guns observed doubtless belonged to Walton's battalion which had been ordered to report to Longstreet's command on August 28.[55]

The unusually large fires and mass cheering, accompanied by sporadic gunfire, are less easy to identify in the general scheme of events at the time. It is likely that the fires and cheering may have been in celebration of the recent engagement at Lewinsville, which was hailed enthusiastically as a victory by the Confederates. On September 11th Brigadier General William F. Smith had sent a reconnaissance in force of about 2,000 men and four guns under Colonel Isaac I. Stevens to examine the ground in the vicinity of Lewinsville.[56] This force had been attacked by Confederate cavalry under Colonel J. E. B. Stuart, and had retreated, leaving eleven dead and captured.[57]   Stuart's

[52] Johnston to Acting Secretary of War Benjamin, September 26, 1861, 1 O. R., V, 881.

[53] Lowe had taken observations of Falls Church on September 9, and had found only scattered pickets and small outpost activity. Lowe to Porter, September 9, 1861, MS, Lowe Papers; also printed in 3 O. R., III, 261.

[54] Beauregard to Jefferson Davis, September 13, 1861, MS Letterbook [Official and Private Correspondence March 1, 1861–March 31, 1863], p. 137, Beauregard Papers; Beauregard to Johnston, September 13, 1861, Roman, I, 478.

[55] Special Orders, No. 274, 1st Corps, August 28, 1861, MS Orderly Book, Beauregard Papers.

[56] Reports of McClellan, Smith, and Stevens, 1 O. R., V, 167-172.

[57] Longstreet's Report, September 12, 1861, 1 O. R., V, 182.

loss was reported as " not a scratch to man or horse." [58] The skirmish was received in the Confederate camps as a brilliant victory,[59] and mass cheering, bonfires and salutes would have been likely and appropriate demonstrations of enthusiasm in the Southern encampments.

Other examples of the accuracy of Lowe's observations are apparent in the records of his operations during the period. On September 24 he reported a considerable force of cavalry visible at Munson's Hill.[60] Stuart's command had occupied this position for some time, and had returned there after the attack on Stevens' column at Lewinsville.[61] On the 25th Lowe observed a line of smokes and camps ten miles distant from Fort Corcoran in direct line with Bailey's Cross Roads.[62] These camps and smokes can be identified as marking the position of Brigadier General Richard S. Ewell's brigade at Annandale, ordered to this ground on September 13,[63] and the units of the Confederate advance guard reported by Johnston to be occupying Springfield Station on the 26th.[64] These locations on the contemporary maps plot roughly from eight to nine miles from Lowe's station at Fort Corcoran, in line with the crossing at Bailey's.[65]

Another important step in the progress of Lowe's successful service was his participation on September 24, in the first artillery fire direction with aerial observation in American history. General Smith had planned to fire on Falls Church from his positions at Chain Bridge, with an aerial observer at Fort Corcoran to sense the shots for range and deflection. By means of telegraph from the balloon car to the gun positions, the effect of each shot was to be communicated to Smith, more than three

---

[58] Stuart's Report, September 11, 1861, 1 *O. R.*, V, 184.

[59] General Orders, No. 119, September 12, 1861, Headquarters Army of the Potomac [CSA], 1 *O. R.*, V, 181. Stuart was recommended for the rank of brigadier as a result of this engagement.

[60] Philadelphia *Daily News,* September 25, 1861; New York *Evening Post,* September 24, 1861; New York *Commercial Advertiser,* September 24, 1861; Providence *Post,* September 25, 1861.

[61] Longstreet to Jordan, September 12, 1861, 1 *O. R.*, V, 182; Stuart to Longstreet, September 11, 1861, *ibid.,* pp. 183-184.

[62] Lowe to Porter, September 24, 1861, 3 *O. R.*, III, 263.

[63] Beauregard to Johnston, September 13, 1861, Roman, I, 478.

[64] Johnston to Benjamin, September 26, 1861, 1 *O. R.*, V, 881.

[65] *Official Atlas,* Vol. I, Plate vii, Sheet 1.

miles distant from the point of observation.[66] In fact the procedure planned called for the first form of " lateral observation
of fire " [67] employing an air observer, a method of fire control
well established in modern American field artillery. Smith cautioned Lowe to note especially the ranges, with particular attention to all shots that fell short.[68] The system was found to work
well and excellent artillery practice was obtained.[69]

After the telegraph had been used, a system of visual flag
signals was also tried for communicating the observer's sens-
ings to the gunners. When shots fell to the right, a white flag
was raised vertically in the balloon car; deflection left was indicated by lowering the flag. Range over was signalled by holding the flag stationary in a horizontal position; and short ranges
were shown by waving the flag in a horizontal arc.[70] Secondary
observers at the Chain Bridge gun positions read these signals
with glasses and the corrections were applied by the gunners
accordingly.[71] Although this first instance of aerial fire direction did not involve the accomplishment of a vital mission,
Lowe demonstrated for the first time in American history that
artillery adjustment by balloon observation was a practical reality.[72] Later in the war field batteries were directed in the same
manner against concealed troops and materiel, and at Island
No. 10 one of Lowe's assistant aeronauts directed mortar fire
from the air with accuracy and effect.[73]

But observation, not artillery fire direction, was Lowe's main
function. His constant surveillance of the enemy lines and
positions during this period, even though the reports of observa-

[66] Smith to Porter, September 23, 1861, 3 O. R., III, 262.

[67] Lowe's observation post at Fort Corcoran was more than two miles to the
flank of the gun position, almost at right angles to the line of fire.

[68] McQuesten to Lowe, September 24, 1861, MS, Lowe Papers; also printed
in 3 O. R., III, 262-263.

[69] Smith to Porter, September 24, 1861, MS, LRPD.

[70] Porter to Lowe, September 24, 1861, MS, Lowe Papers; Smith to Porter,
September 24, 1861, 3 O. R., III, 263.

[71] Smith to Porter, September 24, 1861 [second telegram], MS, LRPD.

[72] It is probable that this instance was the first artillery fire direction from the
air in the military history of any country. The writer has found no evidence in
available sources that this technique had been tried in European armies prior to
Lowe's experiment.

[73] See pp. 322-323, and 396-397, infra.

tions did not reveal information of particular consequence, was important. The panic that had seized Washington after the Federal disaster at Bull Run was slow to disappear, and throughout the weeks that followed, alarms of supposed Confederate movements in force often revived the fear of sudden attack. The presence of the aerial observer helped in alleviating the uneasiness that prevailed after such rumors. When Lowe frequently reported the enemy camps the same as previously observed, or announced that there had been no change in their positions,[74] the natural result of such information was a greater sense of security than would have otherwise been enjoyed. At times scouts or civilians from localities outside the Union lines would come in and announce the approach of large forces of the enemy. On such occasions Lowe went up and scanned the hostile camps and lines, and the roads leading to the Capital, and dispelled the rumors by reporting all quiet in a wide radius.[75] In the same manner he was able to prevent unnecessary preparations and useless " stand to " precautions among the units of the Federal army when false reports of a Confederate advance were circulated. In his report to the Secretary of War he emphasized this element of his early service and declared:

From this time [September 7] to the 27th of September many alarms were given, and the troops called out in line of battle, and in every instance after an examination had been made by means of the balloon the troops were sent back to their quarters and allowed to rest without danger of being surprised.[76]

---

[74] Boston *Transcript*, September 7, 1861; New York *World*, September 7, 1861; Philadelphia *Ledger*, September 7, 1861; Lancaster *Daily Evening Express*, September 6, 1861; Detroit *Free Press*, September 8, 1861; Richmond *Enquirer* [semi-weekly], September 10, 1861; Lowe to Porter, September 8, 1861, MS, LRPD, also MS, Lowe Papers, also printed in 3 *O. R.*, III, 261; Lowe to Porter, September 9, 1861, MS, Lowe Papers, also printed in 3 *O. R.*, III, 261; Detroit *Free Press*, September 11, 1861; Lowe to Porter, September 13, 1861, MS, LRPD; New York *Times*, September 17, 1861; Detroit *Free Press*, September 19, 1861; New York *Herald*, September 7, 1861.

[75] Boston *Tribune*, October 2, 1861; Hartford *Daily Courant*, November 20, 1861; Baltimore *American*, November 27, 1861; Washington *Star*, December 11, 1861; New York *Herald*, October 2, 1861; Providence *Post*, September 16, 1861; Providence *Press*, September 16, 1861.

[76] *Lowe's Report*, 3 *O. R.*, III, 260.

In this way Lowe's constant vigilance from his aerial observatory made possible a measure of security against sudden attack in force, a factor that was of major importance to the Union high command. That the Confederate army at this time was neither in condition nor in sufficient strength to attack Washington, could not be apparent to the national authorities, either civil or military. Hence Lowe, in maintaining continual observation of the enemy and the roads leading to the Capital, performed excellent service even though the face value of his reports seems unexciting and insignificant to a present-day reader.

Lowe's aerial activities also caused apprehension at the headquarters of the Confederate high command. His constant inspection of the lines led the Southern commander to take immediate steps to prevent accurate observation of his dispositions by the Federal balloonist. Three days after Lowe's first ascension from Arlington, Beauregard issued instructions to Longstreet, suggesting methods of camouflage:

The General commanding this Army Corps [wrote Beauregard's adjutant general] wishes every precaution taken to prevent the enemy from discovering by balloons or other means the numbers of our advanced commands or outposts. No lights should be kept at night except where absolutely necessary, and then under such screens as may conceal the lights from observation. Further, tents, if used, ought to be pitched under cover of woods and sheltered in all cases as far as possible from accurate computation.[77]

At the same time, Beauregard had cautioned an engineer officer to conceal field works in process of construction at Triplet's and Powhatan Hills, and admonished him to " use the utmost caution and dispatch in constructing these works without being discovered." [78] Similar precautionary measures were later

[77] Jordan to Longstreet, September 2, 1861, MS Letterbook, 1st Corps, p. 124, Beauregard Papers.

[78] Beauregard to Captain W. H. Stevens, September 1, 1861, Henkels' Catalogue of MSS, p. 10, No. 49, Beauregard Papers. Colonel G. B. Anderson was also instructed to conceal the camps of outposts guarding the fords of Occoquan Creek and Bull Run. " Their tents," Beauregard cautioned, " should be so located and arranged as to be concealed from view, or carefully masked." Beauregard to Anderson, October 15, 1861, MS Letterbook, 1st Corps, p. 200, Beauregard Papers.

ordered as a result of Lowe's continued ascensions and the free flights of La Mountain.[79]

Beauregard, however, does not seem to have relied solely on the success of foiling the aerial observers by these methods, and on occasion ordered the kindling of numerous unnecessary fires, to indicate the presence of much larger force than was actually present.[80] Another expedient adopted by the Confederate commanders was the construction and mounting of dummy guns to conceal their shortage of heavy ordnance, and thus deceive the Union balloonist. To give the appearance of strength to advance posts, the Confederates fabricated a number of "quaker" guns from large logs and stovepipes of various sizes, which were painted black and placed in position to resemble formidable batteries.[81]

On one or two occasions Lowe was taken in by these several deceptive expedients. During one night ascension he reported that he could discover nothing at all, not even the usual bivouac fires,[82] a fact that seems to indicate that in this instance Beauregard's initial instructions ordering the camouflage of bivouac fires were being carried out. Again on the same evening that he had detected the large fires and cheering, Lowe also noted in his report with some curiosity that the picket smokes in the late afternoon seemed to be more numerous than usual, and that they all appeared soon after he had ascended.[83] From these facts it would seem that the Confederate general's scheme of deliberately kindling fires to deceive the Yankee observer was in practice and effective in this case. On the afternoon after the first artillery fire direction, Lowe took observations of the field works near Mason's and Munson's Hills, and recorded that there seemed to be a number of heavy guns in position there, with a strong force in support.[84] In all probability the pieces he

[79] See pp. 135-136, *supra,* and 375, *infra.*
[80] Roman, I, 135.
[81] Longstreeet, p. 60; Roman, I, 155; Report of E. J. Allen [Allen Pinkerton], enclosing testimony of Confederate deserter, January 27, 1862, 1 *O. R.,* V, 737.
[82] Lowe to Porter, September 8, 1861, MS, LRPD.
[83] Lowe to Porter, September 13, 1861 [second telegram], MS, LRPD.
[84] Lowe to Porter, September 24, 1861, 3 *O. R.,* III, 263.

discovered were of the quaker variety, placed for his special benefit. From Lowe's position, they doubtless looked genuine enough. With the exception of these few instances, however, Lowe's observations continued to be correct in substance. The Confederate efforts to vitiate the effect of the aerial observations clearly indicate that Lowe's operations were regarded as a serious threat to the security of the Southern army.

One observation taken during the period shows how some forms of enemy activity, though carefully observed and reported, might not be correctly interpreted. On September 8 Lowe ascended before dawn, and remained aloft until after daylight. At first nothing unusual was seen, but at five o'clock two still lights suddenly appeared, one to the right of Munson's Hill, and the second at Taylor's Corners, near Upton's Hill.[85] The sudden appearance of these two lights from the hills, though not understood by the observer at the time, may well have been part of a secret signal system established at this time between Washington and the Confederate advanced posts. Signal lights displayed on these hills were answered by similar lights from the windows of a house in Washington, in accordance with a prearranged code, by which information of Federal movements was conveyed to the Southern command. Powerful telescopes obtained from Charleston, and the services of an experienced signalman, E. P. Bryan, contributed to the scheme. Captain Alexander, who had served as chief signal officer at Manassas, had established the system.[86] Little could Lowe suspect the possible significance of the lights he had seen and casually reported to the Federal headquarters.

Lowe's operations, as well perhaps as those of Wise and Allen before him, also induced Beauregard to follow the Union example. According to his inspector general, he had endeavored to procure a balloon for the Confederate service through the authorities in Richmond, but had met with no success. He eventually obtained one from " a private source," but defective

[85] Lowe to Porter, September 8, 1861, MS, Lowe Papers; also printed in 3 O. R., III, 261.

[86] Alexander, p. 55; Roman, I, 153. See also dispatch in the New York Herald, December 10, 1861.

construction prevented successful operation.[87] It is evident, however, that this balloon was at least raised, for it was reported hovering over Munson's Hill on September 4.[88] Beauregard, despite the failure of his first aerostat, did not lose interest in military aeronautics, and two years later, employed a balloon in the defense of Charleston.

Possibly the most important result of Lowe's successful operations was the recognition and confidence that he steadily gained among the various officers of the Union army, including several ranking generals and the General-in-Chief. Whipple's complete satisfaction with the workmanship and materials of the new balloon; his impatience over the delay in securing a ground crew; his statement that a balloon for observing the enemy positions was important: all indicate that his skepticism exhibited during Wise's operations was now dispelled and replaced with confidence and interest. Bache's reference to " substantial benefit from aerial ascensions " is also indicative of a similar favorable opinion.

Likewise General McDowell, even though his adjutant general had refused to furnish the ground detachment, took a personal interest in the operations. The first day that the balloon was in the field he specially ordered an ascension before daybreak the following morning to examine the number and extent of enemy campfires.[89] And just as Lowe was about to leave the army on September 30 to begin building several new balloons, McDowell recalled him for further observations of the situation on his front.[90] McDowell himself also ascended with Lowe and made personal observations soon after the balloon had been first taken into the field.[91]

The attention and interest of McClellan also was quickly aroused, and on September 7 the General-in-Chief made his first

[87] *Ibid.*, p. 136.

[88] Detroit *Free Press*, September 6, 1861 [special dispatch].

[89] Whipple to Lowe, August 29, 1861, MS, Lowe Papers; also printed in 3 *O. R.*, III, 260.

[90] Brigadier General R. B. Marcy [C of S, AP] to Porter, September 30, 1861, MS, LRPD.

[91] *Lowe's Report*, 3 *O. R.*, III, 260; Boston *Transcript*, September 5, 1861; Providence *Press*, September 5, 1861; Providence *Post*, September 6, 1861.

ascension in company with Lowe and spent two hours in reconnaissance, studying the Confederate positions and surrounding terrain.[92] His personal participation in the new method of observation provoked widespread comment in the press,[93] and was also announced in one of the Richmond papers.[94] The experience must have impressed McClellan more than casually. More than a year later when preparing his final report after having been relieved of command, he made special note of the ascension in a chronological work-sheet.[95] McClellan again ascended with Lowe several weeks later.[96] He also intervened in Lowe's behalf when General Keyes requested that the men from his brigade serving as the ground crew be sent back to their regiment.[97] The request was returned with the indorsement that until further instructions the detachment was permanently assigned to balloon duty by order of the General-in-Chief.[98]

Other general officers also displayed interest in the balloon. John H. Martindale, an experienced regular, ascended with Lowe [99] and grizzled old Samuel P. Heintzelman, with nearly forty years' service behind him, was so interested at his first sight of the aerostat, that he made special note of the event in his diary the same evening.[100] Heintzelman later made numer-

---

[92] *Lowe's Report*, 3 *O. R.*, III, 260; New York *World*, September 9, 1861; New York *Herald*, September 9, 1861.

[93] Boston *Transcript*, September 9, 1861; Hartford *Daily Courant*, September 9, 1861; Philadelphia *Inquirer*, September 9, 1861; Philadelphia *Ledger*, September 9, 1861; Philadelphia *Press*, September 9, 1861; Lancaster *Daily Evening Express*, September 9, 1861; Baltimore *American*, September 9, 1861; Providence *Press*, September 9, 1861; Providence *Post*, September 9, 1861; " Sidney " [war correspondent with 2nd R. I. Infantry] to editors of the Providence *Post*, September 12, 1861, published September 16.

[94] Richmond *Enquirer* [semi-weekly], September 10, 1861.

[95] " Scheme of Report—Chronological Table of Events," MS 17751, McClellan Papers, Series 1, Vol. LXXXVII.

[96] New York *World*, October 2, 1861.

[97] Keyes to McDowell, September 3, 1861, MS LBNV, II, 69.

[98] McDowell's indorsement, September 4, 1861, MS LBNV, II, 69.

[99] *Lowe's Report*, 3 *O. R.*, III, 260.

[100] Diary of General S. P. Heintzelman, entry of September 9, 1861, MS, Papers of Samuel Peter Heintzelman, Division of MSS, Library of Congress [cited hereinafter as Heintzelman Papers].

ous ascensions with Lowe, both in the Potomac lines and on the Peninsula.

But of all the ranking officers who became interested in Lowe's work, none was more sympathetic or enthusiastic than McClellan's most intimate subordinate, Brigadier General Fitz John Porter. He took a keen interest in the balloon from the very first, and had ascended with McDowell on September 5.[101] Thereafter he manifested an increased interest and frequently requested Lowe to make ascensions and report the results.[102] On one occasion he cautioned Lowe to be sure to report everything he saw, whether it appeared to be " of interest or not." [103]

Within a short time Porter became thoroughly convinced that military balloons offered a valuable means of observation, and had wide possibilities if efficiently managed. He also seems to have developed a strong personal attachment for Lowe. Accordingly he proceeded to use his influence to advance the interests of the aeronaut through the favor of General McClellan and other ranking officers, and to this end suggested to the General-in-Chief that an ascension be made from Chain Bridge, to ascertain the position and movements of the enemy in the vicinity of Falls Church.[104] At the time Lowe was conducting observations at Ball's Cross Roads. From a glance at the map, it will be readily seen that his position there would have been more advantageous for such a reconnaissance, since it was closer to the objective by more than two miles, with the same general terrain features intervening.[105] But McClellan was expected to be at Chain Bridge on the day picked for the observation, and Brigadier General William F. Smith's command occupied this vicinity. Obviously Porter intended to keep the balloon operations before McClellan's notice, and also to attract Smith's

[101] Boston *Transcript*, September 5, 1861; *Lowe's Report*, 3 *O. R.*, III, 260.

[102] See dispatches and correspondence from Porter to Lowe, 3 *O. R.*, III, 261-264, *passim*. Also numerous letters and telegrams in the Lowe Papers, the records of the Army of the Potomac, and Porter's Division.

[103] Lieutenant J. F. McQuesten [ADC, Porter's Staff] to Lowe, September 7, 1861, MS, Lowe Papers; also printed in 3 *O. R.*, III, 261.

[104] Porter to McClellan, September 8, 1861, MS [copy], LRPD; Porter to Lowe, September 9, 1861, MS, Lowe Papers; also printed in 3 *O. R.*, III, 261.

[105] Cf. *Official Atlas*, Vol. I, Plate vii, Sheet 1.

attention at the same time. Porter had already praised Lowe's work to Smith, and here was an opportunity to impress his colleague with a practical demonstration. Also, McClellan's recent ascension would be likely to add to the army commander's interest.

McClellan approved the plan, and ordered the balloon transferred to Chain Bridge as Porter had suggested.[106] Whereupon his calculating subordinate promised to have it there early next morning.[107] In order not to interfere with the technical jurisdiction of Whipple, under whose direction the balloon was nominally assigned, Porter sent the engineer officer a copy of McClellan's request calling for the ascension at Chain Bridge.[108] He then notified Lowe of his action in a letter which clearly disclosed both his desire to advance military ballooning and his personal regard for the aeronaut:

General McClellan [Porter wrote] desires you to transfer your balloon to the Chain Bridge early tomorrow to take observations. I have informed him that you will inflate as early as practicable and move up to Chain Bridge. I am desirous to see you prosper, and think that you are now on the road. I have recommended an increase of two balloons and movable inflating apparatus, and as soon as the utility of the science is made apparent (which will depend on your energy), I have no doubt of success. Strike now while the iron is hot. I suggested your balloon should be sent up to Chain Bridge or its vicinity, and I doubt not that General McClellan will be there, or others who will work for you if they are satisfied of its utility. General Smith is in command, and I promise a good reception for you.

If I can aid you in any manner, don't hesitate to call. I will be pleased to see you before you go over in the morning, and the result of your morning observation, which I beg of you to take.[109]

The results of this carefully planned ascension from Chain Bridge are not disclosed in the records. Smith seems to have been thoroughly interested, and had telegraphed Porter after midnight that the balloon was due at his headquarters at daybreak to take observations. The results, he added would be

---

[106] McClellan to Porter, September 9, 1861, MS, LRPD.
[107] Porter to McClellan, September 9, 1861, MS, LRPD.
[108] Locke to Whipple, September 9, 1861, MS, LRPD.
[109] Porter to Lowe, September 9, 1861, MS, Lowe Papers; also printed in 3 O. R., III, 261.

relayed to Porter's headquarters.[110] A lone dispatch in a western newspaper announced that the reconnaissance disclosed the Confederates in strong force beyond the Potomac, but revealed no change in their position.[111] Other than these items there seems to be nothing to indicate the success of Porter's scheme at the time.

The following day, however, Porter telegraphed Lowe suggesting that he ask Smith if further reconnaissance from Chain Bridge were desired, and intimated that the latter would be pleased to have an ascension made in the morning. He closed his communication with the significant comment, "You are of value now." [112] Evidently Lowe had made a good impression, as Porter had hoped. Smith later evidenced increased interest in further ascensions from his position, and personally ordered troops to assist Lowe and his crew in getting the balloon over the bridge to the Virginia heights for observations.[113] When the balloon had been established in its new position, Smith then ascended with Lowe and made a detailed inspection of the surrounding terrain and carefully noted all visible enemy activity opposite his lines.[113a]

Porter in the meanwhile did not cease his efforts to advance the progress of aeronautics with the army and to enlarge its facilities. On September 15 he wrote to Lowe repeating his conviction that the balloon service could be made even more valuable if it were enlarged, and requested that the aeronaut submit a statement of requirements in materiel necessary to make more frequent ascensions and also to conduct simultaneous observations from different points along the lines.[114]

The following day Lowe replied in detail, advising that two additional balloons, one of 30,000 cubic-foot capacity and a second of 20,000, be built similar in quality and construction to the one then in use. He also urged the manufacture of a port-

[110] Smith to Porter, September 10, 1861, MS, LRPD.

[111] Detroit *Free Press*, September 11, 1861.

[112] Porter to Lowe, September 11, 1861, MS, Lowe Papers; also printed in 3 *O. R.*, III, 261.

[113] Smith to Porter, September 13, 1861, MS, LRPD; Smith to Porter, September 13, 1861 [second telegram], MS, LRPD.

[113a] New York *Herald*, September 15, 1861.

[114] Porter to Lowe, September 15, 1861, MS, Lowe Papers.

able generator for making gas available at any point in the field. Such an apparatus would enable the balloonist to replenish as soon as the lifting power of the envelope should become weakened, and would also remove the necessity of returning to Washington at intervals for reinflation. With such an apparatus, the generation of gas in the field would be cheaper, and since the apparatus he wished to build would make hydrogen available in place of coal gas, the greater buoyancy of the former would permit the observers to attain higher altitudes. The cost of the two balloons recommended, Lowe estimated at $1,500 and $1,200, respectively. The generator was calculated not to exceed $500. Lowe further suggested that the generator might be built at the navy yard at Washington by ship carpenters already in government service. Two weeks would suffice for the manufacture of the entire addition in materiel. With the accession of this equipment, Lowe assured Porter that he would be able to " keep the Government informed constantly of the movements of the enemy, as well as the topography of the country." [115]

With this information at hand, Porter continued to press the matter with McClellan, but no decision was announced at the time. Growing appreciation of the aeronaut's service was again manifested in a message from Porter to Lowe on September 18, while the latter was temporarily in Washington attending to the reinflation of the balloon:

Your balloon can be of great service here, and is wanted every moment. Colonel Sawtelle will aid you.[116]

Two days later General Smith wrote Assistant Adjutant General Williams that he considered it of much importance " that his division be furnished with a balloon and portable gas generating apparatus," to be assigned for his permanent use.[117] Porter's carefully laid scheme designed to arouse Smith's interest had done its intended work, and Lowe's efforts to secure recognition

[115] Lowe to Porter, September 16, 1861, MS, Lowe Papers; also printed in 3 O. R., III, 262.
[116] Porter to Lowe, September 18, 1861, MS, Lowe Papers. Colonel Charles G. Sawtelle, Assistant Quartermaster, McClellan's staff, is the officer mentioned.
[117] Smith to Williams, September 20, 1861, MS S217, LRAP.

of the value of his services were thus meeting with success after success.

An excellent example of this recognition appears in Smith's request for balloon observations as a security measure during his second reconnaissance to Lewinsville on September 25. Because of the cavalry attack sustained during the previous movement to this place, the Union force on this occasion was increased to more than 5,000 men and sixteen pieces of artillery.[118] On the evening before the march Smith telegraphed Porter announcing the proposed movement, and asked that Lowe ascend about eleven in the morning to watch for another such onslaught as Stuart had made on Stevens during the previous movement.[119] Porter was in position, should Lowe report any threatening movement on the part of the enemy, to send out assistance to Smith.[120] Porter gave Lowe his orders, warning him that any heavy firing would indicate Smith's collision with hostile forces.[121] Gravely concerned about the safety of Smith's column, and expecting strong opposition in the path of its advance, Porter later sent word to Lowe to watch for battle at Lewinsville, and to inform headquarters of " anything important." [122] Still anxious, and apprehensive lest the wind then blowing should prevent Lowe's ascension, he again telegraphed the balloonist that he was worried about the movement, and feared that the enemy had thrown a strong force between Smith and Lewinsville, to cut off his return.

I wish [Porter continued] to get as much information of his [the enemy's] movements or what is transpiring, as possible before sundown. I expect the return of the enemy, and if much dust be visible, wish to know it, that I may send out a force.[123]

---

[118] Smith's Report, Reconnaissance to Lewinsville, Va., September 27, 1861, 1 *O. R.*, V, 215-216.

[119] Smith to Porter, September 24, 1861, MS, Lowe Papers; also printed in 3 *O. R.*, III, 263.

[120] Porter to Lowe, September 25, 1861, MS, Lowe Papers [first telegram].

[121] Captain Charles Mundee [AAG, PD] to Lowe, September 25, 1861, 3 *O. R.*, III, 263.

[122] Porter to Lowe, September 25, 1861, MS, Lowe Papers; also printed in 3 *O. R.*, III, 263 [second telegram].

[123] Porter to Lowe, September 25, 1861, MS, Lowe Papers; also printed in 3 *O. R.*, III, 263 [third telegram].

Lowe had ascended according to his orders, and remained up until noon, during which time he had observed part of Smith's column on the road three miles out from Chain Bridge. Two guns had been fired from the vicinity of the bridge as he descended. The wind, he added, prevented further ascension for the present, but he assured Porter that there was no large movement of the enemy afoot, or he would have seen something of it. The visibility was poor, however, because of the thick haze that prevailed, and Lewinsville was a considerable distance away.[124]

The aeronaut's opinion and estimate of the situation were proven correct by Smith's report of the affair two days later. His force encountered no resistance during the movement and saw only a few hostile cavalry scouts. In the late afternoon a single regiment supported by one gun appeared and deployed for action as the heavy reconnoitering column was withdrawing. Some minor skirmishes followed, but by half-past five the enemy force had entirely disappeared.[125] Although Lowe's actual part in the affair was small, Smith's request for his observation as a security measure, and Porter's reliance upon his reconnaissance to determine whether Smith should be sent supports, show clearly how highly the value of the balloon was regarded by these general officers.

Fitz John Porter's conviction that the balloon service should become an essential auxiliary to field operations, and his continued efforts to secure the extension and enlargement of its facilities and benefits, eventually exceeded both his own hopes and those of Lowe. He had early discussed with McClellan the advisability of increasing the balloon establishment; and when Lowe had submitted his recommendations on September 16, he had turned the letter over to the Commanding General for his opinion and decision.[126] Although he gave no definite answer at the moment, McClellan was favorably impressed with Lowe's statements, and Porter's enthusiasm, plus Smith's request for a

---

[124] Lowe to Porter, September 25, 1861, 3 *O. R.*, III, 264.

[125] Smith's Report, September 27, 1861, 1 *O. R.*, V, 216.

[126] MS Letterbook, Secretary of War, Vol. C, Entry M-375, September 17, 1861, Serial " BB," War Department Division, National Archives [cited hereinafter as LBSW].

balloon for his division must have influenced his decision. On the 20th he forwarded Lowe's letter to Secretary of War Cameron with the recommendation that the increase in materiel be made. But rather than asking for the authorization of two balloons, he recommended that the number be increased to four, and also suggested the manufacture of sufficient mobile inflating apparatus for the additional balloons and the one then in the field.[127] When his interest was once aroused, McClellan evidently did not believe in doing things by halves.

Cameron immediately approved the suggestions and issued orders to the Quartermaster General to direct Lowe to begin construction of the new apparatus.[128] As a result, on the same day that Porter and Smith were depending on Lowe for security during the second reconnaissance to Lewinsville, Quartermaster General Meigs informed Lowe that

upon recommendation of Major General McClellan, the Secretary of War has directed that four additional balloons be at once constructed under your direction, together with such inflating apparatus as may be necessary for them and the one now in use. It is desirable that they be completed with the least possible delay.[129]

Lowe had made progress indeed. Exactly two months before, he had been refused admittance to General Scott, even though he presented a card of introduction from the President. A week after this rebuff, Captain Whipple had impatiently declared that " he wished to have nothing more to do " with the experimental balloon. After overcoming these obstacles, Lowe had gained his opportunity to prove his ability and the value of the service he had proposed; and now, in the space of less than two months, he had the General-in-Chief and his ranking subordinates squarely behind his efforts.

The instructions issued by General Meigs marked the second turning point in Lowe's career as a military balloonist. The increase in materiel called for more than the operations of a

---

[127] McClellan to Cameron, September 20, 1861, MS, LBSW, Vol. X [Orders and Indorsements], Serial " BB," p. 615.

[128] Cameron's indorsement, *ibid.*

[129] Brigadier General Montgomery C. Meigs to Lowe, September 25, 1861, MS, Lowe Papers; also printed in 3 *O. R.,* III, 264.

single aeronaut loosely attached to the Corps of Topographical Engineers. In reality Meigs' letter was the initial step in the creation of the first balloon corps in American history, the end to which Lowe had worked from the very beginning. Lowe had now reached the second of his major objectives. Before him lay the task of building the corps from its physical materiel to its organization and operation. But the first line had been taken, and no one had contributed more to this advance than had Lowe's friend and advocate, Fitz John Porter.

CHAPTER VII

THE FIRST AMERICAN AIR FORCE: THE BALLOON SERVICE
OF THE ARMY OF THE POTOMAC

Part One: *Materiel and Personnel*

Although the communication from General Meigs had called for the " least possible delay " in the construction of the new balloons and generators, Lowe's duties in the field prevented him from undertaking the work for several days; and even after he had entered upon the task, he was to be called back to the army for active service long before the project was completed. Two days after he received Meigs' instructions, further observations were ordered from the vicinity of Arlington, during which it was reported that Confederate troops had been observed moving towards Alexandria from the direction of Springfield Station, and that new batteries were being erected on the slope of Ender's Hill.[1] The following day Lowe instructed Sergeant Charles F. Eaton,[2] senior non-commissioned officer of his detachment, to take charge of the balloon and detail, warning him to preserve strict discipline and maintain the balloon and ground crew in readiness for operations at all times.[3] He then prepared to leave the army and begin his new assignment, which was to lead to the creation of a corps of balloonists.

Just as his preparations for leaving were completed, General McDowell requested that ascensions be made from Upton's Hill,[4] a position recently evacuated by the Confederates in their

---

[1] Philadelphia *Press,* September 28, 1861.

[2] Company " D," 22nd Regiment, New York Volunteer Infantry.

Sergeant Eaton continued in service with the Balloon Corps and accompanied Lowe's train to the Peninsula the following spring. During this campaign he contracted typhoid fever and was invalided to a hospital in Washington, where he died later in May, 1862. E. Mason, Jr. [Assistant Aeronaut, USA] to Lowe, June 2, 1862, MS, Lowe Papers.

[3] Lowe to Eaton, September 28, 1861, MS, Lowe Papers.

[4] Brigadier General R. B. Marcy [Chief of Staff] to Brigadier General Fitz John Porter, September 30, 1861, MS, Letters Received and Miscellaneous

228

general withdrawal from their fortified ridges southwest of Washington.[5] On September 30 the balloon was moved out to this former enemy strong-point[6] and several ascensions were made in the afternoon, disclosing only a few scattered companies of infantry and cavalry videttes in the vicinity of Fairfax.[7] McDowell and Louis Philippe d'Orleans, Comte de Paris, who had recently been appointed aide-de-camp on McClellan's staff, accompanied the aeronaut in the course of the observations.[8] The following day the balloon was transferred to the former Confederate stronghold at Munson's Hill, where McClellan himself ascended to reconnoiter the position of the enemy forces.[9]

These operations completed, Lowe was able to take leave from active duty and begin arrangements for building the new equipment. A Philadelphia correspondent had picked up rumors of the expected increase in the balloon establishment more than ten days before,[10] and shortly after these last ascensions, fellow reporters elsewhere declared that the aerial reconnaissances had been " so satisfactory in results " that the General-in-Chief had ordered four more balloons added to his army.[11] Lowe left the *Union* and its crew in charge of Sergeant Eaton, and soon after the first of October proceeded to Philadelphia where the new materiel was to be built. " The balloonist Lowe

Papers, Porter's Division, Army of the Potomac, War Department Division, National Archives [cited hereinafter as LRPD].

[5] General Joseph E. Johnston, *Narrative of Military Operations During the Late War Between the States* (New York, 1874), p. 74 [cited hereinafter as Johnston].

[6] New York *Tribune,* October 1, 1861; Philadelphia *Ledger,* October 1, 1861; Detroit *Free Press,* October 2, 1861; Providence *Post,* October 1, 1861; Providence *Press,* October 1, 1861.

[7] New York *World,* October 1, 1861.

[8] *Lowe's Report,* in *Official Records of the Union and Confederate Armies* (Washington [Government Printing Office], 1880 *et seq.*), Series III, Vol. III, p. 264 [cited hereinafter as *O. R.,* with series number preceding abbreviated title]. New York *World,* October 3, 1861 [special letter]; New York *Herald,* October 2, 1861.

[9] New York *World,* October 2, 1861.

[10] Philadelphia *Inquirer,* September 17, 1861.

[11] Boston *Transcript,* October 2, 1861; Providence *Post,* October 3, 1861; New York *Herald,* October 2, 1861.

is forming a balloon corps for army service," ran several
dispatches in the press.[12]

General Meigs' office had furnished Lowe with an order on
the deputy quartermaster general in Philadelphia, instructing
that officer to pay all bills incurred during the process of construction, the entire expenditure being limited to general conformity to the amounts per balloon specified in Meigs' original
order of September 25.[13] Armed with this blanket authority for
the purchase of materials and for the hiring of assistants, he
began work early in October, and was making good progress [14]
when he was ordered on the 12th to return to the army and
report to General Smith at Johnson's Hill, near Lewinsville.[15]

Arriving in Washington the same day, he reported for orders
to Lieutenant Colonel John N. Macomb,[16] Chief Topographical
Engineer of McClellan's staff, who had recently been detailed in
charge of all balloon operations.[17] That evening Lowe received
instructions from McClellan's headquarters to prepare the balloon then in service for the field, and to march to Smith's headquarters at once.[18] This maneuver was attended by hardships
and difficulties. Night had fallen before the party had started,
and darkness necessitated extremely slow progress because of
the constant danger of running the swaying envelope against the
many obstacles that marked the route. The way through Washington and Georgetown was seriously impeded by the numerous

[12] New York *Evening Post*, October 15, 1861; Boston *Journal*, October 15,
1861; Boston *Transcript*, October 15, 1861; Providence *Post*, October 16, 1861.

[13] Colonel Ebenezer Sibley [Deputy Quartermaster General] to Lieutenant
Colonel G. H. Crossman, October 1, 1861, MS [copy], Lowe Papers; also
printed in 3 *O. R.*, III, 264.

[14] Philadelphia *Inquirer*, October 9, 1861; New York *World*, October 10,
1861; Lowe to Colonel A. V. Colburn [AAG, AP], October 13, 1861, MS,
Lowe Papers.

[15] Colburn to Lowe, October 12, 1861, 3 *O. R.*, III, 264.

[16] Macomb to Brigadier General Seth Williams [AAG, AP], October 12, 1861,
MS M233, Letters Received, Army of the Potomac, War Department Division,
National Archives [cited hereinafter as LRAP].

[17] Macomb to Whipple, October 9, 1861, MS, Papers of Colonel John N.
Macomb, Records of the Bureau of Topographical Engineers, War Department
Division, National Archives [cited hereinafter as Macomb Papers]; "Statement
of Colonel John N. Macomb, CE, USA, in relation to Abstract Charges No.
1923," MS, Macomb Papers.

[18] Lowe to Colburn, October 13, 1861, MS, Lowe Papers.

telegraph wires that had to be dodged and crossed, and when the party reached Chain Bridge, about 3 a. m., the span was jammed with artillery and cavalry columns crossing into Virginia. Rather than wait for the columns to pass, and cross the bridge after it should be cleared, Lowe had his men " mount the trestle-work and walk upon the stringers, only eighteen inches wide and nearly 100 feet above the bed of the river," thereby towing the balloon over while the gun-teams occupied the main roadway across.

Thus [Lowe later described the maneuver], with the balloon above their heads, myself in the car directing the management of the ropes, the men getting on and off the trestle work, with a column of artillery moving below, and still 100 feet lower, the deep and strong current rushing over the rocks, while the sky was dark above, the scene was novel, exciting, and not a little dangerous.[19]

At dawn on the 13th the party, exhausted from the night's exertions, reached a point near Lewinsville within a mile of Smith's headquarters. During the night a stiff breeze had been blowing, and by morning it had increased to a howling gale. Lowe ordered the balloon securely lashed to anchoring stakes, intending to wait until the storm should abate before proceeding on to Smith's camp. Some two hours were spent in making the moorings fast and in keeping the tugging envelope secure, but shortly before six o'clock the tempest reached its height and rendered their efforts useless. " An uncommonly heavy whirlwind swept along," Lowe reported several hours later, " and turned the balloon around like a top, causing the strong cordage to break like pipestems, completely clearing the gas envelope from the network." The free envelope then rose to a high altitude and, taking a northeasterly direction, soon disappeared.[20]

Two hours later the runaway balloon came to earth on a farm at Laurel, near Seaford, Sussex County, Delaware. The owner of the farm was suspected of being a Southern sympathizer, and since several newspapers later published details of

---

[19] *Lowe's Report,* 3 *O. R.,* III, 265.
[20] Lowe to Colburn, October 13, 1861, MS, Lowe Papers. *Lowe's Report,* 3 *O. R.,* III, 265.

the accident,[21] it was soon established that the stray balloon was the same one that had escaped from McClellan's army. Accordingly several ardent Unionists of the neighborhood demanded that the balloon be given up and restored to the authorities in Washington; and likewise, other neighbors who were Secessionist in sympathy sought to get possession of the envelope in order to destroy it.[22] The Unionists finally secured the balloon and turned it over to John C. Aiken, United States Marshal for that district, who notified the War Department on October 18 that he had the balloon in his custody.[23] Thereupon Assistant Secretary Scott telegraphed Aiken to forward the apparatus to McClellan's headquarters.[24] In the meanwhile Lowe had learned of the balloon's whereabouts and had instructed John Toy, superintendent of the Assembly Buildings in Philadelphia, where the new balloons were being built, to proceed to Delaware and claim the envelope. Toy took with him an impressive letter of identification and authority, signed by Lowe over the title of " Aeronaut, Commanding Balloon Department, Army of the Potomac." [25] The marshal thereupon disregarded Secretary Scott's instructions and surrendered the balloon to Toy, who brought it to Philadelphia for repairs.[26] The fabric of the envelope was found to be entirely uninjured, despite the rigors of the storm. Only the valve, and the cordage where it had been severed by the wind, had been damaged. These equipages were replaced and the envelope was given a fresh coat of varnish, thus restoring the *Union* to excellent condition.[27]

[21] Philadelphia *Inquirer*, October 16, 1861; Philadelphia *Ledger,* October 14, 1861; New York *World*, October 14, 1861; New York *Tribune*, October 14, 1861.

[22] Philadelphia *Inquirer*, October 18, 1861.

[23] Aiken to the Secretary of War, October 18, 1861, MS [number omitted], Letters Received, Secretary of War, War Department Division, National Archives [hereinafter cited as LRSW].

[24] Assistant Secretary Thomas A. Scott to Aiken, October 21, 1861, MS Letterbooks of the Secretary of War, Vol. XLVI, p. 367, War Department Division, National Archives [cited hereinafter as LBSW].

[25] Lowe to " Whom It May Concern " [letter of identification, naming John Toy authorized to receive government balloon], October 15, 1861, MS, Lowe Papers.

[26] Philadelphia *Inquirer*, October 18, 1861; *ibid.*, October 19, 1861.

[27] Lowe to Macomb, October 21, 1861, MS, Lowe Papers.

Lowe, now free from further interruption, continued work on the new balloons. The first of the four was completed about the first of November and was shipped to Washington along with the newly overhauled *Union*.[28] The remaining three were finished about a week later, and the press announced that " Professor Lowe has completed his contract for five balloons, to be used for observatory purposes." [29] The increase in aeronautic apparatus fostered by Porter and McClellan did not end with the completion of this materiel. Some three weeks later, after the new aerostats had been placed in service at various points along the Potomac lines, and after McClellan had ordered one sent to General Thomas W. Sherman at Port Royal, South Carolina, Lowe requested permission to build still two more balloons of smaller size, declaring that the interests of the new branch of the service required this additional equipment immediately.[30] After a short delay, McClellan approved the request,[31] and the two extra aerostats were completed shortly after the first of January, 1862.[32] Thus at the turn of the year the equipment under Lowe's direction included seven balloons of varying sizes and capacities.[33]

All the new aerostats were similar in construction to Lowe's first government balloon built during the previous August. Fawn-colored india silk, of the type commonly called " pongee," sewn in double thickness, was used throughout in the envelopes. In the *Union*, fifty-nine bolts of fabric, comprising approximately 1,200 yards, were expended in construction, which had required the labor of fifty seamstresses to complete.[34] The large

[28] Philadelphia *Inquirer*, November 18, 1861.

[29] Boston *Journal*, November 8, 1861; Boston *Transcript*, November 8 and 9, 1861; Hartford *Daily Courant*, November 8, 1861; New York *Tribune*, November 8, 1861; New York *World*, November 8, 1861; Philadelphia *Ledger*, November 8, 1861; Washington *Star*, November 8, 1861; Providence *Post*, November 8, 1861; New York *Herald*, November 8, 1861.

[30] Lowe to Colburn, December 3, 1861, 3 *O. R.*, III, 268.

[31] Lowe to Macomb, January 16, 1862, MS, Lowe Papers.

[32] E. Mason, Jr. [Assistant Aeronaut, USA] to Lowe, January 3, 1862, MS, Lowe Papers; Lowe to Colburn, February 17, 1862, MS, Lowe Papers; Philadelphia *Inquirer*, January 3, 1862; Baltimore *Republican*, January 3, 1862.

[33] Lowe to Macomb, February 17, 1862, MS, Lowe Papers.

[34] Philadelphia *Inquirer*, November 8, 1861; Whipple to Bache, August 7, 1861, MS W938, LRTE.

gas bags consisted of segments of fabric cut from carefully designed patterns, in principle not unlike those employed by a dressmaker.[35] These segments, which somewhat resembled sections of a carefully peeled orange rind, were finished with tapering, biased sides, and were joined with reinforced seams to form the spherical and pear-shaped envelopes of the assembled balloons.[36] In the top of each envelope was fitted a valve, constructed of mahogany with fittings of brass [37] and operated by india rubber springs,[38] carefully cemented into the fabric, which was strengthened by several additional folds at this point of particular strain.[39] The valve shutter and the flanges that received it when closed were treated with a substance called valve cement, prepared from a special mixture of beeswax, brown soap, mutton tallow, and paraffin.[40] The lower part of the envelopes terminated into small, tapering "necks," or appendices, not unlike the trunk of an elephant in shape. These appendices were left open at all times to provide for the escape of gas which might expand to excessive pressure as a result of increased temperatures and also allowed the escape of any moisture that might form within the envelope.[41] Since the gas always expanded upward and outward, the open appendix did not allow the lifting vapor to escape except when the

[35] Mason to Lowe, January 3, 1862, MS, Lowe Papers.

[36] Pencil sketches, with notations, illustrating balloon construction [undated], MS, Lowe Papers.

[37] Lowe to "Whom It May Concern," October 15, 1861, MS, Lowe Papers.

[38] Statement of Equipment Expenditures, 1861, 1862, 1863, Item: "December 2, 1861, India Rubber Springs for Balloon Valves," MS L429, enc., LRSW (1864); "Inventory of Balloon Eagle" [undated], signed by E. S. Allen [Assistant Aeronaut, USA], MS, Lowe Papers; Statement of Accounts, Voucher 17, February 11, 1862, MS, Lowe Papers; Statement of Accounts, Voucher 26, March 17, 1862, MS, Lowe Papers.

[39] Captain Frederick F. E. Beaumont [Royal Engineers], "On Balloon Reconnaissances as Practiced By the American Army," Papers Connected with the Duties of the Corps of Royal Engineers, n. s. (Woolwich, 1863), Vol. III, p. 95 [cited hereinafter as Beaumont].

[40] Statement of Accounts, Subvoucher 30, October 20, 1862, MS, Lowe Papers; Lowe to Brigadier General Rufus Ingalls [QMG, AP], March 6, 1862, MS, Lowe Papers; Lowe to Ingalls, February 28, 1863, MS, Lowe Papers.

[41] During Lowe's flight to South Carolina moisture in the gas was frozen into minute nodules resembling hail, and almost a bushel of these fine particles escaped from the open neck of the balloon Enterprise. Cf. Lowe's MS Memoirs, p. 41.

envelope was filled to capacity. The top section, where the strain was greatest, was reinforced with extra thicknesses of fabric. A control cord for operating the valve passed through the center of the envelope and, issuing from the open appendix, was loosely attached to the netting hoop, within easy reach of the aeronaut.[42]

To insure the retention of gas and for protection against weather, the fabric of the envelopes was treated with a varnish made from raw linseed oil, benzine, and japan drier. The oil was first boiled over a slow fire until it approached the consistency of gum, benzine and japan drier being added to thin the oil and render it applicable with a brush. The envelopes were then thoroughly coated on the outside with the varnish and allowed to dry, after which they were turned inside out and treated on the interior surface with neat's-foot oil to render the fabric soft and pliable.[43] Altogether four coats of exterior varnish were applied to the new balloons before they were commissioned into service,[44] and the effects of sun, rain, and frost frequently necessitated revarnishing in the field.[45] The chief function of the varnish was to render the fabric gas-tight.

[42] Beaumont, p. 95.

[43] Lowe to Brigadier General A. A. Humphreys, March 19, 1863, MS, Lowe Papers; E. Seaver [Assistant Aeronaut, USA] to Lowe, February 23, 1862, MS, Lowe Papers; Statement of Accounts, Aeronautic Corps, Item: November 18, 1862, " French and Richards, for Benzine for mixing the varnish as coating for balloons," MS, Lowe Papers; Item: [undated] " John Woodruff, for Prepared Oil to be used in recoating balloons," *loc. cit.*; "List of Property Purchased by Captain J. B. Howard, A. Q. M., for Use of the Balloon Party," Item: " French and Richards, December 27, 1861, 42 gals. Benzine for varnish," MS [number omitted], in LRSW (1863); Item: " John Woodruff, January 10, 1862, 42 gals. Prepared Oil," MS, *loc. cit.* Some of the details of the varnishing process were furnished by Mr. Roy Knabenschue, of Washington, D. C., a balloonist who was personally acquainted with Lowe in his later life, and used Lowe's process for treating his balloon envelopes.

[44] Philadelphia *Inquirer,* October 14, 1861.

[45] New York *Commercial Advertiser,* September 3, 1861; John H. Steiner [Assistant Aeronaut, USA] to Lowe, January 20, 1862, MS, Lowe Papers; Steiner to Lowe, January 22, 1862, MS, Lowe Papers; Steiner to Lowe, January 25, 1862, MS, Lowe Papers; James Allen [Assistant Aeronaut, USA] to Lowe, July 12, 1862, MS, Lowe Papers; Allen to Lowe, July 25, 1862, MS, Lowe Papers; Lowe to Captain C. B. Comstock [CE, AP], April 21, 1863, MS, Lowe Papers; Starkweather to Lowe, January 13, 1862, MS, Lowe Papers.

It added no tensile strength to the silk, nor did it increase or preserve elasticity, although Lowe referred to the preparation he used as " elastic balloon varnish." [46]

The cordage and rigging consisted of an enveloping network of strong " linen balloon cord " [47] woven into meshes of diamond pattern,[48] reinforced with drawn knots.[49] The whole netting fitted over the envelope and was drawn together just below the appendix, where it terminated in a series of cords attached to a stout ring, technically called a concentration hoop.[50] The cordage originally employed with the *Union* had been designed to withstand a strain of ten tons, but after the accident at Lewinsville Lowe equipped all his balloons with rigging capable of resisting a twenty-five-ton stress.[51]

The wicker car, or basket, was attached to the concentration hoop by a number of strong ropes, which were kept free from interfering with the aeronaut or observer by a second spreading, or stay hoop, placed below the concentration ring.[52] The cars, constructed of willow or rattan, varied in size from small baskets accommodating only one observer, to larger ones capable of carrying four or five men.[53] The dimensions of one of

[46] Lowe to Ingalls, March 6, 1863, MS, Lowe Papers.

[47] Statement of Accounts, Balloon Department, 1861, 1862, Item: " John S. Lee and Co., December 28, 1861, 127½ lbs. linen balloon cord," MS [number omitted], in LRSW (1863) ; Item: June 10, 1862, " 122 lbs. linen balloon cord," *ibid.*

[48] The Brady photograph of the *Intrepid* in process of inflation, in T. S. C. Lowe, " The Balloons with the Army of the Potomac," in George Trevelyan Miller (ed.), *Photographic History of the Civil War* (New York, 1911), VIII, 379, illustrates the pattern clearly.

[49] Mason to Lowe, January 3, 1862, MS, Lowe Papers.

[50] Beaumont, p. 95; Lowe to Macomb, February 13, 1862, MS M289, enc., LRAP; Seaver to Lowe, March 10, 1862, MS, Lowe Papers; " List of Property at Camp Lowe " [undated], signed by John Steiner, Assistant Aeronaut, MS, Lowe Papers.

[51] *Lowe's Report, 3 O. R.,* III, 265.

[52] Steiner to Lowe, March 10, 1862, MS, Lowe Papers. The position and function of the concentration and stay hoops are clearly illustrated in a published Brady photograph in T. S. C. Lowe, " Observation Balloons in the Battle of Fair Oaks," *Review of Reviews* (New York, 1911), XLIII, 187.

[53] Lieutenant W. A. Roebling [ADC, Hdq. 2nd Corps, AP] to Major General A. A. Humphreys [C of S, AP)] February 24, 1863, MS L429, enc., LRSW (1864).

the smaller baskets is given in one of Lowe's dispatches as five feet by three, and two feet deep.[54]

The control cables for captive ascensions and for towing the inflated balloons during marches were of strong manilla rope, varying in length up to 5,000 feet.[55] Lowe made it a rule to employ not less than three cables during ascensions, and sometimes used four when conditions called for added security.[56] Attached to the concentration hoop, they distributed the strain of the buoyant envelope evenly throughout the netting, thus assuring all reasonable safety during ascension.[57] These ascension ropes, as they were technically called, were equipped with pulley blocks [58] of tough *lignum vitae*,[59] for the paying out and recovery of the cables as the balloon was allowed to rise or was lowered by the ground crew. Some of the pulleys were of the snatch block type,[60] for checking the ropes when the desired altitude had been attained. Large windlasses were not carried with Lowe's train. The ropes were controlled by hand through the pulley blocks, which were often attached to trees, and the ground crew was split into several groups operating like tug-of-war teams, each handling a single cable.[61]

[54] Lowe to Assistant Secretary of War Peter H. Watson, April 15, 1863, 3 *O. R.*, III, 274.

[55] Marcy to Secretary of War Stanton, April 12, 1862, MS Letterbooks of the Army of the Potomac, War Department Division, National Archives, Vol. XXXIII [cited hereinafter as LBAP]; Lowe to Macomb, February 13, 1862, MS, Lowe Papers; Lowe to Macomb, February 13, 1862, MS M289, enc. 1, LRAP; "List of Property at Camp Lowe," signed by John Steiner, MS, Lowe Papers.

[56] *Lowe's Report*, 3 *O. R.*, III, 274.

[57] Beaumont, p. 95; Lowe to Porter, September 6, 1861, MS [number omitted], LRPD; see Brady photograph in *Review of Reviews*, XLIII, 187.

[58] Seaver to Lowe, March 10, 1862, MS, Lowe Papers; "List of Property at Camp Lowe," signed by John Steiner, MS, Lowe Papers; Lowe to Ingalls, February 28, 1863, MS, Lowe Papers; Lowe to Ingalls, March 6, 1863, MS, Lowe Papers; "Inventory of Balloon *Eagle*, paraphernalia in charge of E. S. Allen" [undated], MS, Lowe Papers.

[59] "Inventory of Articles received from Lieutenant H. R. Williams [AAQM], August 10, 1863," MS L429, enc., LRSW (1864).

[60] *Ibid.*; Williams to Brigadier General W. Scott Ketchum [AIG, USA], November 6, 1863, MS 429, enc., LRSW (1864).

[61] The method is well illustrated in the Brady photograph in *Photographic History*, VIII, 377; see also Philadelphia *Press*, June 23, 1862; Philadelphia *Inquirer*, March 8, 1862, New York *Tribune*, June 24, 1862.

In size and lifting power the several balloons varied considerably. The largest of the seven were the *Union* and her sister aerostat, *Intrepid*, each of 32,000 cubic-foot gas capacity and a normal lifting power of four cables and five men. In very favorable weather, when the envelopes could be fully inflated, this lifting power was doubled at ordinary altitudes. Next in size were the *Constitution* and the *United States,* not so large or powerful as those of the *Intrepid* class, yet capable of lifting three men and four ropes as a normal load. Their capacity was rated as 25,000 cubic feet. The *Washington* was next in order, containing 20,000 cubic feet when fully inflated, and normally raised a load of two men and four ascension ropes. The latest additions to the corps, the two balloons built in December, 1861, were the smallest to be commissioned into service. Known as the *Eagle* and the *Excelsior*, they had envelopes of only 15,000-foot capacity, less than half of those of the *Intrepid* class. Their normal lifting power was one man and four cables. Like the aerostats of the largest type, the lifting power of the several lesser classes became doubled when the envelopes were filled to capacity, and such maximum inflation could be done only in calm weather.[62]

The dimensions of these balloons did not vary as much as did their gas capacities. Because of their spherical and semi-spherical shape, a few feet difference in diameter and height increased or decreased the volume of the envelope in much greater proportion than the corresponding change in their lineal measurements. The *Intrepid* and *Union* measured 38 feet in diameter at the widest point of circumference, and their height from valve to tapering appendix was 45 feet.[63] The others ranged from 30 to 35 feet in diameter, with corresponding variations in height.[64] The *Eagle* and the *Excelsior* were nearly spherical; the envel-

[62] Roebling to Humphreys, February 24, 1863, MS L429, enc., LRSW (1864); "Report on the Present Condition of the Balloon Corps and the Balloons," June [n. d.] 1863, MS, Gouverneur K. Warren Papers, Letters Chancellorsville, May 1–June 28, 1863, Vol. I, New York State Library, Albany [cited hereinafter as Warren Papers]. Permission to use General Warren's papers was graciously granted by his daughter, Miss Emily B. Warren, of Newport, R. I.

[63] Philadelphia *Inquirer,* August 8, 1861.

[64] Philadelphia *Inquirer,* October 14, 1861.

opes of the larger classes were of the more tapering, pear-shaped type.

The strong nationalist sentiments expressed in the selection of names for the several balloons were also responsible for the decorations that were added to the envelopes. Huge figures of the symbolic American eagle, portraits of George Washington and other distinguished patriots, and brightly painted representations of the national flag were imposed as a final touch on the neutrally colored, varnished fabrics of the envelopes. The wicker cars were likewise painted a bright sky-blue, studded with white stars.[65] Some of the smaller baskets were adorned in representation of the national standard, bearing red and white vertical stripes in addition to the blue field and white stars.[66]

The *Union,* which had been completed in August and hurried into the field as soon as the varnish was dry, had escaped such colorful touches, but during the period of repairs after the accident at Lewinsville, it also received similar embellishments. " It is now stretched out," wrote a Philadelphia reporter who visited the Assembly Buildings where the balloon construction and repairs were in progress, " in order to be decorated with a gigantic spread eagle, the Stars and Stripes, and other devices of the Professor's fertile fancy." [67] The *Intrepid,* in addition to similar graphic expressions of patriotism, also bore its name conspicuously on its side in large serrated letters.[68]

The *Constitution* was likewise floridly adorned, and a huge portrait of George Washington, staring form its bulging side so impressed a Massachusetts chaplain whose regiment was stationed near the balloon at Budd's Ferry that he noted the fact in a regimental history some five years later.[69] A similar portrait displayed on the envelope of the balloon *Washington* also

[65] Philadelphia *Inquirer,* October 22, 1861. The cars shown in the photographs in *Review of Reviews,* XLIII, 189, and in *Photographic History,* VIII, 381, clearly show the star decorations.

[66] See photograph in *Photographic History,* VIII, 377.

[67] Philadelphia *Inquirer,* October 22, 1861.

[68] See photograph of the *Intrepid,* in *Photographic History,* VIII, 375.

[69] Chaplain Warren H. Cudworth, *History of the First Regiment, Massachusetts Volunteer Infantry* (Boston, 1866), p. 99.

drew comment from the historian of another regiment from the same State.[70]

In addition to the major pieces of aeronautic materiel, such as the envelopes, cordage, netting, and cars, the trains of the balloon service were supplied with a large and diverse assortment of auxiliary equipment, including various cordage and netting attachments, spare parts, repair materials, tack and harness for transportation, headquarters property and other miscellaneous items.

Among the several cordage attachments were the sandbags for ballast,[71] and the grappling irons.[72] The bags were of canvas or strong cotton duck, and carried a normal load of 100 pounds of sand ballast. They were fitted at the top with brass grommets, through which the ropes attaching them to the car and netting were passed. The ropes were attached to the car or netting with brass snaphooks, which could be instantly disengaged when it should be necessary to cast loose from the ballast.[73] Thirty or forty of these sandbags were kept available at each balloon station. The grappling irons were simply strong, anchor-like hooks of iron, which were at times fastened to the mooring cables for the purpose of making the balloon fast during ascensions, thus relieving the ground crew. They were also used to anchor the balloons when they were inflated and not in use.[74]

Just as an artillery unit carried spare parts for its ordnance materiel in a battery store-wagon, the several balloon detachments of the corps kept with their bases of operations spare

[70] Captain Joseph Keith Newell, *Annals of the 10th Regiment, Massachusetts Volunteers, in the Rebellion* (Springfield, 1875), p. 194.

[71] Lowe to Ingalls, March 6, 1863, MS, Lowe Papers; Seaver to Lowe, enclosing inventory of balloon equipment at Budd's Ferry, Md., March 10, 1862, MS, Lowe Papers; Lowe to Captain J. B. Howard [AQM], January 24, 1863, MS, Lowe Papers.

[72] Seaver to Lowe, March 10, 1862, MS, Lowe Papers; Lowe to Macomb, February 14, 1862, MS, Lowe Papers; Statement rendered for costs of balloons *Eagle* and *Excelsior,* February 17, 1862, MS, Lowe Papers; Inventory of balloon property received from Lieutenant H. R. Williams [AAQM], August 18, 1863, MS [number omitted], LRSW (1864).

[73] Lowe to Ingalls, March 6, 1863, MS Lowe Papers; Beaumont, p. 97.

[74] Lowe to Macomb, February 14, 1862, MS, Lowe Papers.

parts or fittings for the aerostats and appendages in case of accident, breakage, or expenditure in service. Extra valves, valve-springs, concentration and stay hoops, and other replacement parts were carried with the balloon parties as regular equipment.[75] Hence a worn out or damaged unit could be replaced in the field without serious interruption to the observation facilities of the station involved. Spare ascension ropes, with their complements of pulley and snatch-blocks, were also kept ready for use so that damaged cables and their attachments could be replaced in time of emergency. A reserve supply of additional sandbags, with spare grommets and snaphooks, was also maintained against possible need.[76]

The repair materials and maintenance equipment of the corps present a list of diverse articles that must have seemed extraordinary to the quartermaster and finance officers who were called on to approve and pay the accounts. The items carried in the trains and kept on hand at the several stations varied from brooms to beeswax and from thimbles to pickaxes. A complete set of materials was maintained for the repair of nettings, which included varying quantities of linen balloon cord, sailmaker's twine, mesh sticks for knitting the nets, and other minor articles. Ready repairs to the silk envelopes were assured by the many items of repair equipment that were stocked in the trains. For this purpose the aeronauts kept a supply of spare silk and reinforcing muslin for patching and replacing worn segments of the fabric, balloon thread, sailmaker's needles, thimbles and palms, and sewing boards for the convenience and efficiency of repair work.

Stores of fabric varnish and ingredients for its mixing, and supplies of paint and brushes to keep the wicker-work of the cars safe from weather damage were also kept ready for use.

[75] Lowe to Ingalls, February 28, 1863, MS, Lowe Papers; Lowe to Ingalls, March 6, 1863, MS, Lowe Papers; Inventory of balloon *Intrepid,* signed by John H. Steiner [n. d], MS, Lowe Papers; Inventory of balloon *Eagle,* signed by E. S. Allen [n. d.], MS, Lowe Papers; Statement of Accounts, Balloon Corps, Voucher No. 22, MS L429, enc., LRSW (1864).

[76] Lowe to Ingalls, February 28, 1863, MS, Lowe Papers; Inventory of balloon *Eagle,* MS, Lowe Papers; Lowe to Howard, January 24, 1863, MS, Lowe Papers; Allen to Lowe, July 8, 1862, MS, Lowe Papers.

Extra rattan for repairing the cars was also included. For the valves, raw materials for making the fixative cement that sealed their flanges were also available in the regular supplies. As a rule these materials were carried in a large canvas container called a " property bag," in which other items of repair equipment of small size were also stored. Some of the stations carried ready prepared valve cement, which saved time in preparation.[77]

The maintenance equipment included a variety of carpenter's and engineer's tools, among which were axes, saws, brace and bits, nails, grindstones, shovels, spades, picks, mattocks, and large brooms.[78] Large pieces of heavy canvas or other material, called ground cloths, were spread on the ground and provided a clean, dry area for the balloons during inflation. The heavy brooms served to prepare the ground for these cloths and to keep them clean. The ground cloths were often made from old tentage, and a number of captured Confederate tents, some of those left standing after the evacuation of Yorktown, were

[77] Lowe to Ingalls, February 28, 1863, MS, Lowe Papers; Mason to Lowe, January 3, 1862, MS, Lowe Papers; Inventory of balloon Eagle, MS, Lowe Papers; Seaver to Lowe, February 23, 1862, MS, Lowe Papers; James Allen to Lowe, June 15, 1862, MS, Lowe Papers; Statement, "United States to Edgar G. Steener, Dr." [for paints and brushes], items from September 10 to November 11, 1862, MS, Lowe Papers; Lowe to Macomb, February 15, 1862, MS, Lowe Papers; Lowe to Ingalls, March 6, 1863, MS, Lowe Papers; Inventory of balloon Constitution, in Seaver to Lowe, March 10, 1862, MS, Lowe Papers; Lowe to Humphreys, March 9, 1863, MS, Lowe Papers; Lowe to Howard, November 18, 1862, MS, Lowe Papers; Statement of Accounts, Balloon Corps, July 24, 1862–March 27, 1863, MS, Lowe Papers; Lowe to Howard [enclosing accounts], January 24, 1863, MS, Lowe Papers; Statement of Accounts, Balloon Corps, Voucher No. 22, MS L429, enc., LRSW (1864); Descriptive List, " Property purchased by Captain J. B. Howard, AQM," MS [number omitted], LRSW (1863); Statement, Thain and McKeone, Philadelphia, October 20, 1862, for paraffin, MS, Lowe Papers; Statement, Jackson, Brother & Company, Washington, December 31, 1862, for brown soap, MS, Lowe Papers.

[78] Inventory of Balloon Constitution, MS, Lowe Papers; Statement, "United States to Edgar G. Steener, Dr." [for saws and brace and bits], November 10, 1862, MS, Lowe Papers; Statement of Accounts, Balloon Corps, Voucher No. 22, MS L429, enc., LRSW (1864); Inventory of balloon Eagle, MS, Lowe Papers; Inventory of balloon Intrepid, MS, Lowe Papers; Inventory of balloon property received from Lieutenant H. R. Williams [AAQM], August 18, 1863, MS [number omitted], LRSW (1864); Williams to Brigadier General W. Scott Ketchum [AIG, USA], November 6, 1863, enc. 1 [dated May 12, 1863], MS L429, enc., LRSW (1864); Lowe to Comstock, April 13, 1863, MS, Lowe Papers.

appropriated for this purpose.[79] Lowe also personally contributed the huge envelope of the ill-fated *Great Western* for the same use.[80]

Added to this miscellany was the headquarters property and transportation equipment. The former category included various forms of tentage, camp stoves, field tables, and other items of furniture. Lowe's quarters also housed two field desks for the safe filing of orders, dispatches, and accounts. Stationery, drawing and sketching paper for aerial maps and panoramic sketches, pencils, and similar material were also kept available. Other important items were field glasses, of which four were issued to the corps for the use of the aeronauts.[81]

Since Lowe and his assistants frequently had numerous errands to perform in the field, such as changing observation positions and visiting distant stations and headquarters, saddle horses became requisite to their functioning in the field. The addition of animals to balloon headquarters also added various items of tack, harness, and other items incident to the service of mounted men.[82]

[79] Statement rendered for costs of balloons *Eagle* and *Excelsior*, February 17, 1862, MS, Lowe Papers; Lowe to Brigadier General Edward M. Canby [AAG, USA], January 22, 1864, MS [number omitted], LRSW (1864).

[80] Seaver to Lowe, December 17, 1861, MS, Lowe Papers.

[81] Receipt for Balloon Corps property, signed by James Allen, May 7, 1863, MS [number omitted], LRSW (1863); Lowe to Ingalls, March 6, 1863, MS, Lowe Papers; Lowe to Ingalls, February 28, 1863, MS, Lowe Papers; Inventory of balloon property received from Lieutenant H. R. Williams, August 18, 1863, MS [number omitted], LRSW (1864); Lowe to Comstock, April 13, 1863, MS, Lowe Papers; Lowe to Macomb, February 13, 1862, MS, Lowe Papers; Inventory of Balloon *Eagle*, MS, Lowe Papers; Captain J. B. Howard to Lowe, March 16, 1863, MS, Lowe Papers; Memorandum Receipt [issue] for QM property issued Balloon Department, signed by Captain Daniel G. Thomas [MSK], December 13, 1861, MS, Macomb Papers; Statement, Charles Macarce & Company, Philadelphia, for stationery, December 1, 1862, MS, LRSW (1863); Statement, W. H. Harrower, Washington, for camp stove, December 10, 1862, MS, LRSW (1863); Statement, H. J. Gregory, Washington, for camp stove, November 12, 1861, MS, LRSW (1863); Statement, J. C. Moran, Washington, for pens, paper, and drawing equipment, January 18, 1862, MS, Lowe Papers.

[82] Inventory of balloon *Eagle*, MS, Lowe Papers; Special Requisitions, QMC Form No. 40, November 30, 1861, and March 10, 1862, signed by Colonel J. N. Macomb, MSS, Macomb Papers; Memorandum Receipts, QM Stores, Form No. 27, November 30, 1861, and March 10, 1862, signed by Captain J. J.

Still another classification of equipment in addition to those enumerated was the signal and telegraphic apparatus that accompanied the corps. When the value of balloon observation had become firmly established, General McClellan permitted the addition of a telegraph train to the aeronautic establishment, thus relieving it of dependence upon the Army Telegraph Corps for telegraphic equipment.[83] The train consisted of five miles of wire, insulated with vulcanized rubber, three Beardslee Patent Magneto-Electric Field Telegraph sets,[83a] four wire reels and stands for setting them up in the field, four bearers for carrying the reels, and two crowbars equipped with binding screws for ground connections.[84] With this separate apparatus, the balloon corps was fully independent of other branches of the service with regard to telegraph communication, except that the operators who sent and received for the corps were not permanent balloon personnel. When aerial telegraph was used, operators were borrowed from the Telegraph Corps or from the headquarters of various high commands.

After the Fredericksburg Campaign, Lowe designed a new system of long distance visual signalling by means of small balloons, which bore conspicuous markings for day communication and different colored flares for night operations.[85] He also constructed a powerful oxyhydrogen or calcium light apparatus,

Dana [AQM, AP], MSS, Macomb Papers; Statement, F. A. Lutz, Washington, for horse blankets and surcingles, December 28, 1861, MS, Lowe Papers; "List of QM stores, transferred by Captain J. J. Dana to Lt. Col. J. N. Macomb," November 30, 1861, MS, Lowe Papers.

[83] Lowe to Major General John G. Parke [C of S, AP], November 20, 1862, 3 *O. R.*, III, 293; Acting Master Henry J. Rogers [USN, WNY], to C. Poppenhausen, November 13, 1862, MS, Lowe Papers.

[83a] A technical description and discussion of the functioning and operation of the Beardslee telegraph, accompanied by a detailed illustration, appears in *Scientific American,* December 7, 1861.

[84] Descriptive List, "Signal Telegraph Train, USA," Lowe Papers. This document is a printed regulation list of equipment allotted to a standard signal telegraph train, undated, but probably of the period of the Peninsular Campaign. The items mentioned in the above text as assigned to the balloon corps are those on the printed form, checked off with notes in Lowe's handwriting. It is evident that he selected items for the use of his train from the descriptive list, and checked them off accordingly.

[85] Lowe to Parke, November 20, 1862, 3 *O. R.*, III, 293.

to facilitate night operations of the balloonists when secrecy of position was not essential.[86] It was designed to provide ample light for night inflation of the envelopes and at the same time it removed the danger of igniting the highly inflammable hydrogen. A strong reflector, 18-inches in diameter, was used with this light to throw beams in any desired direction.[87] The usefulness of this extra equipment, Lowe reported to the Chief of Staff, was not confined to the balloon corps, but could be made advantageous for engineer operations at night, such as bridge building, earthwork construction, or the crossing of streams.[88]

Shortly after Colonel Macomb had been detailed to supervise the balloon operations of the army and had examined the various accounts connected with the service, he expressed surprise at the cost of equipment and the expenses of operation. " Ballooning for the army," he wrote to Major Bache in October, 1861, " is a very expensive business." [89] If the normal cost of warfare is considered, then this statement was unwarranted. Doubtless it was the result of Macomb's lack of knowledge and familiarity with the operation and maintenance of balloons. Macomb's opinion, however, raises the question of what the expenses were, and also the cost of the various items of materiel involved.

It is not possible to obtain figures for the cost of all items of equipment, but some of the accounts and vouchers still extant throw light on the subject. The gross cost of the several balloons, complete with netting, cordage, ropes and appendages, ready for service, varied according to size. The largest, the

[86] Receipt for Balloon Corps property, signed by James Allen, May 7, 1863, Item: " One set, Hydro-Oxygen Light Apparatus, complete," MS [number omitted], LRSW (1863) ; Lowe to Howard, January 31, 1863, MS L429, enc., LRSW (1864) ; Lowe to Parke, November 20, 1862, 3 *O. R.,* III, 293.

[87] Inventory of balloon property received from Lieutenant H. R. Williams, August 18, 1863, MS [number omitted], LRSW (1864) ; Williams to Ketchum, November 6, 1863, enclosing list of property, MS L429, enc., LRSW (1864) ; " List of balloon property purchased by Prof. T. S. C. Lowe, Paid for by Captain J. B. Howard," MS [number omitted], LRSW (1863).

[88] Lowe to Parke, November 20, 1862, 3 *O. R.,* III, 293.

[89] Macomb to Bache, October 23, 1861, MS, M1310, Letters Received, Bureau of Topographical Engineers, War Department Division, National Archives [cited hereinafter as LRTE].

Intrepid and the *Union,* required expenditures of $1,500 each. The lesser aerostats, including the *Constitution, United States,* and the *Washington,* averaged $1,200.[90] And the two smallest of the fleet, the *Eagle* and *Excelsior,* cost $1,000 each.[91] Thus the initial outlay in balloons for the corps, each with an original set of fixtures and appendages complete for service, amounted to about $8,600, a sum slightly less than the pay and allowances of a major general of the line for one year.[92]

The cost figures for individual parts, replacement and repair items, are available in some instances. These figures, however, varied with the fluctuating prices of the commodities involved. Even so, they give some idea of the prices of various articles used by the corps. The cost of silk for the envelope of the *Union* amounted to $531,[93] a figure that may serve as a maximum, since this balloon was one of the largest built. A complete balloon net normally cost $200, of which sum approximately $40 was paid for the labor of knitting, unless the work was done by the aeronauts in the field. Linen balloon cord used in the nets and for their repair averaged $1.00 per pound, and was generally ordered in lots of fifty or one hundred pounds. Cars of the larger type were bought for $12; those of the smaller sizes, for two dollars less. Concentration and stay hoops were obtained for $2.00 each, and sandbags, towards the end of 1862, were high at $.70. The brass and mahogany valves of the latest improved pattern averaged from $20 to $30 each. One was listed in the accounts at $26. India rubber springs for their control were rated from $1.00 to $1.50.[94] The various small

[90] Lowe to Porter, September 16, 1861, 3 *O. R.,* III, 262; Lowe to Macomb, December 5, 1861, MS, Lowe Papers; Sibley to Crossman, October 1, 1861, 3 *O. R.,* III, 254.

[91] Statement, "United States to T. S. C. Lowe, Dr., To two superior pongee silk balloons, named *Eagle* and *Excelsior,*" February 17, 1862, MS, Lowe Papers.

[92] *Pay Table, United States Army, 1863-1865* [compiled by J. Lowenthal under direction of the Chief Paymaster, USA], Washington, 1865. When it is recalled that the Federal and State governments poured out the sum of more than $500,000,000 on bounties for enlistments in two years, the expenses of the balloon service seem less than trivial. Cf. Fred Albert Shannon, *The Organization and Administration of the Union Army* (Cleveland, 1928), II, 80.

[93] Whipple to Bache, August 7, 1861, MS W938, LRTE.

[94] The figures on costs given in the above paragraph are taken from the

articles for repairing the envelopes were obtained at trifling expense.[95] Cost figures for the mooring ropes and other articles of equipment are lacking in the available accounts and documents. The two other major items of expense include the payment of personnel and the cost of generators and materials for the manufacture of gas.

The problem of keeping the balloons inflated in the field, in sectors not adjacent to commercial sources of gas, had been apparent to all concerned with the aeronautic operations from the first. To remedy this difficulty, Lowe designed a mobile field generator, adapted to an ordinary army wagon frame, capable of accompanying the army into almost any reasonable position. Both Wise and La Mountain had experimented with apparatus working on the water decomposition principle and had failed to produce a practical field generator. Excess weight, size, complicated machinery, and general unwieldy proportions were all unfavorable factors in these earlier attempts. Discarding the unsuccessful methods previously tried by his predecessors, Lowe adopted the sulphuric acid and iron process, which enabled him to design a relatively light, compact, and wholly efficient apparatus, capable of producing sufficient hydrogen to fill a balloon within a few hours after inflation operations were begun. In fact the machine thus produced was the the first successful mobile field generator ever to accompany an army in the field.

The Lowe Generator was a comparatively simple apparatus.

following documents: Lowe to Ingalls, March 6, 1863, MS, Lowe Papers; Statement of Accounts, Balloon Corps, Voucher No. 22, MS L429, enc., LRSW (1864); Statement of Accounts, Balloon Corps, 1861-1862, MS L429, enc., LRSW (1864); Statement, N. B. Williams, for muslin and sandbags, April 24, 1863, MS, Lowe Papers; Receipt, signed by J. Murray, April 27, 1863, "for labor in knitting balloon net," MS, Lowe Papers; Statement, John F. De Hart, November 3, 1862, for balloon cars, concentration and stay hoops, MS, Lowe Papers; "List of Balloon Property purchased by Prof. T. S. C. Lowe, Paid for by Captain John B. Howard, AQM, 1861, 1862, 1863, MS [number omitted], LRSW (1864).

[95] "List of Balloon Property . . . Paid for by Captain Howard," MS [number omitted], LRSW (1864); Statement, Jackson, Brother & Company, Washington, December 31, 1862, for brown soap, MS, Lowe Papers; Statement, Thain and McKeone, Philadelphia, October 20, 1861, for paraffin, MS, Lowe Papers.

It consisted of a strong wooden tank, five feet high, eleven feet long, and of the same width as an ordinary escort wagon,[96] acid-proofed on the inside, and braced on the exterior with strong longitudinal, vertical, and diagonal braces to protect the wooden walls against pressure from the generated gas. For the same purpose, the tank was further strengthened by a series of copper rods, fitted horizontally from side to side on the interior. The sides and ends of the tank were fitted with a series of shelves to facilitate the more even distribution of the iron filings, which were introduced through an eighteen-inch circular opening in the top, called a "manhole." This opening was fitted with a tight cover, opened and closed with a series of strong wing-knobs, and sealed when closed with gas-proof packing. In the top of the tank was also fitted a copper funnel and tube, attached to the planking with gas-tight flanges, and equipped with a stop-cock. The tube extended vertically into the tank and served to convey the sulphuric acid into contact with the iron and water in the generator. At the forward end, an elbow pipe of copper issued from the top of the tank and served as a connector to the gum hose that carried the newly generated gas to the cooler and purifier. In the rear wall of the tank was cut a tight door or "gateway," fitted and controlled with iron straps and locking keys, to permit drawing off the oxide of iron and other waste material after the gas had been generated. In the wall above the gateway was fixed a brass stop-cock to facilitate the passage of water after the generating process was complete. The tank and all its attached fixtures were mounted on the running gear of an ordinary escort wagon, held firmly in place by a series of iron bars bolted to the wagon frame and turned up at the sides, thus securing the tank to the running gear.

The cooler and purifier were separate from the generator tank and were connected with it by copper couplers and 6-inch rubber hose. This auxiliary equipment consisted of two stout wooden boxes, treated to make them gas-tight, each containing an inner inverted box with inclined planes over which the gas flowed during the cooling and purifying process. The first box or cooler

[96] Lowe to Henry Forrest [Master Ship-joiner, WNY], October [n. d.], 1861, MS, Lowe Papers.

was filled with water through which the gas passed, flowing over the surfaces of the inclined planes of the inner inverted box. The gas, after passing through this water cooling process, escaped through a copper coupler into a second gum hose which conveyed it into the purifier, which was charged with lime and water. The lime solution absorbed the carbonic acid and other extraneous gases, thus purifying the hydrogen before it was fed into the balloon through a 12-inch connecting hose and valve. A hand force pump was often connected between the purifier and the balloon to facilitate inflation.[97]

The process for making gas with this apparatus was simple and rapid. It required, however, a thorough knowledge of the proper mixture of materials, and careful handling of the equipment. For a single inflation, four barrels of fine iron filings or borings, each weighing approximately 834 pounds, or a total of approximately 3,300 pounds,[98] were introduced into the tank through the manhole at the top. This material was spread as evenly as possible. The tank was then filled with water to within about two feet from the top. This done, the manhole was closed and the wingknobs tightly fastened. Ten carboys of sulphuric acid, averaging 161 pounds each, or a total of approximately 1,600 pounds,[99] were poured into the tank by means of a syphon inserted into the copper funnel. The syphon was made of lead, to resist attack by the acid.[100] The acid was fed in according to a prescribed schedule and rate: five carboys at first,

[97] " Description of T. S. C. Lowe's Portable Gas Generator for Balloons and other Purposes," accompanying application for patent, January 2, 1862, MS, Lowe Papers; also Lowe to Forrest, October [n. d.], 1861, MS, Lowe Papers. One war correspondent, after viewing one of the generators, comically described the apparatus as " much resembling the cages used by itinerant showmen for their curiosities." Philadelphia *Ledger*, December 14, 1861.

[98] The average weight of the filings per barrel has been calculated from the mean weight of fifteen barrels listed in a MS invoice of A. Whitney and Sons, Philadelphia, dated April 21, 1862. The heaviest of the lot weighed 969 lbs.; the lightest, 754. The exact mean for all fifteen barrels was 833.8 pounds. Lowe Papers.

[99] The average figure for the weight of the acid per carboy has been calculated from the mean weight of 200 carboys listed in an invoice of Messrs. Savage and Stewart, Philadelphia, dated April 18, 1862, MS, Lowe Papers.

[100] Lowe to Starkweather, November 28, 1861, MS, Lowe Papers; Inventory of Balloon *Intrepid*, MS, Lowe Papers.

followed by a waiting period equal to the time expended in pouring the acid; then three more carboys, followed by a second time interval; and finally the remaining two carboys. The timed delays between pourings were adopted to prevent too rapid generation of gas, which might strain the walls of the tank. The generated gas then passed through the copper elbow coupling at the forward end into the rubber hose which conveyed it into the water cooler, from which it again passed into the lime purifier which absorbed the impurities and foreign gases. As a result, the gas which flowed from the lime solution into the balloon envelopes was almost pure hydrogen.[101]

In filling the balloons, care had to be exercised to prevent impure gas from entering the envelopes. For this reason the connection between the balloon and the purifier was not made until a pure stream of hydrogen should be obtained. The gas passed from the generator tank at a high temperature, and the connecting pipes to the water cooler became so hot that the hand could hardly bear to touch them. But after passing through the cooler and purifier, the gas was barely warm when it entered the balloon.[102]

Barring accidents, and when handled by competent men, the whole process from the time of preparations to the completed inflation of one of the larger balloons was approximately three hours.[103] Normal inflation of a balloon in the field, Lowe wrote to General Hooker's adjutant general in March, 1863, " never required over three hours and fifteen minutes, and since adding my last improvements . . . the gas makes in two hours and thirty minutes." [104]

The construction of the generators for the corps was begun about the same time that Lowe commenced building the four balloons authorized in September, 1861. By October 21, Lowe reported to Colonel Macomb that the generators were under construction and would be ready for service with the completion

---

[101] Instructions for operating field generator, in Lowe to Starkweather, January 26, 1862, MS, Lowe Papers.

[102] Beaumont, p. 96.   [103] *Ibid.*

[104] Lowe to Brigadier General Seth Williams, March 30, 1863, 3 *O. R.*, III, 300; also MS [copy], Lowe Papers.

of the balloons.[105] They were built under Lowe's plans and supervision by naval carpenters at the Washington Navy Yard.[106] The copper pipes, couplers, flanges, and other fixtures were supplied by Megee and Sons of Philadelphia. Goodyear's Rubber, Belting and Packing Company, of the same city, furnished the hose, rubber pipes, packing and other similar materials.[107] A total of twelve of these portable gas works were built for the corps,[108] six of which were in service in the field by the close of 1861.[109] They were painted a pale blue, with bold black lettering bearing the legend, " Lowe's Balloon Gas Generator," and a serial number.[110]

Although the process of generating gas from the action of sulphuric acid and iron was almost as old as the discovery of hydrogen, and had been proposed for army balloons as early as 1793, Lowe considered his application of the process to a mobile field generator with equipment of his own design an original invention, and accordingly entered a patent application in December, 1861.[111] He subsequently filed caveats on January 2, 1862, again on January 27, and still again on January 29, 1863.[112] There is no record, however, of a patent ever having

[105] Lowe to Macomb, October 21, 1861, MS, Lowe Papers.

[106] Commander J. A. Dahlgren [Comdt., WNY] to Commodore Joseph Smith [Chief of Bureau of Yards and Docks, USN], October 23, 1861, MS, Letterbook, Commandant, Washington Navy Yard, Navy Department Division, National Archives [hereinafter cited as NDNA]; Smith to Dahlgren, October 24, 1861, MS, Letterbooks, Bureau of Yards and Docks, Vol. XVIII [Letters to Commandants], pp. 323-324, NDNA.

[107] Lowe to Macomb, February 13, 1862, MS, Lowe Papers.

[108] George N. Beale [Naval Storekeeper] to the Navy Agent, Washington, indorsed by Commander Dahlgren, November 26, 1861, MS L852, enc. (1863), LRSW.

[109] Lowe to Macomb, February 13, 1862, MS, Lowe Papers.

[110] Statement of Accounts, Balloon Corps, 1861, 1862, Item: " blue paint for generators," MS, Lowe Papers; anon., " Three Months with the Balloons," St. James's Magazine (London, 1863), p. 102; Statement, " United States to Edgar G. Steener, Dr.," Item: September 17, 1862, 50 lbs. blue paint," MS, Lowe Papers; Statement of Accounts, Balloon Corps, Voucher No. 22, Subvoucher No. 8, Item: Lettering Gas Generators, MS L429, enc., LRSW (1864). Also see photograph from Brady print in Signal Corps Collection, Plate XXV, infra.

[111] Application for Patent, December 6, 1861, MS, Lowe Papers.

[112] Receipts for Caveat Fees, signed by John L. Hayes for the Commissioner of Patents, January 2 and 27, 1862, and January 29, 1863, MSS, Lowe Papers.

been granted him for this apparatus. It is possible that the disbanding of the balloon corps in June, 1863, led Lowe to abandon his claims for patent rights. On the other hand, the commissioners may have refused to grant the claims for the new application of a known process.[113]

Each generator was drawn by a team of four horses.[114] When the balloon trains were on the march, loose fixtures, such as pipes, hose, couplers and flanges were carried in the empty generator tanks.[115] Additional wagons were required for the transportation of iron and acid supplies, as well as for the balloons and miscellaneous equipment that accompanied the corps. Lowe's normal train for one balloon generally carried enough iron and acid for one inflation, which would sustain a balloon for several days. Additional gas materials were brought up from rear bases by quartermaster trains or by wagons assigned to the corps for this purpose.[116] Because of the quartermaster department control of most of Lowe's transportation, it is almost impossible to determine the exact number of wagons that were required for each balloon in the field when they operated separately. Lowe gives a figure of seven wagons (not including generators) for a train including four balloons, their appendages, and inflating materials, as well as the camp and garrison equipage of the personnel.[117]

The satisfactory mobility of the generators was proven successfully in several major campaigns. They accompanied the army through the sloughs and mudholes of the Peninsula's impossible roads; through the snow and mire from Washington to the slaughter field at Fredericksburg in mid-winter; over the mountain roads of western Virginia after Antietam; and over the wagon tracks of the wilderness in the Chancellorsville campaign.[118] Only once, during the critical action of the Seven

---

[113] A search of the records of the Patent Office fails to reveal any patent issued in favor of Lowe for these generators.

[114] Beaumont, p. 98.

[115] Lowe to Starkweather and Jacob C. Freno [n. d.], MS, Lowe Papers.

[116] Lowe to Humphreys, May 30, 1862, MS, Lowe Papers.

[117] Lowe to Williams, March 30, 1863, MS, Lowe Papers; also printed in 3 O. R., III, 300.

[118] See below, Volume II.

Days when McClellan withdrew to the James, was it necessary to abandon any of this equipment to the enemy. In this instance, three generators were left on the field, comprising the only aeronautic equipment ever captured by the Confederate forces.[119]

Like the balloons that they serviced, the field generators also necessitated the addition of auxiliary equipment, replacement parts, and repair materials to the trains of the balloon corps. Various tools used in the process of making gas included wrenches of several sizes for the connection of syphons, couplers and adapters, and conveyer pipes; also pumps to expedite the inflation of the envelopes and for filling the generator tanks with water when conditions permitted their use; water hose, and gutta-percha folding buckets; and an assortment of small tools for tightening fixtures and general maintenance. A soldering outfit for use in fitting flanges, joints, and other generator parts was also provided in the equipment.[120] Replacement parts included copper couplers, funnels, and acid conveyer tubes; extra footage of gum connecting hose, a supply of packing for various connections and valves; and spare copper flanges, bolts, and screws.[121] In general only a few of these extra units were carried in the field, since apparatus of this type withstood wear better than the more fragile balloons and their equipages. Barring accidents and damage from hostile gunfire, the generator parts resisted the strain of field service well, and replacements were required less frequently than those necessary for the balloons.[122]

[119] Lowe to Humphreys, March 19, 1863, MS, Lowe Papers.

[120] Lowe to Howard, January 24, 1863, MS, Lowe Papers; Williams to Ketchum, November 6, 1863, MS L429, enc., LRSW (1864); Inventory of Balloon *Eagle*, MS, Lowe Papers; Inventory of Balloon *Intrepid*, MS, Lowe Papers; Statement, Charles Burnham, for pumps, December 9, 1862, MS, Lowe Papers; Lowe to Macomb, February 13, 1862, MS, Lowe Papers.

[121] Statement, " U. S. Government, per Prof. T. S. C. Lowe, to Goodyear's Rubber, Belting, and Packing Co., Dr.," October 24, 1862, MS, Lowe's Papers; Inventory of Balloon *Constitution*, March 10, 1862, MS, Lowe Papers; " List of Balloon Property Purchased by Prof. T. S. C. Lowe, Paid for by Captain J. B. Howard," 1861, 1862, 1863, MS [number omitted], LRSW (1864).

[122] This statement is based on miscellaneous correspondence in the Lowe Papers, and on the fact that only a small number of items for such materials appear in the accounts as against a large number of items for balloon materials over the same period of time.

The cost of the generators is difficult to determine. Exact statements of the expenses involved in the construction of a complete unit are not available. Lowe advised General Porter before the first set was built that the cost of the entire apparatus would not exceed $500,[123] and this figure appears to be as definite as can be found. Since the labor was supplied by government employees in the Washington Navy Yard, this item of expense must be disregarded in the total estimates. The costs of individual parts are listed in some of the balloon corps accounts. One voucher shows the sum of $508.45 for complete sets of couplers, flanges, bolts, pipes, valves, and other metal work used in building three generators,[124] or a total of approximately $170 for each. Another item in one of the accounts reveals $109 expended on hose, gum packing, and other similar materials for one generator.[125] These figures, with $41 added for the cost of one of the pumps,[126] give a total of $320 for known expenditures for parts. Estimated expenses of lumber for the tanks, coolers and purifiers, paint and other incidental materials [127] bring the total relatively close to the limit of $500 submitted by Lowe in his original statement.

A few figures for individual replacement parts and repair materials appear in some of the vouchers. The 6-inch connecting gum hose was listed at $1.45 per foot; the larger 12-inch at $2.75. A cheaper hose of oiled canvas, which Lowe treated with balloon varnish, was later obtained for 35 cents per foot and substituted for the expensive gum. Packing averaged $1.50 per pound, and miscellaneous pipes and hose attachments varied from 50 cents to $1 per foot.[128]

[123] Lowe to Porter, September 16, 1861, 3 *O. R.*, III, 262.

[124] Lowe to Howard, January 24, 1863, item: Statement of Joseph Oates and Son, *c.* December, 1862, MS, Lowe Papers.

[125] Lowe to Howard, November 18, 1862, item: Statement of Goodyear's Rubber, Packing, and Belting Co., MS, Lowe Papers.

[126] Lowe to Howard, November 14, 1862, item: Statement of Charles Burnham, $125 for 3 pumps, MS, Lowe Papers; also original statement, Charles Burnham, for pumps, December 9, 1862, MS, Lowe Papers.

[127] The estimates for lumber, paint, and other incidental expenses are naturally only approximations, but it appears that these expenses could not have been much more than $100.

[128] Lowe to Ingalls, March 6, 1863, MS, Lowe Papers; Statement, "U. S.

A major item of expense that must be considered in connection with the generators was the cost of materials for making the gas. Lowe reported to the assistant adjutant general after the apparatus had been used for more than a year that a single inflation, when the iron filings were bought commercially, amounted to about $75. When the iron was furnished free by the gun factories at the Washington Navy Yard, this sum was pared down to about $60.[129]

An examination of the available accounts, in conjunction with the amounts of iron and acid used for one inflation, verifies these statements reasonably well. The iron, when purchased from commercial foundries, averaged $10 per ton.[130] Sulphuric acid, a more expensive item in the formula for hydrogen, was obtained at an average price of $3 per carboy, or $1\frac{7}{8}$ cents per pound. An extra charge of $1.50 per carboy for the containers was added to the cost.[131] Thus at these rates, a single inflation requiring approximately one and two-thirds tons of iron (four barrels) and ten carboys of acid would involve an expenditure amounting to about $61. Adding in freight charges, packing, and the cost of lime for the purifier, the total reasonably approached $75. When the iron was supplied free from the Navy Yard, about $16 must be deducted, leaving a total that agrees well with Lowe's statement.[132]

Government, per Prof. T. S. C. Lowe, to Goodyear's Rubber, Belting, and Packing Company, Dr.," for packing, February 14, 1862, MS, Lowe Papers; Statement, Goodyear, October 24, 1862, for misc. hose, packing, etc., MS, Lowe Papers; Statement, Goodyear, April 19, 1862, for hose and couplings, MS, Lowe Papers.

[129] Lowe to Williams, March 30, 1863, MS, Lowe Papers; also printed in 3 O. R., III, 300.

[130] Invoice, Matthews and Moore [Bush Hill Iron Works], Philadelphia, April 19, 1862, MS, Lowe Papers; Invoice, Merrick and Sons [Southwark Foundry], Philadelphia, April 19 and 22, 1862, MS, Lowe Papers; Invoice, Whitney and Sons [Philadelphia Car Wheel Works], Philadelphia, April 19 and 21, 1862, MS, Lowe Papers; " List of Balloon Property purchased by Prof. T. S. C. Lowe, Paid for by Captain J. B. Howard," AQM, 1861, 1862, 1863, MS L429, enc., LRSW (1864).

[131] Invoices, Savage and Stewart [Frankford Chemical Works], Philadelphia, April 18, 1862, MS, Lowe Papers; Invoice, Savage and Stewart, May 7, 1862, MS, Lowe Papers; " List of Balloon Property . . . Paid for by Captain J. B. Howard," AQM, MS, L429, enc., LRSW (1864).

[132] See p. 249, supra, for amounts used in an inflation. The incidental

With the balloons in good condition, an inflation would provide lifting power for about two weeks in fair weather and without accidents. Hence the average cost of keeping a large balloon inflated under favorable conditions was about five or six dollars per day.[133] In the case of the small envelopes, the cost decreased in proportion to the reduced capacities. With these figures as a base, some interesting speculation as to the cost of gas for operating the balloon corps might be set down. But many extraneous factors, such as weather, varying numbers of balloons in service over a given period of time, damage, and inexperienced handling, prevent any really accurate estimate of operating costs over long periods. It may be said with a fair degree of accuracy, that under favorable conditions one large balloon in constant service must have cost from $150 to $175 per month for hydrogen. Beyond a calculation of this type, it is not safe to go.

Aeronautic operations from the various positions occupied by the Federal army along the Potomac soon disclosed the advantage of ascensions from the river and its adjacent tributaries. Likewise the establishment of widely separated stations of observation at different points on the extended Potomac lines led to the need of water transportation of supplies between balloon stations. As a result the balloon-boat USS *G. W. Parke Custis* was added to the equipment of the balloon service in November, 1861.[134] This vessel had originally been purchased by the Navy Department in August of the same year for service as a coal barge in the Washington Navy Yard.[135] She had been

expenses ran high. In a shipment of 15 barrels of filings, the extra costs amounted to $17, including $7.50 for barrels, $4.50 for weighing and packing, and $6.00 for cartage. Invoice of Philadelphia Car Wheel Works, April 19, 1862, MS, Lowe Papers.

[133] Based on the above calculations of $60 and $75 for a single inflation. See also Lowe to Williams, March 30, 1863, 3 *O. R.*, III, 300, in which the per diem cost of gas is estimated at $5.30. See also Lowe to Watson, April 15, 1863, 3 *O. R.*, III, 306.

[134] Washington *National Republican*, November 8, 1861.

[135] Commander John A. Dahlgren to Commodore Joseph Smith, August 8, 1861, MS, Letterbook, Commandant, Washington, Bureau of Yards and Docks, NDNA; Smith to Dahlgren, August 9, 1861, MS, Letters to Commandants, Bureau of Yards and Docks, Vol. XVIII, 237, NDNA; Smith to Dahlgren, August 12, 1861, MS, *ibid.*, p. 239.

assigned to Lowe for the aeronautic service by personal instructions of Secretary Gideon Welles,[136] to whom Lowe's application for a vessel had been referred. This balloon-boat, which may properly be called the first aircraft carrier in history, was approximately 8 years old at the time of her conversion to the aeronautic service. Her initial cost to the Navy had been $150.[137] She had an overall length of 122 feet, a 14½-foot beam and was 5½ feet in depth of hold. Of very shallow draft, she had a carrying capacity of 75 tons in 2½ feet of water, which was increased in deep water to 120 tons. The vessel was not equipped with engines, but was dependent upon an attending tug or small steamer for propulsion. In waters too shallow for towing, she could be propelled by large oars or heavy poles.[138]

Under the direction of Lowe, the superstructure of the vessel was remodelled by naval carpenters to suit the requirements of her new service. The entire hull was covered over with a single flat deck, which provided a large level area for the inflation and ascension of the balloons.[139] A small deckhouse on the stern was also added for the headquarters of the aeronaut in charge of the boat.[140] The shallow draft and roomy hold made the vessel admirably suited to balloon operations. She could be towed into any position navigable by her attending tug, and by means of her poles and oars could be maneuvered through more shallow

[136] Welles to Dahlgren, November 6, 1861, quoted in Dahlgren to Welles, December 10, 1861, MS 195, Commandant's Letters, Washington Navy Yard, Vol. III, Naval Records and Library, Navy Department, Washington.

[137] " Abstract of Disbursements on Account of Bureau of Yards and Docks, by Navy Agent at Washington, D. C., for the Month of August, 1861," MS, NDNA; "Abstract of Triplicate Bills Accompanying Requisition of S. P. Brown, Navy Agent, Washington, on Bureau of Yards and Docks, August 17, 1861," MS Requisition 1244, NDNA.

[138] Dahlgren to Smith, August 12, 1861, MS, Letterbook, Commandant, Washington, Bureau of Yards and Docks, NDNA.

[139] Lowe to Dahlgren, November 12, 1861, MS [copy], Lowe Papers. The hull of the vessel, according to a correspondent who visited the boat on the lower Potomac, was " widened on the deck by means of braces, like the New York ferry boats." New York Herald, November 13, 1861.

[140] Lowe to Dahlgren, November 16, 1861, MS, Papers of Admiral John A. Dahlgren, Division of MSS, Library of Congress [hereinafter cited as Dahlgren Papers].

waters into positions of advantage for observation that were otherwise inaccessible. In addition to her usefulness in providing ascension areas on her deck, her hold also furnished ample space for the stowage and transportation of iron turnings, acid, and other materials and equipment. With her accompanying tug, she provided a rapid means of communication and transportation between distant balloon stations along the Potomac. During the fall of 1861 the balloon-boat was particularly useful in connection with the operations of the aeronautic station with Hooker's Division at Budd's Ferry and with the unit assigned to General Stone's corps near Edwards Ferry at the other end of the Potomac lines. In the spring of 1862, the vessel accompanied the balloon corps to the Peninsula, and was operated to advantage in this first major campaign.

*   *   *

The large-scale expansion in aeronautic equipment and materiel for military purposes naturally led to a corresponding increase in personnel. The balloon service, as envisioned in Lowe's plans with the approval of the Commanding General, called for the immediate establishment of observation stations at different points along the Potomac lines, each to be in charge of an aeronaut operating tactically under the officer commanding the particular sector and technically under Lowe as chief of the corps. When the army should leave its comparatively inactive lines and engage in combat operations the balloons were to be stationed in positions designated by the Commanding General or by officers of his staff, under the charge of the several aeronauts directed by Lowe under the supervision of the various corps and division commanders. Lowe, as head of the balloon department of the army, was to supervise the technical operation of the equipment, see to its maintenance, and direct the administration of personnel and materiel. He was also to act as observer wherever his services should be needed.[141]

These plans of operation called for the services of a number

[141] These generalizations are based on subsequent developments of the balloon service and its operation, which are described in later chapters on the winter period of 1861-1862 and the campaigns of the succeeding two years.

of competent, trained balloonists, capable of taking charge of a station, superintending the manufacture of hydrogen in the field, and conducting observations as desired by the officers to whose command they should be assigned. In addition to these experts, there would be needed certain assistants not so highly trained, yet possessed of sufficient knowledge to render the aeronauts technical assistance, to help with the administration and maintenance, and to supervise numerous details attending the balloon service in the field. From time to time there would also be required the services of telegraphers to man the aerial and ground stations of the balloon telegraph lines. Teamsters to take charge of the transportation facilities and care for the draft animals would also form part of the personnel. And finally, each observation station would require its own separate ground crew of enlisted men detailed from line regiments of volunteers for special duty with the balloon establishment. Such, in general, were the personnel requirments of the newly created aeronautic service.

Lowe's pre-war experience in civilian ballooning had brought him into contact with many of the more prominent aeronauts of his day, and the enrollment of assistants in this specialized field presented no difficulties. Professional jealousies and antagonisms, however, seem to have eliminated the possible services of at least two of his widest-known contemporaries. Lowe's openly avowed distrust of the ability of Wise, despite the latter's long experience and high reputation,[142] evidently prevented his inviting the veteran aeronaut to join the corps.[143] La Mountain, now operating independently with General Franklin's division, had become Lowe's *bête noir* by the time the balloon service assumed any semblance of organization, and any understanding or cooperation between the two was impossible. For reasons undisclosed, the nationally known Samuel A. King of Philadelphia, former partner of James Allen, was not asked to serve. King later applied for service in the summer of 1862, and was supported by flattering testimonials from Mayor Alexander Henry,

[142] *Lowe's Report, 3 O. R.,* III, 256; Lowe's MS Memoirs, pp. 77-78.

[143] On the other hand, Wise might also have refused to serve under Lowe, had the invitation been given.

of Philadelphia, but apparently no action was taken on his application.[144]

In the summer of 1862, the French aeronaut Alexander J. B. De Morat, who had been active in the South before the war and had attracted attention by his balloon race with the subsequent Confederate military balloonist Wells,[144a] applied for an appointment in the Balloon Corps. According to his statements given to Secretary Stanton and to the Bureau of Topographical Engineers, he had served for three years as " Chief Engineer of Aeronauts " in the 4th Division of the army of the Second Republic, and was well acquainted with topographical survey operations and reconnaissance. De Morat also asserted that he had made one hundred and eighty-five ascensions, for which he could produce documentary evidence. Signing his communication as " Graduate of the École St. Cyr," he asked to be allowed to take examinations for appointment in the Union balloon service.[144b] Receiving no reply to his application, De Morat then applied directly to General McClellan, but was again ignored.[144c] Whether Lowe knew of this application and for reasons of his own did not want De Morat in his organization, or whether he was not informed of the French officer's offer of service and hence could not take steps to secure his services: these questions are not answerable from available sources.

[144] Lieutenant Colonel Ebenezer S. Sibley [DQMG] to Colonel Stephen S. Long [Chief of Top'l. Engrs.], August 2, 1862, MS S477, LRTE; Alexander Henry to Brigadier General Montgomery C. Meigs [QMG], July 27, 1862, MS K67, enclosed with MS S477, LRTE. There is no documentary evidence that Lowe ignored King when enrolling aeronauts, but since Lowe makes no mention of him in his letters or memoirs, and King later applied independently for service, the assumption is in favor of this conclusion.

[144a] New Orleans *Picayune,* February 16, 1858; clippings in James Allen's Scrapbook, II, 29.

[144b] De Morat to the Bureau of Topographical Engineers, August 3, 1862, MS D536, LRTE.

[144c] De Morat to McClellan, September 6, 1862, MS LBAP, XXII, 72. Failure of the Bureau and of McClellan to answer his letters seems to have dissuaded De Morat from his purpose. But after the Balloon Corps was abandoned in 1863 and rumors circulated that the service was to be reorganized and continued, he again appealed to the Topographical Bureau and offered himself as a " candidate for immediate examination," calling attention to his thirteen years' experience in civil and military aeronautics. De Morat to the Chief of the Bureau of Topographical Engineers, July 20, 1863, MS D571, LRTE.

Lowe nevertheless filled his vacancies with a number of able men, several of whom were nationally prominent in their profession. Available sources disclose the names of some nine aeronauts and three general assistants who were enrolled during the period of Lowe's direction of the balloon department. One of the first to receive an appointment was William Paullin, of Philadelphia, an aeronaut of some twenty-eight years' practical experience, whose reputation and standing ranked well with those of Wise and King. His most outstanding work had been done in South America, Mexico, and the West Indies, and he had gained a wide reputation in the western States of this country.[145] Paullin was also well known in the East, having made several public ascensions from Philadelphia and other seaboard cities.[146] He had been associated with John Wise in 1840,[147] and was well known to Lowe before the war, having assisted in the exhibition for the Japanese Embassy in 1859. Appointed to the Balloon Corps on October 11, 1861,[148] Paullin was assigned to Brigadier General Joseph Hooker's division at Budd's Ferry on the Lower Potomac. Although at first hindered by lack of materials and by weather interference, he did good service at this point, and was cited for his efficiency and skill in a dispatch of Colonel William F. Small [26th Pennsylvania Infantry], under whose command a number of ascensions were made.[149] Paullin's military service, however, was of short duration. Evidently intending to supplement his army pay, he attempted to carry on an ambrotype business as a sideline to his military occupation, and was dismissed by Lowe in January, 1862, for neglect of duty.[150]

James Allen, whose early attempts to perform independent service with Colonel Burnside's 1st Rhode Island Regiment had proven ineffective, was another of the more outstanding aero-

[145] Appleton's *Cyclopaedia of American Biography* (New York, 1888), IV, 682.

[146] Philadelphia *Gazette* [day and month missing], 1844, clipping in Scrapbook of Charles F. Durant, New York Public Library.

[147] John Wise, *Through the Air* (Philadelphia, 1873), pp. 322-323.

[148] Lowe to Paullin, October 11, 1861, MS, Lowe Papers.

[149] Small to Hooker, December 9, 1861, MS H479, enc. 2, LRAP.

[150] Lowe to Howard, January 31, 1863, MS L429, enc., LRSW (1864).

nauts who joined the corps. In response to Lowe's invitation to serve, Allen enrolled in the latter part of March, 1862,[151] and remained in service until the corps was disbanded in June of the following year. He participated in the Peninsular Campaign, in the campaigns of Fredericksburg and Chancellorsville, and became acting chief of the organization when Lowe was stricken with fever after the battle of Malvern Hill. After Lowe's final resignation following the battle of Chancellorsville, Allen became head of the corps and functioned in this capacity until the balloon establishment was suppressed. In the fall of 1862, his brother, Ezra S. Allen also entered service with the corps and remained with the organization until the close of its operations.[152]

One of the more colorful characters among the several aeronauts was the German balloonist, John H. Steiner, of Philadelphia, whose letters and dispatches, penned in curious, phonetic broken English, provide a source of amusement and difficulty for the investigator.[153] Steiner enjoyed a wide reputation before the war, particularly in the Middle West, and is credited with having made in August, 1859, the first balloon ascension ever performed at Toronto, Canada.[154] The previous year he had engaged in a free balloon race with Eugène Godard, subsequent military balloonist of Louis Napoleon's Italian campaign, an event which attracted much attention.[155] A free ascension from Erie, Pennsylvania, the year before, nearly ended in his drowning in Lake Erie, a fate from which he was rescued by a passing steamer.[156] Like La Mountain, Wise, and Lowe, Steiner had also been inspired with ambition to attempt a trans-Atlantic flight, and had published his intention in the

---

[151] Providence *Journal*, March 23, 1862; Providence *Daily Post*, March 28, 1862; unidentified clipping in James Allen's Scrapbook, I, 45. Lowe had written to Allen on February 8 inviting him to join the balloon service. Allen replied favorably on the 14th, and again on the 24th. Their plans seem to have been delayed, however, for Allen did not take up active duties until the latter part of March. Allen to Lowe, February 24, 1862, MS, Lowe Papers.

[152] Ezra S. Allen to Lowe, April 1, 1863, 3 *O. R.*, III, 301.

[153] See Appendix, Volume II, for an example of one of Steiner's dispatches.

[154] Toronto *Globe*, August 4, 1859.

[155] Cincinnati *Gazette*, October 20, 1858; *Harper's Weekly*, October 23, 1858.

[156] *Harper's Weekly*, July 4, 1857.

summer of 1858.[157] Steiner was enrolled in the balloon corps in December, 1861, and was assigned to Brigadier General Charles P. Stone's Corps of Observation on the Upper Potomac. He remained in this assignment until February, 1862, when he was selected to take charge of a balloon ordered to the Western Department for service at Cairo and Island No. 10. Relieved of duty with the Western Department in July, 1862, he rejoined Lowe's organization with the Army of the Potomac and served until December, when he resigned because of the continual delays in receiving his pay.[158]

Among the aeronauts who received early appointments in the corps was John B. Starkweather, of Boston. Very little is known of his civilian career, though it is evident that he knew Lowe before the war. When Lowe was engaged in building the four new balloons authorized in September, Starkweather arrived in Philadelphia on October 15,[159] doubtless for the purpose of applying for service. Sometime before, the Boston papers had published notices of Lowe's activities, and on the day of Starkweather's arrival on the scene of construction, both the *Journal* and *Transcript* announced that the aeronaut was forming a balloon corps.[160] Starkweather appears to have joined the corps towards the middle of November. He was detailed to take charge of a balloon assigned to the expeditionary forces under Brigadier General Thomas W. Sherman at Port Royal, South Carolina, and served with this command until the mid-summer of 1862. He resigned from the service at this time, and was later recommended to assume charge of a balloon ordered to

[157] *Harper's Weekly,* July 10, 1858.

[158] Lowe to Howard, January 31, 1863, MS L429, enc., LRSW (1864). After the war, Steiner seems to have turned inventor, and patented several improvements in gas generators and other gas appliances. Patent 85972, issued January 19, 1862, Gas Generator; Patent 92391, issued July 6, 1869, Apparatus for generating and carbureting gas for railroad cars; Patent 98442, issued December 28, 1869, Gas Generator; Patent 125496, issued April 9, 1872, Improvement in Gas Apparatus; Patent 126652, issued May 14, 1872, Improvement in Gas Generators and Carburetors.

[159] Arrival List, St. Louis Hotel, Philadelphia, in Philadelphia *Press,* October 16, 1861.

[160] Boston *Transcript,* October 2, 1861; Boston *Transcript,* October 15, 1861; Boston *Journal,* October 15, 1861.

General Quincy A. Gillmore's forces before Charleston in 1863.[161]

Another aeronaut, Ebenezer Mason, Jr.,[161a] of Troy, New York, perhaps an earlier associate of La Mountain, joined the balloon corps in December, 1861. Mason appears to have been a sort of jack-of-all-trades, having been employed for some time before his army appointment as an agent for an Albany showman known as " Wyman the Wizard." The latter gave up the show business in September, 1861, and Mason, who seems to have known Lowe personally, appealed to him for employment with the army, declaring that he was willing to do any form of work connected with Lowe's department.[162] Mason was also a competent balloon-maker as well as an aeronaut, and his first duty with the corps involved the building of the *Eagle* and *Excelsior*,[163] the last of the balloons constructed for army service. He was also detailed as an observer at several points along the Potomac lines during the winter and spring of 1861-1862, and accompanied Lowe's train to the Peninsula. The delays in receiving pay, a common annoyance borne by all during this campaign, led to Mason's refusal to perform duty until his salary, more than a month and a half over-due, should be forthcoming.[164] In fairness to the man, there was real cause of complaint in his case for his family was practically destitute.[165]

[161] Lowe to Sherman, November 27, 1861, 3 *O. R.*, III, 267; Lowe to Comstock, April 20, 1863, *ibid.*, p. 308.

[161a] The name is given by General (then Captain) William A. Glassford, who had correspondence with the aeronaut after the war, as " E. Locke Mason." The habitual signature on letters and dispatches of the war period was " E. Mason," but the writer has found one MS dispatch signed with the given name of " Ebenezer." The city directory for Philadelphia, in 1862 lists an " Ebenezer L. Mason " as a harness-maker, and this name does not appear in previous issues of the same directory. Possibly after leaving the army in 1862 Mason settled in Philadelphia and entered the harness making trade. He had engaged in all manner of trades before the war. Cf. William A. Glassford, " The Balloon in the Civil War," *Journal of the Military Service Institution of the United States*, XVIII (Governor's Island, 1896), 261, 266; *McElroy's Philadelphia City Directory, 1862* (Philadelphia, 1862), p. 451.

[162] Mason to Lowe, September 5, 1861, MS, Lowe Papers.

[163] Mason to Lowe, January 3, 1862, MS, Lowe Papers.

[164] Lowe to Howard, January 31, 1863, MS L429, enc., LRSW (1864).

[165] Mason to Lowe, June 2, 1862, MS, Lowe Papers. " I make a final appeal for aid from you to get my pay to relieve my family who have been

But Lowe, feeling that private matters must be disregarded in the interests of discipline and the objectives of the critical campaign, indignantly discharged his " striking " assistant, and thus ended Mason's military career in the first major campaign.[166]

About the middle of November, 1861, Ebenezer Seaver, another balloonist whose previous career is practically unknown, was added to the personnel under Lowe's direction.[167] This individual, whose name was curiously similar to that of a contemporary Confederate balloonist,[168] served with Porter's division in the vicinity of Washington and at other points along the Potomac lines. He was later sent to replace Paullin at Budd's Ferry, where he served until the movement of the army to the Peninsula.[169] Accompanying the balloon corps in the Richmond campaign, Seaver joined his fellow aeronaut Mason in the strike for pay and was summarily discharged from the service.[170] After the war he was associated for a time with James Allen in civilian ballooning.[171]

Nothing can be found in the records concerning John R. Dickinson, an aeronaut employed by Lowe in October, 1861. He appears in one dispatch as an assistant in charge of the *Union* at Washington.[172] Beyond this one mention, his name is missing from the sources for the balloon corps.

With the appointment of Jacob C. Freno in January, 1862,[173] the roll of Lowe's aeronauts appears to be complete. Although Lowe had experienced some difficulty with several of his assistants over the matter of delayed pay, Freno alone seems to have been the one black sheep to enter the aeronautic fold of the army. He was a resident of Philadelphia, and was a mixture of

ordered into the street for non-payment of rent," was Mason's pathetic appeal to Lowe sometime after he was dismissed.

[166] Lowe to Howard, January 31, 1863, MS L429, enc., LRSW (1864).

[167] Seaver to Porter, November 21, 1861, MS, Lowe Papers.

[168] Charles Cevor, balloonist with the Confederate Army.

[169] Porter to Lowe, November 17, 1861, MS, Lowe Papers; Lowe to Macomb, January 16, 1862, MS [copy], Lowe Papers; Lowe to Macomb, January 17, 1862, MS M106, enc. 2, LRAP.

[170] Lowe to Howard, January 31, 1863, MS L429, enc., LRSW (1864).

[171] James Allen's Scrapbook, II, 158.

[172] Lowe to Paullin, October 11, 1861, MS, Lowe Papers.

[173] Lowe to Meigs, March 5, 1863, MS, Lowe Papers.

lawyer and balloonist.[174] Lowe had employed him in civilian work about two years before the war, and had again engaged his services as an assistant in the preparations for the Cincinnati-Unionville flight in April, 1861.[175] In need of aeronauts for the expanding corps, Lowe again employed him as an assistant aeronaut for duty on the Potomac lines in January, 1862. Freno remained in service with the corps until December of the same year, when his conduct became such that Lowe dismissed him. As an aeronaut, the man seems to have been able enough. Lowe reported to Macomb at the time of his appointment that he was " fully competent to take charge of a balloon in my absence." [176] But as an individual, Freno's record shows him to have been a thoroughly shabby character. He had originally obtained a commission in the 66th Pennsylvania Infantry, and was mustered in as first lieutenant on September 27, 1861.[177] According to Lowe's testimony, he was dismissed from the service for conducting a gambling hell in the camp of his regiment,[178] but closer investigation reveals that he was charged with cowardice before his men in the presence of the enemy in an outpost engagement in the vicinity of Harper's Ferry. Ordered held for court-martial, Freno escaped the grim consequences of such a trial through the leniency of his commanding officer, who allowed him to resign in order to prevent regimental disgrace.[179] Freno then applied for service with Lowe, stating that he had voluntarily resigned his commission in order to serve with the Balloon Corps. Lowe took him at his word, and employed him. For several months he did creditable service, but after several outbreaks of insubordination and breaches of discipline, he was summarily dismissed " for repeated absence without leave, for expressing disloyal sentiments, open-

[174] *McElroy's Philadelphia City Directory, 1861* (Philadelphia, 1861), p. 327. Freno is listed here as " attorney-at-law, 116 Spruce [Street]."

[175] Lowe to Macomb, January 17, 1862, MS M106, enc., 2, LRAP. See also *supra*, p. 167.

[176] Lowe to Macomb, January 17, 1862, MS M106, enc. 2, LRAP.

[177] MS Muster Roll, 66th Pennsylvania Infantry, War Department Division, National Archives; Service Records, Jacob C. Freno, Old Records Division, Adjutant General's Office, Florida Avenue, Washington, D. C.

[178] Lowe to Howard, January 31, 1863, MS, L429, LRSW (1864).

[179] Confidential Source No. 1. (See bibliography, in Volume II.)

ing a faro bank for the purpose of gambling, and for the demoralizing effect which he had upon the subordinates of this department." [180] Following his dismissal, he gained access to the building in Washington where the balloon *Constitution* had been temporarily stored after a general overhauling, and deliberately damaged the aerostat by ripping a large piece from the envelope.[181] Lowe preferred charges against him, and the Provost Marshal General's office made a futile attempt to apprehend him.[182] Several months later, when the affair had blown over, Freno wrote out a set of malicious charges against Lowe's character, ability, and integrity as chief of the corps, which he had the nerve to send in to the office of the Provost Marshal of the district of Washington.[183] Lowe indignantly refuted the charges to the full satisfaction of his superior officers, but the insult rankled.[184] The employment of the lawyer-balloonist Freno had indeed been a poor choice of personnel.

In all, the total number of trained aeronauts attached to the corps during its existence does not seem to have exceeded nine. Because of gradual enrollment, dismissals, and resignations, the greatest number on duty in the field at one time was seven. During the spring and early summer of 1862, Lowe, James Allen, Mason, Seaver, and Freno were serving with McClellan's army on the Peninsula. Starkweather at this time was with Sherman at Port Royal, and Steiner was with the Western Department at Island No. 10. As the war progressed, the number of aeronauts active in the field diminished. Mason and Seaver were dismissed during the campaign before Richmond,

---

[180] Lowe to Meigs, March 7, 1863, MS, Lowe Papers.

[181] Testimony of James Allen, E. S. Allen, James Starkweather, John O'Donnell, J. H. Hall, William A. Hodges, Francis A. Barringer, L. M. Chickey, L. F. VanNatter, and G. F. Gibson, enclosures with Lowe to Meigs, March 7, 1863, MS, Lowe Papers.

[182] Lowe to HDQ, AP, February 28, 1863, MS, LBAP, XII, 666; indorsement March 3, 1863, *ibid.*; Brigadier General M. R. Patrick [PMG] to Assistant Secretary Watson, March 4, 1863, MS, LBAP, LXX [Letters Received, Office of Provost Marshal General], p. 179.

[183] Statement of Jacob C. Freno, sworn to December 24, 1862, MS F26, LRTE.

[184] Lowe to Howard, January 31, 1863, MS L429, enc., LRSW (1864); Lowe to Brigadier General Edward M. Canby [AAG], January 22, 1864, MS [number omitted], LRSW.

leaving Allen, Lowe, and Freno with McClellan. Starkweather resigned at the expiration of his duties at Port Royal. Steiner came east and rejoined McClellan's army in time for the operations following Antietam, but resigned before the Fredericksburg campaign. In this campaign Ezra Allen performed his first duty, along with his brother and Lowe. From this time to the end of the Chancellorsville campaign, Lowe and the two Allens made up the entire aeronaut personnel of the corps. With Lowe's resignation, the final number was reduced to two.

In addition to the aeronauts, Lowe enrolled several general assistants to aid with the details of administration, maintenance, and sundry other matters relating to the operation of the aeronautic department. Among these was his father, Clovis Lowe, who had become thoroughly familiar with the technique of balloon handling, repairing, and general operation. The elder Lowe proved himself a valuable addition to the personnel of the corps and relieved his son of many details attending the numerous duties he was called on to perform as chief of the department.[185] He enjoyed the full confidence of Brigadier General Humphreys, who assumed military command of the balloon department during the Peninsular campaign, and was placed in charge of all materiel and equipment for the movement to Washington when McClellan's army withdrew from Harrison's Landing.[186] Although his presence in the corps later caused Captain Cyrus B. Comstock, who succeeded Humphreys, to charge the younger Lowe with nepotism and arbitrarily to dismiss the elder assistant,[187] it appears from the records that Comstock's action was unreasonable and unnecessary.[188]

The same curt dismissal at this time was meted out by Comstock to John O'Donnell, another assistant serving with the

[185] It is only natural that Lowe's father should have been particularly helpful. He had taken a keen interest in his son's aeronautic activities before the war, and was thus able to render technical assistance as well as routine help with small details. Misc. correspondence of Clovis and T. S. C. Lowe, Lowe Papers.

[186] Humphreys to Clovis Lowe, August 13, 1862, 3 O. R., III, 291; Humphreys to T. S. C. Lowe, June 25, 1862, MS, Lowe Papers.

[187] Comstock to T. S. C. Lowe, April 12, 1863, MS, Lowe Papers; also printed in 3 O. R., III, 303.

[188] See below, Volume II.

corps.[189] Of O'Donnell's background and career, there is no trace in available sources. He seems to have been employed as a general assistant to Lowe at his headquarters, detailed to help with paper work, records, supply and maintenance, and in fact with all duties from filing dispatches to repairing balloons and arranging for forage for the teams.[190] Still another non-technical member of the corps was Richard Brown, who was employed on March 20, 1862, to take charge of and manage the balloon-boat *G. W. Parke Custis* during the movement to the Peninsula.[191] He served in this capacity for the duration of this campaign, after which his name fails to appear in the sources.[192]

None of the aeronauts or assistants, from Lowe down, enjoyed any military status. They were all rated as civilian employees and had no command functions of a military nature. Even Lowe, who was technically chief of the corps,[193] had no rating as an officer, although he was held responsible for the functioning of the corps and was vested with administrative authority over its personnel. The British observer with the Union army, Captain Beaumont, later remarked that he could never quite make out the chief aeronaut's rank, concluding at length that he was a brigadier general.[194] In reality, however, he was not even a lieutenant of volunteers. The "courtesy title" of colonel was accorded him on occasion,[195] but this dis-

---

[189] Comstock to Lowe, April 12, 1863, MS, Lowe Papers; also printed in 3 *O. R.*, III, 303.

[190] Statement of Mrs. Henry M. Brownback [daughter of Lowe], to the writer, June, 1938. O'Donnell signed himself as "clerk and assistant aeronaut," in his testimony against Freno. Lowe to Meigs, March 7, 1863, MS, Lowe Papers.

[191] Lowe to Macomb, April 2, 1862, MS, Lowe Papers.

[192] Lowe to James Allen, June 29, 1862, MS, Lowe Papers.

[193] Lowe was accorded a number of titles by the officers under whom he served, and his dispatches are signed as: "Chief Aeronaut"; "Chief Aeronaut, U. S. Army"; "Aeronaut, Commanding Balloon Department, Army of the Potomac"; "Chief Aeronaut, Army of the Potomac"; "Chief of Aeronautics, &c"; "Chief of Aeronautics, Army of the Potomac." See dispatches in *Lowe's Report*, 3 *O. R.*, III, 270-316, *passim*.

[194] Beaumont, p. 96.

[195] Lowe's MS Memoirs, p. 166; Anon., "Three Months with the Balloons," *St. James's Magazine* (London, 1863), p. 97; Park Spring to Lowe, May 25, 1862, MS, Lowe Papers; Robert B. Allen to Lowe, October 24, 1862, MS, Lowe Papers; Cincinnati *Commercial*, March 12, 1862; Philadelphia *Press*, May 3, 1862; *ibid.*, May 30, 1862; Boston *Transcript*, June 4, 1862; New York

tinction, so far as being of any value in his relations with the interlocking branches of the service, such as the quartermaster corps and finance office, amounted to about as much as that of a modern "Kentucky Colonel." James Allen, Mason, Seaver, and Steiner all received similar courtesy titles of captain.[196]

Civilian status was later to produce a series of serious difficulties when the balloon corps had to work in cooperation with other branches of the military service. Other aspects of this status were also objectionable to the individual aeronauts. In the first place, since the balloonists were only civil employees, their families would receive no pension in the event of their deaths in line of duty, and the operation of military balloons, often under fire, and at times controlled by details of inexperienced volunteers, was not devoid of unusual risk.[197] Also, there was the possibility of the balloons accidentally escaping their moorings and carrying the aeronauts over the enemy lines. In the event of capture under such circumstances, none of them would have been able to claim the protection of the uniform of his service, and civilians engaged in aerial observation might technically be charged with being spies and arbitrarily treated as such.[198]

Tribune, June 24, 1862; Philadelphia *Press,* June 23, 1862; Joel Cook [Special Correspondent of Philadelphia *Press*], *The Siege of Richmond* (Philadelphia, 1862), p. 33.

The Boston *Journal* of November 25, 1861, referred to the chief aeronaut as "Major Lowe." It appears that Lowe fully expected to receive a regular commission as colonel, and had promised Ebenezer Seaver a majority when this hope should be realized. "I am told," Seaver wrote his chief, "that you are now Colonel by appointment; if this is so, where is my commission as Major you promised me last November?" Seaver to Lowe, March 17, 1861, MS, Lowe Papers.

[196] Lowe to Seaver, March 8, 1862, Lowe Paapers; New York *Tribune,* March 27, 1862; Philadelphia *Press,* March 29, 1862; Baltimore *American,* March 28, 1862; New York *Tribune,* March 28, 1862; Philadelphia *Press,* March 14, 1862; *ibid.,* April 2, 1862; *ibid.,* April 9, 1862; Richmond *Enquirer,* April 12, 1862; Philadelphia *Inquirer,* April 25, 1862; New York *Illustrated News,* May 17, 1862; Philadelphia *Inquirer,* October 10, 1861; *ibid.,* October 24, 1862; Providence *Daily Post,* November 1, 1862.

[197] Lowe stressed these points in a letter to McClellan, in which he urged the General to arrange for his military status. Lowe to McClellan, June 16, 1862, MS, Lowe Papers.

[198] Such at least, was Lowe's conviction. Whether or not the Confederates would have hanged Federal aeronauts as spies because they were civilians is a

In addition to the aeronauts and general assistants, another important group of the Balloon Corps personnel comprised the telegraph operators who manned the aerial and ground stations at various points of observation. Since the use of telegraph was not always needed, these technical specialists were not permanently attached to the corps, but were temporarily assigned from the regular Telegraph Corps whenever circumstances demanded their services. Often regular operators attached to the various division and army corps headquarters were detailed to operate with the balloons, and then resumed their regular duties when their services were no longer required at the aerial stations.[199] Full reliance could not always be placed on the division and corps headquarters for telegraphers. In one instance Lowe appealed to the chief of staff for an operator, and General Porter's headquarters was designated to furnish the man. Porter reported with regret that he could not spare the operator.[200] On the other hand, there were times when telegraphers were available in number. In reply to a request for a man, on one occasion one of the chief telegraphers of the Army Telegraph Corps advised Lowe that a " fresh invoice " of operators was due the next day, and that from these men there would be trained personnel to meet all requirements.[201] Although the assignment of telegraph men to the balloon corps seems to have followed the loose organization characteristic of the aeronautic establishment, the permanent assignment of telegraphers would have been wholly unnecessary. The records reveal no case in which a competent operator was not available in time of emergency, and in general, the relations between the aeronautic department and the telegraph corps seem to have been cooperative and friendly.

Very little concerning the work of individual operators can be gleaned from the records, and the names of only a very few

matter of speculation. They might have done so under the cloak of legality. Lowe believed that they would have done so. *Lowe's Report,* 3 *O. R.,* III, 318.

[199] This statement is based on a number of letters and orders in the Lowe Papers.

[200] Lowe to Marcy, June 13, 1862, MS, Lowe Papers; Porter to Marcy, June 13, 1862, MS, McClellan Papers, MS 12820, and MS 12847, Vol. LXIII, 1st Series.

[201] Park Spring to Lowe, June 23, 1862, MS, Lowe Papers.

appear in available sources. Herbert C. Robinson, telegrapher of the American Telegraph Company, who accompanied Lowe in his first experimental ascension from the Columbian Armory and transmitted the first aerial dispatch on June 18, 1861, remained in service with Lowe for a short while after the *Union* had been built and commissioned into service.[202] He does not seem to have worked with the corps after its organization, and his name does not appear in records relating to balloon service after the first of September, 1861.

The most distinguished of the several telegraphers who manned the wires in Lowe's train was Park Spring, of Philadelphia, one of the chief telegraphers of Colonel Eckert's Army Telegraph Corps. He was attached to the balloon corps at intervals during the Peninsular campaign, and took a keen interest in aerial telegraphy. His work with the balloon corps was conspicuous during the battle of Seven Pines [Fair Oaks], when direct telegraphic communication was established between the balloon car and the offices of the War Department in Washington.[203] He was particularly cordial in his relations with Lowe, and the two men remained on friendly terms throughout the campaign.[204] A close friend of Spring, Charles Fonda, was also assigned to duty with the balloon corps shortly before the Seven Days' Battles.[205] Fonda seems to have remained on duty with the corps until shortly before the repulse of Lee's army at Malvern Hill.[206]

The only other operator who can be identified with the balloon establishment was D. D. Lathrop, a telegrapher from

[202] Robinson to Lowe, June 25, 1898, MS, Lowe Papers.

[203] N. Y. *Illustrated News,* July 5, 1862; Spring to Lowe, May 25, 1862, MS, Lowe Papers; N. Y. *Tribune,* June 5, 1862; Park Spring to Lancaster [Pennsylvania] *Express,* reprinted in the Boston *Transcript,* June 13, 1862; Philadelphia *Press,* June 2, 1862; Spring to Lowe, June 6, 1862, MS, Lowe Papers; Caldwell to Eckert, June 1, 1862, MS, LBAP, Vol. XXXVI; N. Y. *Herald,* June 2, 1862; Washington *National Republican,* June 2, 1862; New York *Times,* June 2, 1862.

[204] *Ibid.*; Lowe to Eckert, June 2, 1862, MS, Misc. Box, Papers of the Army of the Potomac; Caldwell to Lowe, April 30, 1862, MS, LBAP, Vol. XXXIV.

[205] Spring to Lowe, May 25, 1862, MS, Lowe Papers; Lowe to Eckert [n. d], 1862 (Telegram), Misc. Papers, "Balloon Corps"; Spring to Lowe, June 23, 1862, MS, Lowe Papers.

[206] Spring to Lowe, July 5, 1862, MS, Lowe Papers.

Eckert's organization assigned to general headquarters. He was, so far as can be ascertained, the only civilian member of the corps' personnel, permanent or temporary, who was killed in line of duty. Entering Yorktown after the Confederate evacuation, Lathrop stepped on the primer of a concealed Rains torpedo planted at the base of a telegraph pole. The explosion tore off both his legs and Lathrop died in a few minutes.[207]

Still another civilian group forming part of the balloon corps included the teamsters who were attached to the various balloon trains and observation stations. Since each separate balloon required several wagons for the transportation of generators, camp equipment, and other baggage, the drivers of these vehicles became an important cog in the machinery of the corps' mobility. The men assigned to this duty were not soldiers attached to the balloon service on special duty, but were civilians employed by Lowe along with the other civilian personnel.[208] Their duties, in addition to handling the teams in draft, included the care of the animals attached to the trains, the drawing of forage, and the general care and repair of the mobile materiel. A head teamster, or wagon master, was in charge of this group of personnel and supervised their work. On the Peninsula, the wagon master was Robert Collins, who seems to have done efficient work, and was commended by General Humphreys.[209]

The rates of pay for the various classifications of civilian members of the corps personnel varied with individual ratings. Lowe, as head of the organization, received $10 per day, or slightly more than the pay of a full colonel.[210] The two assistant aeronauts who were detailed with separate responsibility to other military departments, Starkweather and Steiner, were next

[207] Lowe's MS Memoirs, p. 135; Diary of General Samuel P. Heintzelman, MS, entry of May 6, 1862, Heintzelman Papers; Brigadier General William F. Barry [Insp. of Arty.], to Brigadier General G. W. Cullum, August 25, 1863, 1 *O. R.*, XI, i, 350; " W. D. B." to editors of the Cincinnati *Commercial*, May 7, 1862, published May 16; New York *Illustrated News*, May 24, 1862.

[208] Humphreys to Clovis Lowe, August 3, 1862, MS, Lowe Papers.

[209] Humphreys to T. S. C. Lowe, August 3, 1862, MS, Lowe Papers; Memorandum Receipt, for oats and forage, signed by Robert Collins, Wagon Master, May 28, 1862, MS, Lowe Papers.

[210] Whipple to Lowe, August 2, 1861, 3 *O. R.*, III, 259.

in the scale with daily rates of $5.75, or somewhat less than captain's pay.[211] The same rate was paid Seaver, although he does not seem to have enjoyed any higher status than the other assistant aeronauts. James Allen, senior assistant aeronaut received $4.75, practically the equivalent of a second lieutenant. The other assistant aeronauts, Mason, Ezra Allen, Freno, Paullin, and the general assistants, Clovis Lowe and O'Donnell, each drew $3.75.[212] The teamsters received $25 per month, nearly twice as much as the private soldiers of the line, and the wagon master drew $30.[213] Since the telegraph operators were paid by their own organization, they need not be included in this connection.

The remaining personnel of the balloon corps consisted of the various details of troops temporarily assigned for duty as ground crews and general utility groups at the balloon stations. These detachments generally consisted of parties from 30 to 50 enlisted men including several non-commissioned officers, commanded by one commissioned officer, usually a lieutenant.[214] The details were drawn from regiments adjacent to the balloon stations, or from those which were serving in the same general locality. The system of temporary assignment, a practice already proven to be inefficient in the earlier experience of Allen and Wise, was continued after the formation of the corps. It produced a number of unnecessary difficulties in administration and operation.

Despite the difficulties attending the system of providing troops for the balloon corps, Lowe made the best of the avail-

[211] Lorenzo Thomas [AG, USA] to Brigadier General Thomas W. Sherman, December 30, 1861, MS, Letters Sent, Headquarters of the Army, War Department Division, National Archives [cited hereinafter as LSHQA]. Certificate of the status of John H. Steiner, Aeronaut, USA, stating rate of pay, April 1, 1863, signed by T. S. C. Lowe, MS, Lowe Papers. For the relative army pay scale, see Pay Table, United States Army, 1863-1865 [compiled by J. Lowenthal, under direction of the Chief Paymaster], (Washington, 1865).

[212] Balloon Corps Payroll, Quartermaster Corps Form No. 15, February, March, and April, 1862, MS, Lowe Papers.

[213] Humphreys to Lowe, August 3, 1862, MS, Lowe Papers.

[214] This statement is based on the average number of men in the detachments mentioned in the mass of correspondence relating to these details in the Lowe Papers, and various collections of the War Department Division, National Archives.

able arrangements and carried on with creditable success. To offset the loss of trained men when his detachments were transferred back to their regiments, he was able to secure the permanent assignment of several men particularly adept at balloon handling and repair.[215] These men he used as key personnel to assist in training green troops as the new detachments were detailed to him for duty. He also assigned them to permanent, specific duties in connection with the balloons, generators, and other equipment. During the Peninsular campaign, the enlisted personnel attached to the balloon corps headquarters contained the following specialist ratings: one man detailed to varnish envelopes; two generator assistants to help with gas manufacture; two repair assistants, assigned to the repair of envelopes, nettings, and other appendages; one machinist, for mechanical repairs to the generators; and one permanent orderly or dispatch bearer for the delivery of written messages.[216] In time of emergency, such as during major action, additional dispatch bearers were attached to Lowe's headquarters and to the other balloon stations in the field. During the battle of Seven Pines, for example, General Humphreys especially requested that four extra orderlies from the general headquarters escort be detailed to Lowe's station, and four more to the station operated by James Allen.[217]

Some of the men permanently assigned to Lowe's corps deserve special mention. Among them was Corporal James Starkweather, of Company " K," 19th Massachusetts Infantry, a brother of the aeronaut who served at Port Royal.[218] Starkweather, who had been a sailmaker in civil life,[219] was doubtless

[215] Macomb to Williams, November 15, 1861, MS [number omitted], LRAP; Special Orders, No. 147, Army of the Potomac, November 18, 1861, MS Records of the Army of the Potomac, Vol. XXVII [Special Orders], p. 294 [cited hereinafter as SOAOP].

[216] Assignment list, undated, but evidently of the period of the Peninsular campaign, in Lowe's hand, MS, Lowe Papers.

[217] Humphreys to Brigadier General R. B. Marcy [C of S], 6:50 a. m., June 1, 1862, MS H453, LRAP.

[218] Special Orders, No. 159, Hdq., Corps of Observation, December 30, 1861, MS, Records of the Corps of Observation, War Department Division, National Archives [cited hereinafter as RCO].

[219] Descriptive List, James Starkweather, Company " K," 19th Massachusetts Infantry, MS, Lowe Papers.

familiar with balloons through his brother's profession, and his own occupation rendered him proficient in balloon and netting repairs. He served in the campaigns of the Peninsula, Fredericksburg, and Chancellorsville, and remained with the corps until it was disbanded.[220] Another valuable soldier was Private William A. Hodges, Company " I," 14th New York Infantry, whose term of service with the balloons equalled that of James Starkweather. Hodges became proficient at many details pertinent to this service and in time was entrusted with numerous assignments of responsibility.[221] Privates Albert Trumbull, W. H. Welch, James F. Case, Robert Wardwell, and Francis Barringer, of the 22nd New York Infantry, also served with the corps from its organization until its end.[222] Three other soldiers who earned special commendation from Lowe were Privates John H. Hall, Lawrence M. Chickey, and George W. Fisher, 29th Massachusetts Infantry. These men were assigned permanently to the balloon service in March, 1862, and served until the spring of 1863.[223] Sergeant Charles F. Eaton, 22nd New York Infantry, a member of one of the first detachments assigned to balloon service in September, 1861, proved himself especially efficient and responsible and served until stricken with typhoid fever which caused his death in May, 1862.[224]

Among the enlisted men who served in the temporary detachments detailed to balloon service Lowe was fortunate enough to find two competent draftsmen capable of making aerial maps

[220] Payrolls, " Non-Commissioned Officers and Privates Employed on Extra Duty," Quartermaster Corps Form No. 3, MSS, Lowe Papers; Muster Roll of Balloon Corps, May 1 to June 30, 1863, MS, War Department Division, National Archives.

[221] Payrolls, Quartermaster Corps Form No. 3, MSS, Lowe Papers; Special Orders No. 193, Hdq., Army of the Potomac, December 20, 1861, MS, Lowe Papers; Lowe to Hodges, April 17, 1863, MS, Lowe Papers.

[222] Special Orders, No. 147, Hdq., Army of the Potomac, November 18, 1861, SOAOP, XXVII, 294; also MS, Lowe Papers; Payrolls, QMC Form No. 3, MSS, Lowe Papers.

[223] Muster Roll of Men on Detached Service with Professor Lowe, by Order of Major General Wool, dated March 22, 1862, MS, War Department Division, National Archives; Payrolls, QMC Form No. 3, MSS, Lowe Papers; Lowe to Humphreys, May 21, 1862, MS, Lowe Papers.

[224] Special Orders, No. 147, Hdq., AP, SOAOP, XXVII, 294; Lowe to Eaton, September 28, 1861, MS, Lowe Papers; Mason to Lowe, June 2, 1862, MS, Lowe Papers.

and panoramic sketches. The first of these was Sergeant William Bancroft, 4th Michigan Infantry, who was attached to the corps during the Peninsula campaign, and who was specially detailed as a draftsman to Lowe's headquarters by Brigadier General George W. Morrell.[225] The other was Sergeant Edward S. Dewey, 10th Massachusetts Infantry. Dewey had been an artist before the war and did excellent work as a draftsman with the balloon corps in the spring of 1863, preceding the battle of Chancellorsville. He remained in service with James Allen after Lowe's resignation and proved to be a valuable asset to the dwindling organization at that time.[226]

Shortly after the creation of the balloon corps, several war correspondents announced that troops enrolled for balloon service were equipped with special distinguishing insignia, and described the device as a cap ornament of metal in the shape of a balloon, surmounted by the letters " B. C." These accounts in some instances were embellished with humorous details of comments from fellow soldiers who speculated on the remarkable age of the balloon troops.[227] Despite these press notices, there is no evidence that a uniform was ever adopted by the members of Lowe's corps, either officially or otherwise. It will be remembered that the French *compagnies des aérostiers* of the Eighteenth Century were distinctively uniformed.[228] In the American forces nearly seventy years later, there appears to have been no imitation of the custom. The aeronauts wore ordinary civilian clothes, generally suited to field service. Lowe usually wore semi-military dress, consisting of trousers stuffed into high riding boots, a dark coat, and in cold weather a long dark coat equipped with military frog fasteners. A black slouch hat, not unlike those worn by some field officers of his day, completed his field attire.[229]

---

[225] Morrel to Lowe, June 19, 1862, MS, Lowe Papers.

[226] Colonel Henry L. Eustis [10th Massachusetts Infantry] to Lowe, April 22, 1863, MS, Lowe Papers; Lieutenant Daniel G. Macnamara, *History of the Ninth Regiment, Massachusetts Volunteer Infantry* (Boston, 1899), p. 291.

[227] New York *World*, December 7, 1861; Washington *Star*, December 4, 1861; Baltimore *American*, December 9, 1861.

[228] See Chapter i, *supra*, p. 8, n. 32.

[229] Anon., " Three Months with the Balloons," *St. James's Magazine* (Lon-

In general, the Chief Aeronaut presented a striking and distinguished appearance. Well-proportioned, with broad, massive shoulders and vigorous, athletic frame, Lowe stood six feet one inch—as tall almost as his early advocate and Commander-in-Chief, the towering Lincoln. His features were regular, strongly but delicately cast, and indicated gentle birth and refinement. His many portraits reveal a straight, aquiline nose, a firm but kindly mouth, and a strong, determined chin. His eyes were a brilliant, piercing blue, deeply set below a high forehead above which his heavy black hair, slightly wavy and worn long, as was the custom of many officers of his time, fell back and away to provide a softening frame for his rather proud and aristocratic countenance. During the war period he wore a heavy, curling mustache, and its bright auburn color contrasted sharply with the intense coal-black of his hair. His skin was fine and delicate, and his complexion, so his daughter later wrote, was " as fair and flawless as a child's,—quite the handsomest man I have ever known." In profile, some of his portraits show a curiously strong resemblance to his friend and commander, General McClellan.[229a]

Only one of his assistants ever affected a uniform. This individual was Seaver, who seems to have designed his own finery and as Clovis Lowe humorously wrote to his son, " General Sickles could not have put on so many things or as many shoulder straps as he did." From the elder Lowe's comments, Mr. Seaver became the unwilling butt of many jests directed at his unofficial regalia.[230] Evidently the aeronaut felt the need of a " courtesy uniform " to go with his unofficial title of captain.

don, 1863), p. 97. See contemporary photographs of Lowe, Plates XXXIV and XL, *infra*.

[229a] This personal description of Lowe at the time of the war is drawn from the several available portraits of him, from a family miniature, and from several descriptive letters from his daughter, Mrs. Henry M. Brownback. See Plates XIII, XXIX, and XXXII, *infra*. One of General Porter's aides-de-camp, Lieutenant Stephen M. Weld, met Lowe near Cloud's Mill, and made special note of the Chief Aeronaut's appearance in a letter to his father. Lowe, he wrote, " is a very good looking man. . . ." Weld to his father, March 20, 1862, *War Diary and Letters of Stephen Minot Weld, 1861-1865* (privately printed, New York, 1912), p. 88.

[230] Clovis Lowe to T. S. C. Lowe, January 25, 1862, MS, Lowe Papers. " I met him [Seaver] with a great deal of cordiality," the elder wrote to his

An estimate of the number of enlisted men who served with the corps during its existence cannot be made with any degree of accuracy. Many of the muster rolls of permanently detached troops are lost or missing from the records. The temporary details were rarely carried on separate rolls, and thus men comprising them cannot be distinguished on the regular rolls of their regiments. From materials available, however, it is possible to record at least a large number of the regiments which furnished troops for this special service. Men from six states, three New England, two Middle Atlantic, and one Western, served at one time or another with the various stations of Lowe's corps. From these six states, seventeen infantry regiments furnished balloon detachments, including the 9th, 14th, 21st, 22nd, 72nd, 81st, 84th, and 92nd New York; the 10th, 19th, and 29th Massachusetts; the 7th Connecticut; the 2nd and 4th Michigan; the 26th and 118th Pennsylvania; and the 4th Maine.[231] How large a percentage of all units furnishing men for this service these regiments represent, is impossible to tell. Likewise beyond calculation is the number of men who served with the balloon corps. When the service was in process of organization, several newspapers stated that it would contain about 600 men.[232] The basis for this statement was not revealed, and thus the figure means little.

son, " and introduced him to a large number of officers. . . . In the course of the evening you would hear one officer inquire of another ' What do all those stripes on his shoulder mean, and what right has he to uniform?' Finally one more bold than the rest inquired of Mr. Seaver what the ' A. D.' on his straps meant, and he answered ' Aeronautic Department,' and he was told that they were not aware that there was any such department. He wished he had had the money for his uniform before they got through with him."

It is rather amusing that the papers should have reported the enlisted men's insignia as " B. C.," when one of the aeronauts insisted on placarding himself with the legend " A. D."

[231] This list of regiments furnishing troops to the Balloon Corps has been compiled from muster rolls, payrolls, correspondence in the Lowe Papers, orders, letters, and dispatches in the various collections of the War Department Division, National Archives, and from lists of men detailed to balloon duty published in the contemporary press. Several regimental histories were also contributing sources.

[232] New York *Times*, November 9, 1861; Philadelphia *Press*, November 11, 1861; Detroit *Free Press*, November 15, 1861.

CHAPTER VIII

THE BALLOON SERVICE OF THE ARMY OF THE POTOMAC

Part Two: *Administration*

The extensive, complicated materiel, and the numerous diverse duties imposed by the balloon service upon its personnel suggest the necessary complement of a well-organized, efficient, and coordinated unit for field operations. Nothing could be farther from the truth. Despite the obvious need of efficient administration and closely-knit organization, the balloon corps was a loosely organized, poorly administered, and decentralized mixture of civilians and military personnel, often hindered by the administrative regulations and red tape of the army, and at the same time frequently denied the benefits of military efficiency. In relation to the other branches of the service it was an orphan, imposed as an unwanted ward upon the Bureau of Topographical Engineers, the Quartermaster Corps, and the Corps of Engineers. At the close of its existence, the Signal Corps was also selected as its unwilling guardian, but the chief signal officer refused to accept the added responsibility.

In addition to being juggled from one branch of the service to another, the balloon corps was subject to several separate channels of command; and it was largely dependent for its physical maintenance on various officers who had practically no knowledge of its requirements and necessities, and in some cases, little interest. In the strict military sense, the balloon service could hardly be called a corps at all.

At first its administration was an application, on a larger scale, of the methods that had been adopted for Lowe's operations with his first government balloon. Originally Lowe had been employed as a civilian to build and operate the *Union* under the supervision of Captain Whipple, of the Bureau of Topographical Engineers. In all technical matters he had been given a free hand, and was authorized to purchase what might

280

be needed for the balloon and its effective operation. The expenses thus incurred were examined, approved, and settled by Whipple from funds at his disposal from the Bureau. At the same time Whipple was assigned to the staff of General McDowell for topographical duty, and field operations with the balloon were carried out in accordance with instructions through channels from McDowell. Thus the administrative affairs of the balloon service became the province of the Bureau; the tactical operations were directed through the chain of command from McDowell. Whipple, in his dual capacity as an officer of the Bureau and a member of McDowell's staff, thus became the supervisor of Lowe's activities with respect to administration and operation alike.

During the period when the Army of the Potomac was being organized under the direction of McClellan as commander-in-chief, the situation was somewhat changed. McClellan and several of his subordinate generals, notably Fitz John Porter, developed a keen interest in the balloon and began to assume personal direction of its operations. Lowe consequently found himself in the position of being responsible to a multiplicity of superiors. Nominally responsible to Whipple, he frequently received orders from Porter and other officers, and carried them out without reference to Whipple's jurisdiction.

Early in September, 1861, when McClellan ordered the balloon moved to Chain Bridge, Whipple inquired of the adjutant general's department whether this action were intended to relieve him of " further responsibility and charge in regard to the balloon and the operations connected therewith." [1] After considerable delay, the inquiry was referred to Lieutenant Colonel John N. Macomb, chief topographical officer on McClellan's staff,[2] who replied that

within a few days past I was told by General Marcy [Chief of Staff] to take charge of the balloon; but I have as yet no further written evi-

---

[1] Whipple to Major Seth Williams [AAG, AP], September 17, 1861, MS W137, enc., Letters Received, Bureau of Topographical Engineers, War Department Division, National Archives [cited hereinafter as LRTE].

[2] Indorsement of Lieutenant Colonel James A. Hardie [AAG, AP], October 7, 1861, MS W137, LRTE.

dence of this fact than the endorsement at Headquarters here, on the letter above named, referring the same to me.[3]

In this casual manner Macomb superseded Whipple in charge of the balloon then in operation, and became the supervising officer of the balloon service which was expanded shortly thereafter. He never did receive any official order to assume this special duty. General Marcy's verbal statement that " it was the General's [McClellan's] wish that I should take charge of the balloon department " was all that authorized his assignment.[4]

The appointment of Macomb did not effect much change in the administration of the balloon corps which was established soon after he superseded Whipple. The tactical direction passed from McDowell to McClellan, who assigned balloons to various stations with the subordinate commands of his army. Macomb took a less active part in the field than had Whipple,[5] and during his administration Lowe's activities were directed by various staff officers in accordance with orders from McClellan.[6] But the methods of maintenance that had been followed in the operation of the one balloon remained unchanged when the service was expanded. Since the personnel of the organization other than the temporary details of ground troops included only civilians and had no military quartermaster attached, a major function of the supervising officer was still to authorize the purchase of supplies and equipment, and to approve and settle the various accounts presented by the chief aeronaut.[7]

[3] Macomb to Whipple, October 9, 1861, MS, Letters and Papers of Colonel John N. Macomb, Records of the Bureau of Topographical Engineers, War Department Division, National Archives [cited hereinafter as Macomb Papers].

[4] Macomb to Major Israel C. Woodruff [Chief, BTE], February 23, 1863, MS [number omitted], LRSW.

[5] All the available correspondence from Macomb relates to matters of supply, maintenance, and finance. None of his letters contains any instructions or directions relating to field operations. Materials examined include the Lowe Papers, the Macomb Papers, the Letters and Letterbooks of the Army of the Potomac, and the Bureau of Topographical Engineers.

[6] Lieutenant Colonel A. V. Colburn to Lowe, October 12, 1861, 3 *O. R.*, III, 264; Lowe to Brigadier General Joseph Hooker, November 10, 1861, *ibid.*, p. 265; Colburn to Lowe, November 16, 1861, *ibid.*, p. 266; Colburn to Lowe, November 16, 1861 [second letter], *ibid.*; Colburn to Lowe, November 22, 1861, *ibid.*, p. 267; Lowe to Hooker, November 30, 1861, *ibid.*, pp. 267-268.

[7] Macomb to Woodruff, February 23, 1863, MS [number omitted], Letters

A number of circumstances combined to create difficulties in this phase of administration. Macomb had no knowledge of the physical requirements of the balloon service. He was unfamiliar with the problems involved, with the materiel and supplies that were essential to the maintenance of the corps, and with the nature of the many operations that military ballooning entailed. Aeronautics had never been practiced in the army before, and hence there were no precedents to follow, and no past experience on which to rely. Also, the unusual character of many necessary materials and expenses peculiar to aeronautic operations prevented the functioning of routine military procedure in matters of supply and expense. For example the ordinary quartermaster stores of the army did not contain for issue materials for the construction and repair of balloons, sulphuric acid in quantity, and other articles of common necessity for balloon operation. Consequently the responsibility for maintaining the corps active in the field devolved upon Lowe. He was authorized " to purchase such materials as are necessary for his operations, to make such journeys as might be required and to direct his assistants to make such journeys." [8] In conformity with this authorization he set his corps in operation, a task that involved details and expenses completely unforeseen and unexpected by his military superior. He travelled the length of the Potomac lines, establishing balloon stations as ordered by McClellan and his staff. He went back and forth from the lines to Washington, performing numerous errands; from Washington to the Navy Yard to supervise the construction of the generators; from the Navy Yard to the balloon stations, to establish the generators in operation. He directed his assistants to go to Philadelphia and elsewhere to obtain necessary supplies. He

Received, Secretary of War, War Department Division, National Archives [cited hereinafter as LRSW]; Macomb to Brigadier General W. Scott Ketchum [AIG], February 11, 1864, MS L429, enc., LRSW; " Statement of Colonel John N. Macomb, C. of E., in Relation to Abstract of Charges No. 1923. . . . Third Auditor's Office," [n. d.], MS, Macomb Papers.

[8] " Verbal direction and authority of the Major General Commanding," quoted in Assistant Adjutant General [signature omitted], to Brigadier General Stewart Van Vliet [CQM], June 2, 1862, MS Letterbooks and Bound Records of the Army of the Potomac, Vol. II [Letters Sent, July 1861–July 1862], p. 177 [cited hereinafter as MS LBAP].

purchased numerous articles for the corps, incurred expenses for the transportation of materials, lodging and rail fare for his assistants during the performance of duty away from the military lines, and numerous other items of legitimate expense in the prosecution of his duties as chief aeronaut.[9]

When the accounts first began to come in for payment Macomb was surprised and disturbed. His previous experience had never encompassed the problems of operating a corps of balloonists, with stations extending from the upper reaches of the Potomac to the shore of southern Maryland. Early in January he reported to the adjutant general that

> the bills which are presented to me for their [the balloon details'] expenses are so great and of such unusual character as to threaten the exhaustion of the appropriation for " Surveys for Military Defenses " [10] and also to make me feel doubtful as to their being passed to my credit when I pay them, and I shall therefore feel under the necessity of asking you to submit them to the General-in-Chief for his sanction.[11]

Authority for the approval of Lowe's accounts was granted by McClellan's headquarters,[12] and for a while the administration of the balloon corps, though irregular in terms of orthodox military procedure, functioned with relative smoothness. Subsequently Macomb again raised the question of certain " irregular " items in Lowe's accounts, despite the general authorization previously issued.[13] The items included several trifling expenses, particularly a bill for the stabling and care of one of Lowe's horses in a livery stable during one of his numerous official errands.[14] No government stable was available in the area in

---

[9] Accounts and vouchers, listing nature of expenses listed, covering period November–December, 1861, MS, Lowe Papers. These vouchers enumerate charges for supplies, materials, field and headquarters equipment, rail fare, porterage, meals for enlisted men and civilian assistants, steamboat fare, etc.

[10] " Surveys for Military Defenses " was the account under which all expenses for balloon operations were carried during Macomb's administration.

[11] Macomb to Williams, January 4, 1862 [Weekly Report No. 13], MS M24, Letters Received, Army of the Potomac, War Department Division, National Archives [cited hereinafter as LRAP].

[12] Williams to Macomb, January 6, 1862, MS, Macomb Papers.

[13] Macomb to Williams, January 25, 1862, MS M145, LRAP.

[14] Macomb to Williams, February 13, 1862, MS M230, LRAP.

which Lowe was working at the time.[15] The regulations, Macomb pointed out, did not provide for the shoeing and stabling of government animals in private livery establishments.[16] The items questioned, however, were returned approved, with the indorsement that " all bills of the balloon establishment are to be audited under the regular and usual system of expenditure and accountability in the Topographical Service." [17]

This statement of policy, though the specific accounts in question were approved, created an anomalous situation. General headquarters approved certain accounts that did not conform to the procedure of the topographical service, yet at the same time that procedure was ordered applicable to all balloon expenses. Thus if Macomb followed his instructions, he was bound to disapprove expenses after the precedent for their approval had been set by general headquarters. Such a policy could provide no satisfactory basis for administering the balloon corps accounts. Strict application of the Topographical Bureau's system of accounts created serious difficulty. Since many of the expenditures involved in military ballooning were not common to the topographical service, irregularities continued to appear in the chief aeronaut's vouchers. The conflict of precedent and orders arising from the policy at first adopted often led to much official haggling over the approval of Lowe's expenses. In some instances items were approved; others of almost identical nature were disallowed.[18]

Further difficulties arose from Lowe's lack of familiarity with the intricacies of military forms and accounting systems. His accounts and vouchers were frequently returned for correction and redrafting to conform with details and precedents required by the auditors and finance officers.[19] Sworn affidavits were

[15] Lowe to Macomb, February 11, 1862, MS, Lowe Papers.
[16] Macomb to Williams, February 13, 1862, MS M230, LRAP.
[17] Indorsement, February 17, 1862, on MS M230, LRAP, MS, LBAP, Vol. XXIII [Indorsements, August 1861–August 1862], p. 253.
[18] Macomb to Lowe, July 25, 1862, MS, Lowe Papers.
[19] " Memorandum in Respect to Prof. T. S. C. Lowe, Chief Aeronaut, Army of the Potomac " [n. d., but from context evidently of the period of spring, 1862], MS, Lowe Papers; Macomb to Lowe, July 25, 1862, MS, Lowe Papers; Humphreys to Lowe, August 6, 1862, MS, Lowe Papers; Humphreys to Wil-

sometimes required to support the validity of his accounts.[20] Much time was thus occupied in clerical exactions which added to Lowe's pressing duties and responsibilities in the field.

One of the most unsatisfactory features of the administrative system appeared in cases when Lowe could not obtain necessary supplies and services on the government credit at his disposal. He could pledge the Government's credit only when the sellers of materials saw fit to accept his authority to bind the Government. When it was necessary to secure services, such as labor, porterage, or transportation, cash payment was often required.[21] Individuals furnishing such service could not be expected to wait for the long process of approval, audit, and eventual payment entailed by the regular military system of accounts. Thus when cash was demanded for supplies or services, Lowe had two alternatives: he could suspend his operations until the administrative machinery could be set in motion to provide official credit or funds, or he could pay for needed materials and services from his own money and present vouchers for personal reimbursement. Often the tactical requirements of the balloon service did not admit the delays attendant on the former course of action. When ordered to establish a station at some corps or division headquarters, or to arrange for observations at some distant point in the lines, it became the chief aeronaut's responsibility to do so without delay. If certain supplies were needed for this purpose, or if porterage, transportation, or other services not furnished at the moment were not available, it became his duty to secure them and execute his orders in the least possible

liams, May 28, 1862, MS, Lowe Papers; Memorandum concerning Balloon Corps accounts, MS LBAP, Vol. XXIII [Indorsements, August 1861–August 1862], p. 359.

[20] Affidavit, supporting validity of accounts, sworn to by Lowe before Fred C. Thomas, notary public, Washington, February 24, 1862, MS, Lowe Papers.

[21] Items on which Lowe expended personal funds and later sought reimbursement included such items as

" November 23, 1861, Express Company, freight on pumps, hose, &c., $10.28
November 23, 1861, Cartage of hose, &c., armory to Navy Yard,   1.50
November 25, 1861, Fare to Old Point, Va., self and assistants,   3.00
November 28, Fare, carriage hire, and porterage, Baltimore to
   Washington,   2.50 "
Cf. Balloon Corps accounts and vouchers, MSS, Lowe Papers.

time. Commanding line officers are not interested in administrative excuses for delays or failure to execute tactical operations.[22] Nor was Lowe as a civil employee in position to dodge responsibility in case the balloon service failed to accomplish a given mission. The balloon corps was a new branch, an untried innovation. To convince the general officers of its utility, Lowe had to produce results in the field, not excuses. Irregularities in accounts and petty expenses were secondary considerations. Hence when time was an essential factor in Lowe's operations, or when needed equipment or personal services were required that could not be secured on the credit at his disposal, he paid such bills with his own money and took receipts as vouchers for reimbursement.[23] In some instances his vouchers were accepted and his claims allowed. In others, items of almost identical nature were refused. The existing records exhibit little consistency in the policy thus adopted towards these irregular accounts. In pressing for reconsideration of one of these disallowed claims, Lowe later declared:

Had I been possessed of Government funds or had some Quarter Master been attached to my department to do this business, I should then have had no occasion to make these accounts; but as it was, my former accounts being paid, and receiving the same orders to keep my department in working order, I was encouraged to pay further small sums as

---

[22] A clear idea of what confronted Lowe in this respect may be obtained from an incident of more modern vintage. On one occasion the present writer was serving as Plans and Training Officer on a battalion staff of a National Guard field artillery regiment during field training. He was given orders to take a detail and erect targets in the impact area of the range for service practice the following day. The regimental supply depot had neither white paint, cloth, or other facilities of any kind to issue for the purpose. To report this deficiency as an excuse for not having the targets in position in time for firing would have resulted only in severe censure from higher command and disruption of the firing schedule. The only alternative was to go to the nearest town, buy cheese cloth, nails, and other necessary materials and pay for them personally. The targets were put in place that afternoon, but the writer has yet to collect the money personally expended. (The incident described occurred in 1931.)

[23] Voucher No. 34, " To refund Mr. Lowe sundry amounts for which all of the sub-vouchers could not be procured. These seem to be for various small outlays incidental to the prosecution of his duties." Macomb to Williams, January 25, 1862, MS, M145, LRAP; also Macomb to Williams, January 30, 1862, MS M145, enc., LRAP.

it was necessary, and to this moment I am unable to understand why one should be paid and the other rejected.[24]

The assumption of payment for sundry items thus expended for the corps eventually resulted in the loss of more than five hundred dollars from Lowe's personal funds. Although continual applications were made for reconsideration, the claims were repeatedly disallowed.[25]

On March 31, 1862, the administration of the balloon corps was removed from the jurisdiction of the topographical engineers and placed directly under the quartermaster department, subject to the direction of Lieutenant Colonel Rufus Ingalls, Chief Quartermaster of the Army of the Potomac.[26] Little change in the handling of accounts was effected by this transfer of control. Lowe was still authorized to make necessary purchases on the credit of the Government, now subject to the approval of the Chief Quartermaster and his subordinates rather than to that of Macomb and the Topographical Bureau.[27] One important change, however, lay in the transfer of all issued property to the accounts of the quartermaster officers, and the inclusion of Lowe and his civilian assistants on the employee rolls of the quartermaster department.[28] Thenceforward, until another change of control took place, all issued equipment, such as animals, wagons, etc., became subject to the orders of the quartermaster officers, a circumstance that frequently disrupted

[24] Lowe to Brigadier General Edward R. S. Canby [AAG, USA], January 22, 1864, MS L429, enc., LRSW.

[25] Lowe to Secretary Stanton, January 22, 1864, MS, LRSW; Lowe to Stanton, April 20, 1864, MS L429, with enclosures, LRSW; "Reconsideration of the Accounts of Mr. T. S. C. Lowe," enclosure to MS L429, April 26, 1864, LRSW; "Summary of the Accounts of T. S. C. Lowe, Aeronaut," enclosure to MS L429, LRSW.

[26] "Endorsement on Accounts of Prof. T. S. C. Lowe," signed by Brigadier General W. Scott Ketchum, December 1, 1863, MS [number omitted], LRSW (1864).

[27] Humphreys to Clovis Lowe, July 15, 1862, MS, Lowe Papers; Humphreys to T. S. C. Lowe, August 3, 1862, MS, *ibid.*

[28] "Endorsement on Accounts of Prof. T. S. C. Lowe," signed by Ketchum, December 1, 1863, LRSW [number omitted], 1864; Certificate of Captain John B. Howard [AQM], transferring Lowe to his rolls, May 7, 1863, MS L429, enc., LRSW (1864); Brigadier General Rufus Ingalls [CQM], to Ketchum, October 28, 1863, MS L429, enc., LRSW (1864).

the balloon service and imposed difficulty upon its operation in the field.

When the army arrived on the Peninsula, the tactical direction of the corps was assumed by the Commanding General, his staff, and the various corps and division commanders to whose units balloons were assigned for use. Lowe received his orders from all these officers, and directed his assistants accordingly.[29] After the fall of Yorktown, the need of more centralized control was recognized. Accordingly Brigadier General Andrew A. Humphreys, Chief Topographical Engineer on McClellan's staff, was appointed to take charge of the balloon corps for purposes of tactical operation.[30] Humphreys had no direct control of the administrative affairs of the corps, but assumed the responsibility of approving Lowe's accounts and assisted him in the use of forms and in other routine quartermaster regulations. He appears to have had a liberal understanding of Lowe's problems, and often acted with consideration and wisdom as a buffer between his civilian subordinates and the exactions of the quartermaster officers.[31] The regime under Humphreys marked one of the better administrative periods of the balloon service. Had he continued in this assignment, it is likely that eventually an efficient maintenance system might have been developed. After the Peninsular campaign, Humphreys was transferred to command a line division, and his able support was thus removed.[32] The balloon corps continued under Ingalls for administrative

[29] Captain Fred T. Locke [AAG, AP], to Lowe, April 6, 1862, 3 *O. R.*, III, 273; Colburn to Lowe, April 6, 1862, *ibid.*, p. 274; Porter to Lowe, April 13, 1862, *ibid.*, pp. 274-275; Colonel Delos B. Sackett [IG, AP], to Lowe, May 26, 1862, *ibid.*, p. 278.

[30] Special Orders, No. 157, Hdq. AP, May 25, 1862, MS, Lowe Papers. See also Humphreys to Lowe, May 26, 1862, 3 *O. R.*, III, 278.

[31] Humphreys to T. S. C. Lowe, May 28, 1862, MS L429, enc., LRSW; Humphreys to Clovis Lowe, July 15, 1862, MS, Lowe Papers; James Allen to T. S. C. Lowe, July 20, 1863, MS, Lowe Papers; Humphreys to Clovis Lowe, August 3, 1862, MS, *ibid.*; Humphreys to T. S. C. Lowe, August 3, 1862, MS, *ibid.*; Humphreys to T. S. C. Lowe, August 6, 1862, MS [number omitted], LRSW (1863); Humphreys to T. S. C. Lowe, August 6, 1862, MS, Lowe Papers.

[32] Special Orders, No. 204, HQA, August 23, 1862, 1 *O. R.*, XII, iii, 640; Henry H. Humphreys, *Andrew Atkinson Humphreys* (Philadelphia, 1924), 167-168.

purposes without noticeable change. When the aeronautic train rejoined the army after the battle of Antietam, Ingalls reaffirmed Lowe's authority to continue his customary practice of maintenance and gave orders that

Professor Lowe is authorized to purchase all articles necessary for his balloons and party. He is also authorized to send an agent for them [articles of supply] who will be entitled to transportation at public expense on any railroad. The bills will be paid on presentation to Captain Howard on being properly prepared, certified, &c.[33]

This policy was soon to be ended. Less than three months later Lowe was advised by the assistant quartermaster under Ingalls that " all articles for the use of the balloon corps must be procured by requisition from this office. In future no bills will be paid for articles purchased . . . unless by written authority of the Chief Quartermaster of the Army of the Potomac." [34] Evidently Lowe found it necessary to make purchases without waiting for the Chief Quartermaster's sanction. Several weeks later Ingalls reiterated Howard's instructions in curt terms:

You were notified some time since by Captain Howard by my order, that you would make no more purchases of articles required for balloon purposes. Hereafter when you require any materials for the purpose above mentioned, you will make out your requisition for the same, which will be presented at this office for approval. I will then see that the articles are furnished you.[35]

The change was certainly not conducive to any added efficiency. The requirement of advance approval before the purchase of supplies and equipment, when the quartermaster officers were ignorant of the needs of the balloon service, led to serious delays and lengthy explanations from Lowe before the materials were forthcoming. Such delays, added to the time required for the ordering of articles not contained in the quartermaster stores, sometimes caused several weeks' delay from the time of requisition to the receipt of the equipment.

[33] Order of authorization, October 7, 1862, signed by Ingalls, MS L429, enc., LRSW (1864). The officer referred to was John B. Howard, Assistant Quartermaster, Army of the Potomac, Ingalls' chief assistant.
[34] Howard to Lowe, December 29, 1862, MS, Lowe Papers.
[35] Ingalls to Lowe, February 2, 1863, MS, Lowe Papers.

A case in point may be useful by way of illustration. Ingalls issued his second order on February 2. On February 28 Lowe sent in a duly prepared requisition for sundry necessities, including lime, acid, iron turnings, sandbags, net cord, and other materials essential to the operation of the corps.[36] On March 5 Ingalls returned the requisition asking for estimates of the cost of the articles enumerated and for what purposes they were needed.[37] Lowe thereupon had to write out a detailed explanation of the items, to the effect that lime was necessary for purifying the gas; that the acid and iron were the ingredients used in making the gas; that the sandbags were used for ballast and for anchoring the balloons; and that the net cord was used for the purpose of repairing a net that had become worn through use and exposure, and so on.[38] Obviously the new administrative regulations did not facilitate matters of supply.

On April 7, 1863, the balloon department was transferred from the immediate supervision of the quartermaster department and placed under the Corps of Engineers,[39] subject to the direction of Captain Cyrus B. Comstock, Chief Engineer of the Army of the Potomac.[40] Although supply accounts, civilian salaries, and other items of expense were still paid from quartermaster funds, approval of all vouchers and requisitions, and general administrative control was exercised by Comstock. The changes he effected and the resulting difficulties with Lowe which ended in the chief aeronaut's resignation are treated in Volume II. The jurisdiction of the Corps of Engineers continued until the final disbandment of the balloon train in June, 1863. Shortly before the close of the balloon service, Comstock

[36] Lowe to Ingalls, February 28, 1863, MS, Lowe Papers.

[37] Ingalls to Lowe, March 5, 1863, MS, Lowe Papers.

[38] Lowe to Ingalls, March 6, 1863, MS, Lowe Papers.

[39] "Endorsement on Accounts of Prof. T. S. C. Lowe," signed by General Ketchum, December 1, 1863, MS [number omitted], LRSW (1864); Lieutenant Colonel John B. Howard [CQM, 3rd AC] to Brigadier General Montgomery C. Meigs [QMG, USA], September 9, 1863, MS L429, enc., LRSW (1864); Lieutenant Colonel Ebenezer Sibley [AQM, USA], to Ketchum, October 12, 1863, MS L429, enc., LRSW (1864).

[40] Special Orders, No. 95, Hdq., AP, April 7, 1863, MS Papers of Brigadier General Cyrus B. Comstock, Division of MSS, Library of Congress [cited hereinafter as Comstock Papers]. See also 3 O. R., III, 302.

was transferred from the Army of the Potomac, and was replaced by Brigadier General Gouverneur K. Warren, the last officer to exercise supervision over the aeronautic department of the Federal Army.[41]

One final change in the administration and direction of the balloon corps was attempted. During the movement north to intercept Lee's invasion of Pennsylvania, General Hooker ordered the Signal Corps to take over the balloon department and its administration.[42] Colonel Albert J. Myer, Signal Officer of the Army of the Potomac, whose part in Wise's attempt to reach the army at the first Manassas may be recalled, then protested that he had neither the men nor the appropriation with which to operate the additional branch of the service.[43] Thereupon the balloon train was ordered back to Washington and was disbanded. Altogether, the aeronautic department had been placed under three separate branches of the army in the space of less than three years. It would have been made subject to a fourth, but for the objection of the officer commanding the branch involved in the last proposed change. Lowe's organization was an unwanted orphan indeed.

An examination of the internal administration of the corps under all these various branches shows much to be desired in efficiency. The ordinary quartermaster property issued to Lowe's train was at first handled under the same system that applied to all branches of the army. Items of regular equipment were issued on memorandum receipt on request of the supervising officer of the corps, who took them up on his property debit accounts.[44] Lowe and his assistants were in turn required to submit inventories and receipts covering all such property in their possession.[45]

[41] See Volume II.
[42] Ingalls to Ketchum, October 28, 1863, MS L852, LRSW; Hooker's indorsement, June 24, 1863, on Lowe to Hooker, June 18, 1863, MS L429, enc., LRSW (1864).
[43] Myer's indorsement, June 26, 1863, on Lowe to Hooker, June 18, 1863, MS L429, enc., LRSW (1864).
[44] Memorandum Receipts, Quartermaster forms No. 40 and No. 27 [for various articles of issued property], Macomb Papers; Macomb to Lowe, December 6, 1862, MS, Lowe Papers.
[45] Macomb to Williams, February 8, 1862 [Weekly Report No. 18], MS

Although the balloon establishment should have fared reasonably well under this system, numerous difficulties arose, of which the worst was the matter of transportation. Wagons and draft animals issued to the balloon trains were rarely sufficient for their demands. When several observation stations were required in the field, Lowe was often forced to appeal to the corps or division commanders in whose sectors the stations were established, to provide the necessary transportation.[46] In many cases these officers had no additional wagons or teams to spare, or needed what they had for their own operations.[47] " Professor Lowe asks me for six wagons to bring up his balloon," wrote General Keyes, commanding the 4th Corps in the advance on Yorktown. " I cannot furnish one until I get up forage and provisions." [48] In such a case Lowe would have to appeal to the Commanding General before he could procure this vital equipment.[49] By and large he received more cooperation from the high command than from the quartermaster department, although that branch was charged with the administration of his corps. On one occasion immediately preceding the fall of Yorktown, after Lowe had been unable to secure wagons or horses from any source and had carried the matter to McClellan,[50] the adjutant general notified Brigadier General Stewart Van Vliet, then Chief Quartermaster, that

the Commanding General directs that you cause to be issued to Professor Lowe, Chief of the Balloon Department of this Army, such means of transportation and Q. M. supplies as may be necessary to enable him to perform the duties with which he is charged.[51]

M288A, LRAP; Inventories of property with balloons *Eagle, Constitution* and *Intrepid,* MSS, Lowe Papers; Ingalls to Lowe, February 2, 1863, MS, Lowe Papers; Macomb to Lowe, December 6, 1862, MS, *ibid.*

[46] Lowe to Colburn, April 20, 1862, MS, Misc. Papers of the Army of the Potomac, War Department Division, National Archives; Lowe to Keyes, April 20, 1862, MS, Lowe Papers; Marcy to Lowe, April 8, 1862, MS, *ibid.*; Howard to Lowe, April 15, 1863, MS, *ibid.*

[47] Marcy to Lowe, April 8, 1862, MS, Lowe Papers; Porter to Lowe, April 16, 1862, MS, *ibid.*; Lowe to Van Vliet, April 13, 1862, MS, *ibid.*

[48] Keyes to Marcy, April 8, 1862, 1 *O. R.*, XI, iii, 78.

[49] Lowe to Marcy, April 24, 1862, MS, Lowe Papers.

[50] *Ibid.*

[51] Williams to Van Vliet, April 29, 1862, MS, Lowe Papers; see also 3 *O. R.*, III, 275.

McClellan had intervened in Lowe's behalf on the issue of transportation on previous occasions,[52] but the necessity of such intervention in such matters by the General-in-Chief hardly speaks well for the administration of supplies in the army under his command.

When the quartermaster department replaced the Topographical Bureau in administrative control of the balloon corps, the question of transportation went from bad to worse. After the wagons and teams had been issued, there were times when they were claimed and taken back for other purposes by officers of the quartermaster corps, regardless of the pressing needs of the balloon trains. During operations near Pohick Church, prior to the movement to the Peninsula, a wagon master attached to the quartermaster corps commandeered three teams assigned to Lowe's generators and escort wagon. An urgent telegram to Captain Dana, of the Quartermaster Office in Washington, stating the absolute necessity of retaining the teams was of no avail. "It is very necessary," Lowe explained, "that these teams remain here for a day or two until I can come in and make arrangements for others, *as we have got to move.*"[53] The appeal fell on deaf ears. The teams were taken away, and Lowe's generators were left stranded. An application for assistance from General S. P. Heintzelman,[54] in command near the position, led to an order from General McClellan's headquarters directing Van Vliet to furnish additional transportation.[55]

Again, after the return from the Peninsula, when McClellan's army was pushing through Maryland to parry Lee's first thrust across the Potomac, the balloon trains were unable to accompany the army in time to be of service in the battle of Antietam because the quartermaster general had requisitioned all

[52] Instructions to General Van Vliet, March 7, 1862, MS LBAP, II, 43, letter 67.

[53] Lowe to Captain J. J. Dana, March 6, 1862, MS, Lowe Papers.

[54] Lowe to Heintzelman, March 6, 1862, MS, Papers of General Samuel Peter Heintzelman, Division of MSS, Library of Congress [cited hereinafter as Heintzelman Papers].

[55] Instructions to General Van Vliet, March 7, 1862, MS LBAP, II, 43, letter 67.

Lowe's transportation for other purposes.[56] Only when orders were issued by McClellan's chief of staff was Lowe provided with requisite teams and wagons. The delay thus occasioned resulted in Lowe's arrival after the battle was over.[57] Such examples as these serve to illustrate the obstacles with which Lowe had to contend with regard to one of the most vital administrative functions of his corps.

Nor were the difficulties arising over issued property confined to transportation equipment. There were frequent delays in obtaining forage.[58] On one occasion it was impossible to secure supplies because the office of the quartermaster in the field could not provide the necessary requisition forms required for the issue of stores. " Four blanks are all I can obtain. . . . The clerk tells me that he has been expecting blanks from Washington for some time," was the reply to Lowe's requisition two days before the battle of Seven Pines.[59] During the siege of Yorktown Lowe had to go directly to General Van Vliet in order to draw tents for the men of his temporary detail.[60] And this was necessary when the administration of the corps was a quartermaster department function. Translated into terms of civilian affairs, one might as well imagine the head of the shipping department of a huge corporation reduced to applying to the president for a ball of wrapping string. When Ingalls inaugurated his new policy in the maintenance of the balloon corps in 1863, he reduced the tentage assigned to Lowe's establishment to two wall tents for five aeronauts and the headquarters of the corps, one tent for the seven permanently detailed enlisted men, and a small hospital tent for the storage of all the supplies. Every other scrap of canvas issued to the corps was ordered turned in.[61]

[56] Lowe's MS Memoirs, p. 166.

[57] *Ibid.*; *Lowe's Report, 3 O. R.,* III, 292.

[58] Lowe to Ingalls, May 27, 1862, MS, Lowe Papers; Lowe to Keyes, April 24, 1862, MS, *ibid.*; James Allen to Lowe, April 22, 1862, MS, *ibid.*; Humphreys to Clovis Lowe, July 15, 1862, MS, *ibid.*; Lowe to Van Vliet, April [date missing], 1862, MS, *ibid.*

[59] F. W. Vaughn to Lowe, May 28, 1862, MS, Lowe Papers.

[60] Lowe to Van Vliet, April 13, 1862, MS, Lowe Papers.

[61] Howard to Lowe, March 16, 1863, MS, Lowe Papers.

296

If the administration of physical property and supply was poorly handled, then the pay of the aeronauts and other civilian personnel was even worse administered. In the midst of the Peninsular campaign two of the aeronauts, Mason and Seaver, refused to perform service until their salaries, long over-due, should be paid. Their subsequent discharge by Lowe, regardless of the merits of the case, was a direct result of the continual delays in this phase of administration.[62] Frequently the aeronauts did not receive their salaries for months on end, although payrolls were submitted on time. " I have plendy off all things except *monny*. I can't even buy tobacco," complained the German aeronaut Steiner, in his broken, phonetic English after his pay was long over-due.[63] He later begged Lowe to send his wife two dollars.[64] With the Western Department, he found matters worse. After repeated delays, he wrote his chief, " I was paide after a grade deal of trubel, and the lord knowes wen I will be paide again." [65] Steiner, too, eventually left the army because of the continual wait for his pay.[66] James Allen, Lowe's most loyal assistant, also voiced similar complaints. " It is now nearly three months since I received my salary, and I am destitute—with no means of sending my family any support," he wrote. ". . . I must have it, as my family are suffering for want of it." [67] The civilian teamsters fared no better. In August, 1862, when Lowe was on sick leave, General Humphreys wrote to him that he was astonished

to learn that the teamsters had not been paid for four months—some for even a greater length of time. They left the service in a body at the end of July, and I understand chiefly on this account. Some of their families have doubtless suffered.[68]

While delays in the paying of troops in the field during active

[62] Lowe to Howard, January 31, 1863, MS L429, enc., LRSW; also copy in Lowe Papers.
[63] Steiner to Lowe, January 22, 1862, MS, Lowe Papers.
[64] Steiner to Lowe, February 11, 1862, MS, Lowe Papers.
[65] Steiner to Lowe, June 9, 1862, MS, Lowe Papers.
[66] Lowe to Howard, January 31, 1863, MS L429, enc., LRSW (1864); also copy in Lowe Papers.
[67] Allen to Lowe, August 2, 1862, MS, Lowe Papers.
[68] Humphreys to Lowe, August 3, 1862, MS, Lowe Papers.

service are common and usual incidents to wartime administration, continual failure to pay the civilian employees for month after month would seem to constitute rank mismanagement somewhere. The responsibility for this chronic situation is impossible to determine. In the case of the teamsters, Humphreys specifically denied that the trouble was in his headquarters.[69] This statement implied that either Lowe or the quartermaster department was at fault. The former implication was inadmissible, since Lowe as a civilian directly under the administrative jurisdiction of the quartermaster department could not have delayed the teamsters' payrolls unless his superiors allowed him to do so. This reasoning might possibly justify the condemnation of Ingall's headquarters as responsible for the mismanaged payrolls. But there may have been other circumstances over which Ingalls and his staff had no control, which contributed to the deficiency. Thus the root of the trouble over pay cannot be definitely traced. With the numerous channels of command, responsibility, and interdependence which made up the structure of military administration and finance, any number of individuals might have been responsible, and official buck-passing, particularly under the stress of field campaigning, is an old and familiar story. Certainly Lowe could not have prevented mismanagement in the matter of the teamsters' pay. As the aeronauts' salaries, it is only reasonable to suppose that he did his best to prevent the delays that were common occurrences in the corps. Here his own pay was involved, and he, too, had a family dependent on him for support.

The administrative management of the separate balloon trains sent to other departments were by and large no better than those of the corps with the Army of the Potomac. In the case of Steiner's service with the Western Department at Cairo and Island No. 10, things were decidedly worse. The details of this phase of the corps' activity, and John Starkweather's experience with the Department of the South, are treated in a subsequent chapter dealing with the detached service of the corps.[70]

Another source of difficulty that continually plagued the administration of the balloon service was the system of assign-

---

[69] *Ibid.*   [70] See Chapter xi, *infra.*

ing temporary detachments for duty as ground crews. The evils of this procedure are readily apparent and have been touched on in earlier chapters. The system was condemned by Lieutenant Abbot during the early experiments with James Allen's balloon in June, 1861; its deficiencies again came to the fore during the short-lived operations of John Wise; and La Mountain did not escape its shortcomings. Finally when the aeronautic service was greatly expanded, the same old problem arose and was permitted to continue.[71] The system of temporary assignment caused periodic removals of trained men from the various balloon stations. Since the technique of managing a captive balloon was new to the details of volunteer troops, each detachment had to be carefully instructed in the duties peculiar to the service before its members could become proficient in the work required. Then the sudden transfer of such a detail immediately left the aeronauts just where they were in the beginning, with new platoons of untrained men. The tour of this special duty cannot be estimated satisfactorily. Transfers generally took place when regimental commanders demanded the return of their troops, or at times when a balloon station was moved to a considerable distance from its original position. Since seventeen regiments are known to have furnished details, with the possibility that many others for which records are now lacking did likewise, it is reasonable to assume that transfers took place every few weeks. The detachments that were assigned to La Mountain's balloon were relieved every fifteen days.

The commanding officers of regiments cannot be blamed for demanding the return of their detachments assigned to this special duty. No unit commander cares to lose men indefinitely

[71] Ebenezer Seaver to Brigadier General Charles P. Stone, December 28, 1861, MS S42, Letters Received, Corps of Observation, War Department Division, National Archives [cited hereinafter as LRCO]; Indorsement, July 15, 1862, on Captain N. P. Hollowell [20th Maine Infantry] to Hdq., AP, MS LBAP, Vol. XXIII, 443; Indorsement, October 21, 1861, on Colonel Walter Phelps [22nd N. Y. Infantry] to Hdq. AP, MS LBAP, Vol. XXIII, 41; Brigadier General I. N. Palmer [3rd Brigade, Casey's Division] to Captain H. W. Smith [AAG], May 14, 1862, with indorsements of Generals Silas Casey and Erasmus D. Keyes, MS, Lowe Papers; and other similar communications and indorsements, in the Lowe Papers and records of the Army of the Potomac.

from the effective strength of his command, and it is only natural that field officers of regiments furnishing Lowe's ground crews should have made efforts and have brought pressure to bear to effect the recall of their detached platoons. On such occasions their demands were generally supported through the channels of indorsement on up to general headquarters.[72] For example a regimental commander who furnished a platoon during the advance on the Peninsula applied to his brigadier to secure the recall of his men.[73] The general commanding the brigade promptly approved the request and indorsed it on to the division commander,[74] who concurred and forwarded it with an indorsement of approval to the commander of the army corps to which the division was attached.[75] From corps headquarters the request went forward to general headquarters, having gathered three favorable indorsements during the process.[76] In this particular case the request with its battery of indorsements was then referred to Lowe, with a further indorsement from the adjutant general's office, inquiring whether or not the balloon service could be carried on properly with temporary details from adjacent regiments. As a matter of policy, the assistant adjutant general continued, " it is preferable that the details should not be permanently drawn from one body of troops." [77] On another occasion when a regimental commander made a similar request, General Burnside's adjutant general referred it to Lowe with the comment that it was of utmost importance that the detachment in question rejoin its regiment.[78] In one case the demand of even a company commander

[72] The statement of the general approval of the request of regimental commanders to have detachments relieved is based on an examination of the various indorsement books of the Army of the Potomac, and its component corps.

[73] Colonel J. S. Belknap to Brigadier General I. N. Palmer, May 13, 1862, MS, Lowe Papers.

[74] Palmer to Captain H. W. Smith [AAG, Casey's Division], May 14, 1862, MS, Lowe Papers.

[75] Indorsement of Brigadier General Silas Casey, May 14, 1862, MS, *ibid.*

[76] Indorsement of Brigadier General Erasmus D. Keyes [CG, 4th Army Corps], May 15, 1862, MS, *ibid.*

[77] Indorsement of Lieutenant Colonel James A. Hardie [ADC, AAG, AP], May 19, 1862, MS, *ibid.*

[78] Indorsement, November 14, 1862, on Lieutenant Colonel P. Powers to Hdq. AP. MS LBAP, Vol. XXIV, p. 122.

was upheld, and General Humphreys was ordered to relieve a ground crew and arrange for a new one.[79] Once the transfer of a detail on balloon duty was postponed until a new platoon could be trained.[80] But this instance seems to have been an exception to the general rule. Perhaps long experience with the faulty system finally induced the commanding officer to modify the routine procedure; this exceptional case did not occur until the middle of May, 1863. Little good resulted. Like many reforms of this nature, if indeed it was a reform, it came too late to be of value. The balloon corps was disbanded less than a month later.

The difficulties incident to the temporary assignment of details did not affect the aeronauts alone. The men often suffered in the process. These troops were carried on the rolls of their own regiments and were paid by their own regimental paymasters.[81] When on detached service they had to be mustered in every month in order to qualify for their pay, a cumbersome procedure when the details were scattered at stations separated from their regiments.[82] The system imposed an additional problem in the matter of messing the balloon detachments in the field. When their regiments were closely adjacent, they drew their rations from their own quartermasters. But when the distance was at all great between their regiments and the balloon stations, or when they were assigned to duty which required their presence away from both their regiments and the stations, severe hardship was imposed on the men in the matter of getting their food. " The balloon party at Cloud's Mill complain that their rations are not sent out regularly," reported a brigade officer of the day on duty with Gen-

[79] Captain H. P. Hollowell [Company " D," 20th Maine Infantry] to Hdq., AP, with indorsement, July 15, 1862, MS LBAP, Vol. XXIII, p. 443.

[80] Indorsement of Colonel S. F. Barstow [AAG], June 1, 1863, on Major General John Newton to Hdq. AP, MS LBAP, XXIV, p. 347.

[81] This statement is based on the fact that none of the men assigned for temporary duty appear on the payrolls of the balloon corps. Hence they must have been carried on the rolls of their own regiments. Further, since their duty with the balloons was only temporary, transfer from their own regimental rolls would not have been made, as they would have only to be picked up again at the expiration of their special duty.

[82] Howard to Lowe, February 11, 1863, MS, Lowe Papers.

eral Philip Kearney's brigade in December, 1861.[83] A week later Colonel Macomb advised the adjutant general's office that

the men detailed from the army for extra duty with the balloon parties are employed to work for Mr. Lowe and other aeronauts sometimes at the Columbian Armory, sometimes at the Navy Yard, and are frequently called on short notice for distant service, and are thus prevented from having any systematic arrangements for cooking their rations or messing generally.[84]

To remedy this defect, Macomb asked permission to commute the rations of the men thus assigned.[85] Several days later official approval, quoting sundry paragraphs of Army Regulations, was forthcoming.[86] This one difficulty, at least, was remedied early in the war.

The temporary assignment of enlisted men also produced other minor problems. Records of duty assignment were sometimes lost or confused, and in some cases commanding officers of regiments actually lost track of their men assigned to Lowe's corps. In one instance Private William A. Hodges of the 14th New York Infantry was reported missing while in fact he was on balloon duty, and inquiries concerning him were made through McClellan's headquarters.[87] The inquiry drew the vague comment that " there seems to be some mistake about the regiment to which W. A. Hodges belongs. If he be in Mr. Lowe's party, it would be desirable to ascertain his company and regiment." [88] All this time Hodges had been performing faithful duty with Lowe, and had been definitely assigned to the corps by Special Orders from general headquarters.[89] A worse

[83] Major William Birney [Brigade OD] to Kearney, December 10, 1861, MS, Letters Received, Department of Northeastern Virginia, Package No. 9 [Reports of Brigade Officers of the Day], War Department Division, National Archives.

[84] Macomb to Williams, December 18, 1861, MS M566, LRAP.

[85] Ibid.; Macomb to Williams, January 18, 1862, MS M83, LRAP.

[86] ADC [no signature] to Macomb, December 21, 1861, MS LBAP, I, 507; AAG [no signature] to Macomb, January 18, 1862, MS LBAP, I, 577.

[87] Col. James M. McQuade to Williams, December 13, 1861, MS M537, LRAP.

[88] Indorsement of Lieutenant Colonel Hardie, December 14, 1861, MS LBAP, XXIII, 135.

[89] Special Orders, No. 147, Hdq., AP, November 18, 1861, MS LBAP, Vol. XXVII [SO, July 1861–March 1862], p. 294.

case was that of Corporal James Starkweather, who was detailed permanently on balloon duty from the 19th Massachusetts Infantry. Starkweather's colonel subsequently reported him as a deserter when he did not return with the detachment when it was relieved. The man was fortunately cleared from this serious charge by a timely letter from Lowe to his colonel, in which Starkweather's true status was explained.[90]

On one occasion the transfer of troops from balloon duty during Lowe's temporary absence from his headquarters resulted in the loss of most of his headquarters furniture and equipment. The detail when relieved casually made off with all Lowe's bedding, stove, tables, mess equipment, and even his tent. The senior non-commissioned officer of the new detail succeeded in retrieving the tent, but the rest of the property was gone beyond recovery. The officer commanding the previous detachment had resigned and gone home, and the field officer of the day reported after a perfunctory search of his regimental area that the property could not be found.[91] In this case Lowe had personal reasons to condemn the system of temporary assignments.

A great many of the evils and deficiencies that occurred in the administration of the corps may be traced to the civilian status of Lowe and his assistants. The juggling of the department from one military branch to another was probably a direct result of this status. Since the aeronauts possessed no military rating, traditional military procedure required that they be attached to some branch of the army, under officers who could be held responsible for their activities and operations. Needless to say, these officers, with routine duties of their own corps to perform, often found the additional organization something of a white elephant. But had Lowe and his assistants been commissioned, and the balloon department created as a separate branch, there would have been no need for such an arrangement as was followed. The chief aeronaut, as an officer in charge of his own department, could have exercised all the administrative author-

---

[90] Lowe to Commanding Officer, 19th Massachusetts Infantry, December 19, 1861, MS, Lowe Papers.

[91] Sergeant Thomas Boyd [4th Michigan Infantry], to Lowe, March 4, 1862, MS, Lowe Papers.

ity and supervision that had been, because of his civil status, thrust on other officers, such as Macomb, Ingalls, and Comstock. Because of his intimate knowledge of the requirements and technique of the service, such administrative supervision by Lowe would have been far more effective than that of individuals who did not know that net cord was for making nets. Also, the relations between the corps and other interdependent military branches would have been smoother and less detrimental to the balloon service. For example, had Lowe been commissioned in commensurate rank, no wagon master with a requisition from a quartermaster captain would have summarily seized his teams and left his generators stranded.

In reviewing the problems of administration, Lowe later expressed himself clearly on the question of civil status:

And right here [Lowe was describing his difficulties in getting his train to Antietam] I would like to say that what delays, annoyances, &c. fell to my lot while connected with the army came from my being a civil employee. From the first I had hoped to be allowed to organize a balloon corps as a military branch, and applied for a commission to command it. But all that the " powers that be " would do was to grant me the courtesy title of colonel and accord me the courtesies and not the authority of that rank. Consequently I was subject to every young and inexperienced lieutenant or captain who for the time being was placed in charge of the balloon corps.[92]

These officers, Lowe continued, " had no knowledge whatever of aeronautics and were often a serious hindrance rather than a help."

Lowe had done his best to secure proper military status. During the Peninsular campaign he appealed directly to McClellan to remedy the defect.

Permit me [Lowe wrote] respectfully to present for your consideration a succinct statement showing the anomalous position held by me, having *civil* charge only, of the balloon corps acting under your command. . . . I deem it not improper in me to request your kind consideration in the bestowal of such military rank or its equivalent as shall be correlative with my civil advantages at the present time.[93]

---

[92] Lowe's MS Memoirs, p. 166.
[93] Lowe to McClellan, June 16, 1862, MS, Lowe Papers.

A few days later, Lee launched his onslaught that opened the bloody Seven Days. McClellan had to think more of the safety of his army and the change of base than of the propriety of bestowing a commission on his chief balloonist. Hence nothing came of the request. Later, Lowe writes, when he arrived on the field too late to be of service in the battle of Antietam and gave as his reasons the seizure of his transportation by the quarter-master department, McClellan promised him a commission with full command of the aeronautic department, which was to be established as a separate, independent branch.[94] The promise, though evidently made in good faith, was never carried out. McClellan was shortly thereafter relieved of command and his successor continued the former policy.

Lowe did not abandon his hope for a commission and an independent balloon corps even after he had resigned following his troubles with Comstock. When his former assistants, James and Ezra Allen, who carried on after he left, wrote complaining of the continued poor management and general ineffective condition of the then decadent balloon corps, he addressed a communication to the Secretary of War urging the reorganization of the corps along military lines, separate from the other branches of the service, under the control of a chief aeronaut who should have " the power to provide all the apparatus necessary for the successful operation of this branch of the service wherever it may be employed, under the direction of . . . the Commander-in-Chief." [95] Grimly remembering his troubles over accounts, and the disallowed claims for personal reimbursement which he was even then pressing for reconsideration, Lowe further recommended that

for defraying the expenses of the corps and for avoiding the great inconvenience and delay heretofore experienced in settling the accounts, a sufficient fund should be assigned to the service, and placed in charge of an officer, who would make payment of bills approved by the Chief of the Corps.[96]

---

[94] Lowe's MS Memoirs, p. 166.

[95] Lowe to Stanton, June 9, 1863, MS L356, Letters Received, Headquarters of the Army, War Department Division, National Archives [cited hereinafter as LRHQA]; also MS [copy], Lowe Papers.

[96] *Ibid.*

In reply to these suggestions, the chief clerk of the War Department notified Lowe that his letter had been referred to the General-in-Chief for consideration.[97] It went no farther. The letter has remained in General Halleck's file to this day.

A final difficulty as a result of the civil status of the balloonists arose after Lowe resigned. While he remained in service, the company officers commanding the temporary details respected his knowledge and authority in all matters of technical operation, regardless of his lack of military rank. This recognition may have been the result of his wide reputation before the war, or his personality and bearing may have commanded the respect of these junior officers assigned to his service. When Lowe left, however, this respect and recognition were not accorded the Allens. No better picture of the conflict between military and civilian personnel can be found to illustrate this defect in the corps than is shown by the letter of James and Ezra Allen, written to Lowe a fortnight before the balloon service was abandoned:

Since you left the Army of the Potomac, we have endeavored as you requested, to make this branch of the service as efficient as possible, but to accomplish what was done when you were in charge is simply impossible under the present management, and I fear that things will soon be in such a condition that when an important observation is wanted, it cannot be had. This comes from one or two aspiring young officers without any knowledge or experience whatever, assuming command of the business you have so long and so well managed. We, as *civilians* are not in the least respected by these officers, and the men whom we are obliged to direct (when using the balloons) do not obey our orders. We are often ordered to do things against our better judgment; on obeying one of these peremptory orders on Monday last, an accident occurred which might have proved fatal to the aeronaut.[98]

This friction between civil and military personnel was also described in an official report in stronger language, which stated that the relations

[97] Chief Clerk, War Department [signature omitted] to Lowe, June 12, 1863 [copy], MS L356, enc., LRHQA.

[98] James and Ezra Allen to Lowe, June 6, 1863, MS, Lowe Papers; also copy, MS L356, enc., LRHQA.

between the Messrs. Allen and the two lieutenants is of the most unfriendly character; their language . . . being a series of bitter recriminations and personal insults. . . .[99]

Such were the results of trying to mix oil and water: in the field, civil and military authority cannot be made to harmonize, and these examples are the fruit of such an attempt.

As viewed through the perspective of time by the present observer, there seems to have been little excuse for the slip-shod methods and chronic bungling that prevailed throughout the administrative affairs of the balloon service. No past experience, no precedents to follow, the new branch of the service could be operated only on trial and error: these might be reasons offered by the apologist of the system, or lack of system, adopted. Such reasoning cannot convince a practical critic. The French, nearly seventy years before, in the turmoil of revolution, counter-revolution, foreign invasion, and violent internal discord, had designed a system of military aeronautics that shames the Union procedure by contrast. Their *compagnies des aérostiers* contained troops recruited for that service alone, men carefully selected because of special qualifications. They were trained in their duties before being sent into the field. Their companies were well organized, with their own quartermasters to handle the question of supply, and they were commanded and directed by their own officers, subject to the tactical direction of the army commanders. A school for training the men and for perfecting materiel was also set up, equipped with a depot of stores and supplies for the aeronautic companies.[100] The members of the Committee of Public Safety who directed military affairs of the unsteady new Republic had had no precedent, no past experience to follow: the balloon had been invented only a decade before. And they were civilians, without military experience. Yet they had the vision and foresight to provide an organization and administrative system that should have been a precedent for the Federal army, which if adopted in Lowe's

[99] " Report on the Present Condition of the Balloon Corps," June, 1863, MS, Gouverneur K. Warren Papers [Letters, Chancellorsville, May 1–June 28, 1863, Vol. I], New York State Library, Albany [cited hereinafter as Warren Papers].
[100] See Chapter i, *supra*, p. 8, ff.

corps would have removed most, if not all the major difficulties and inefficiencies that plagued its operations month after month. Whipple, in condemning the failure that attended Wise's experiments, had declared that American aeronauts were less successful than the French had been at the close of the last century.[101] If judged by the same criteria, the American ranking commanders and administrative officers deserve far greater censure.

In the final analysis, however, judgment should include consideration of army administration as a whole. To single out the balloon corps as the single example of maladministration would be scarcely worthy of creditable reasoning. At no time, and in no branch was the Federal army a paragon of efficient administrative achievement. Considering the problems faced by those in command and in charge of the many departments, the graft and corruption, the attempt to create a great army from raw volunteers, the pernicious appearance of military politicians and political soldiers, and the ingrained red tape of army administration, one must conclude to a certain extent that the maintenance of the balloon corps was not far out of mesh with the general pattern of the Federal army.

[101] Whipple to Bache, August 6, 1861, MS W920, LRTE; see also Chapter iii, *supra*, p. 79.

CHAPTER IX

THE BALLOON SERVICE OF THE ARMY OF THE POTOMAC

Part Three: *Operation*

The tactical mission of the balloon service varied. The primary function was observation: to obtain and transmit to the various division, corps, and army headquarters accurate information of enemy activity. Other duties included continual vigilance against possible surprise, the direction of artillery fire by aerial observation, the making of maps and sketches, and the securing of information relating to distant friendly commands. In fulfilling this mission the aeronauts and the military officers serving with the balloons developed a technique of operation which enabled them to perform dependable service and to meet the varying situations arising as a result of field conditions. Lowe and his assistants applied their civilian experience to the requirements of military service, with necessary modifications in methods as circumstances demanded. Initially ignorant of military affairs, they learned much by experience, and their technique improved and expanded as their field duty continued.

In the field the basic unit of the balloon corps was the observation station. Its location was determined by the commanding general, members of his staff, or by the officer supervising the tactical operations of the corps.[1] When balloons were assigned to several sectors of the line, the corps or division commanders in whose sectors the balloons were used often designated the stations of observation. The station generally consisted of the headquarters of the aeronaut in charge, including his tent and various items of headquarters equipment, furniture, and personal equipage; a small supply depot for the storage of mainte-

---

[1] Lowe to Hooker, November 10, 1861, 3 *O. R.,* III, 266; Colburn to Lowe, November 17, 1861, *ibid.,* p. 267; Lowe to Colburn, December 3, 1861, *ibid.,* p. 268; Colburn to Lowe, April 6, 1862, *ibid.,* p. 274; Sacket to Lowe, May 26, 1862; *ibid.,* p. 278; Humphreys to Lowe, May 26, 1862, *ibid.*

nance equipment and materials; the bivouac area for the ground crew; and the ground station of the balloon and its attending generators. A British observer with the Federal army once described his impression of Lowe's headquarters near the Chickahominy:

I had ridden forward from the main body and joined General Stoneman's command, then occupying for the first time, the west bank of the Chickahominy. I found the balloon snugly ensconced in a hollow, protected from view by a hill in front, from the top of which a convenient position for ascent was gained. The Professor's tent and those of the rest of the balloon corps were scattered around, forming a small distinct encampment.[2]

The degree of improvement of the ground and planning of the station naturally depended on the operations of the army at the moment. When the main forces occupied fixed lines or were comparatively inactive, the balloon camps and stations were improved and laid out with elaboration. During battles and rapidly moving maneuvers, the observation stations were stripped of everything not vitally essential to rapid and frequent ascensions and to the transmission of intelligence. Whenever possible, the ground stations of the balloons were selected to afford cover from artillery fire and concealment from enemy observation. On the Peninsula, positions permitting defilade or concealed by patches of woods were occupied.[3] A ravine in rear of Burnside's headquarters at the Phillips House, later designated as Lowe's Ravine, was selected for one of the balloons at Fredericksburg.[4] Near the Chickahominy, a convenient

[2] Captain Frederick F. E. Beaumont, " On Balloon Reconnaissances as Practised by the American Army," *Papers Connected with the Duties of the Corps of Royal Engineers*, New Series (Woolwich, 1863), XII, 97-98 [cited hereinafter as Beaumont]. See Plates XXXVIII and XXXIX, *infra*.

[3] Letter of " R. S. T.," Philadelphia *Press*, May 6, 1862; New York *Tribune*, June 4, 1862; Philadelphia *Inquirer*, June 25, 1862; Frederick Milnes Edge [Spec. Corres., London *Star*], *Major General McClellan and the Campaign on the Yorktown Peninsula* (London, 1865), p. 75; Captain Eugene A. Nash, *History of the 44th Regiment, New York Volunteer Infantry, in the Civil War, 1861-1865* (Chicago, 1911), p. 68; Captain Frederick C. Floyd, *History of the 40th Regiment New York Volunteers* (Boston, 1909), p. 137.

[4] Lieutenant S. Millett Thompson, *Thirteenth Regiment of New Hampshire Volunteer Infantry in the War of the Rebellion: A Diary Covering Three Years and A Day* (Boston, 1888), p. 89 [diary entry of December 23, 1862].

hollow behind a range of hills was similarly employed.[5] Before Yorktown, the ground station of one of the balloons was described by an officer as being " in a cavern, which seemed to have been prepared by nature for the purpose." [6]

When the envelopes were deflated at the commencement of operations at a station, care was exercised in selecting positions for inflation that were sheltered from the wind. " Be sure you have good protection against the wind, even if you have to build one by planting a circle of trees," Lowe instructed one of his assistants.[7] It was also essential to conduct inflation operations well away from the camp and its bivouac fires, to prevent the hazard of igniting the highly inflammable hydrogen.[8] On one occasion when the army was inactive before the Chancellorsville campaign, a strong enclosure was built around the area where the balloon *Washington* was anchored.[9] The ground on which the balloons were inflated was cleared off, swept, and then spread with large ground cloths, in order to keep the envelope free from dirt or from damage through contact with stones or debris.[10] At Poolesville, the assistant aeronaut Seaver had the ground carefully levelled, and then covered with clean straw.[11]

On calm days the sending of a balloon aloft was a comparatively simple operation, and the preparations required were relatively few. The car, which was generally unhooked from the

[5] Beaumont, p. 96; Anonymous war correspondent, " Three Months with the Balloons," *St. James's Magazine* (London, 1863), p. 102.

[6] Captain Henry N. Blake [11th Mass. Vols.], *Three Years with the Army of the Potomac* (Boston, 1865), p. 55.

[7] Lowe to Starkweather, January 14, 1862, MS, Lowe Papers. Where natural protection was not available, Lowe sometimes had tall saplings cut, preferably evergreens, and planted in trenches around the area occupied for inflation, to serve as a windbreak. See contemporary war correspondent's drawing, Plate XXXVII, *infra*.

[8] Steiner to Lowe, January 11, 1862, MS, Lowe Papers.

[9] Alfred S. Roe, *The 10th Massachusetts Volunteer Infantry, 1861-1865* (Springfield, 1909), p. 174; Correspondence of the Springfield *Republican*, April 17, 1863, clipping in James Allen's Scrapbook; Captain Joseph Keith Newell, *Annals of the 10th Regiment, Massachusetts Volunteer Infantry, in the Rebellion* (Springfield, 1875), p. 194.

[10] Beaumont, p. 96; Seaver to Lowe, December 17, 1861, MS, Lowe Papers; Lowe to Canby, January 22, 1864, MS [number omitted], LRSW; Steiner to Lowe, January 22, 1862, MS, Lowe Papers.

[11] Seaver to Lowe, December 17, 1861, MS, Lowe Papers.

envelope and cordage when the balloon was anchored and inactive, was attached to the concentration ring. The mooring cables were then made fast to the same ring, with pulleys and snatch-blocks attached to trees or other fixed objects. Telescopes, maps, and other equipment for the observers' use were placed in the car, along with the telegraph apparatus when that form of communication was to be employed. Anchoring ballast was removed from the netting and ascension ballast hooked to the car. The aeronaut and those to accompany him took their places in the car, and the command " Stand by your cables," was given to the ground crew. At this order, the crew, divided into several parties, manned the cables and paid them out slowly and evenly, thus allowing the balloon to rise to any desired elevation. When the proper altitude for observation was attained, the aeronaut signalled the crew to make fast their cables. The operations of the crew were supervised by the officer in charge of the detail, who received instructions from the aeronaut during the process of elevating the balloon. After the desired observations had been taken, the crew hauled in the ropes on signal from the aeronaut and lowered the balloon to the ground.[12]

At times when fixed anchoring points were not available, the pulley blocks were dispensed with, and the men in charge of the cables paid out their ropes directly and held the balloon in position by hand.[13] In windy weather the operations were practically the same as those above described, except that the crews in charge of the ropes had to control their lines with great care and skill in order to keep the balloon as steady as possible, and counteract the effect of the wind on the elevated envelope. Effective manipulation of the ropes at all times, and especially during windy periods, was accomplished only through training and practice and attention to the aeronaut's instructions. In

[12] Beaumont, p. 97; " Three Months with the Balloons," p. 102; A War Correspondent, " A Balloon Adventure on the Potomac," *Ballou's Monthly Magazine* (August, 1887), p. 152; Philadelphia *Press,* June 23, 1862; New York *Tribune,* June 24, 1862; Hartford *Daily Courant,* September 2, 1861; New York *Tribune,* August 31, 1861.

[13] Louis Philippe d'Orleans [Comte de Paris], *Histoire de la guerre civile en Amérique* (Paris, 1874), I, 511 [cited hereinafter as Comte de Paris].

heavy winds the balloons could not be sent up, regardless of the skill of the crew and balloonist. At times of severe storm or gale, it became necessary to discharge gas in order to lessen the strain on the cordage and insure the safety of the envelope.[14]

Every effort was made to keep the balloons in readiness for ascension at all times. Lowe once assured the adjutant general that his inflated balloons were kept ready to ascend within five minutes from the time orders were received.[15] That such a standard of proficiency could be maintained at all times is highly improbable. The balloons and generators required constant attention and repair. In very hot weather the valve cement often melted, causing leakage of gas, with the result that the valves often required re-cementing.[16] The generator tanks were also subject to strain from the pressure of the hydrogen, and seams thus started necessitated caulking with white lead.[17] After each generation of gas, the tanks were washed out thoroughly, as a safeguard against undue deterioration.[18] And the envelopes required revarnishing at intervals to insure the retention of gas.[19] A thorough recoating process included turning the envelopes inside out in order that the interior surfaces might be treated with neat's foot oil to preserve flexibility.[20]

Although the construction of mobile generators eliminated to

[14] Hooker to Lowe [Telegram, Budd's Ferry Md., n. d.], MS, Lowe Papers; Lowe to Porter, September 11, 1861, MS, Letters Received and Miscellaneous Papers, Porter's Division [cited hereinafter as LRPD]; Lowe to Brigadier General Gorman, February 21, 1862, MS, Lowe Papers.

[15] Lowe to Williams, March 30, 1863, MS, Lowe Papers; see also 3 *O. R.,* III, 300.

[16] James Allen to Lowe, June 15, 1862, MS, Lowe Papers; James Allen to Lowe, July 8, 1862, MS, *ibid.*

[17] James Allen to Lowe, July 8, 1862, MS, Lowe Papers.

[18] Lowe to Seaver, March 4, 1862, MS, Lowe Papers.

[19] Lowe to Paullin, December 1, 1862, MS, Lowe Papers; see also 3 *O. R.,* III, 268; Lowe to Starkweather January 14, 1862, MS, Lowe Papers; Lowe to Comstock, April 21, 1863, MS, *ibid.*; see also 3 *O. R.,* III, 309; Steiner to Lowe, January 22, 1862, MS, Lowe Papers; Steiner to Lowe, January 25, 1862, MS, *ibid.*; Steiner to Lowe, January 20, 1862, MS, *ibid.*; James Allen to Lowe, July 12, 1862, MS, *ibid.*; James Allen to Lowe, July 25, 1862, MS, *ibid.*; Starkweather to Lowe, January 13, 1862, MS, *ibid.*; New York *Commercial Advertiser,* September 13, 1861.

[20] Lowe to Comstock, April 21, 1863, MS, Lowe Papers; see also 3 *O. R.,* III, 309; Interview [Washington, D. C., April 5, 1939] with Mr. Roy Knabenschue, former balloonist to whom Lowe gave his varnishing formula.

a large extent the necessity of towing the inflated balloons, the aeronauts frequently found it expedient to continue the old method of moving the balloons by hand. Often the balloons were kept inflated at their ground stations and then towed to positions some distance away for ascension, so that artillery fire directed against them might not fall in the vicinity of the balloon camps.[21] The envelopes were also towed to positions permitting more advantageous observation, even though they might be some distance from the original bases of operation. When the army lay opposite Fredericksburg in the spring of 1863, one of the balloons was towed for some thirteen miles, with an engineer officer in the car, taking observations at 1,000-foot elevation throughout the march.[22] Sometime later one of the assistant aeronauts reported to his chief that " Lieutenant Libby took the balloon in tow along the [Rappahannock] River bank." [23] On another occasion James Allen and an engineer officer took observations over a course of two miles from a balloon towed from its position at Banks Ford.[24] And Lowe once stated that he had towed one of the smaller balloons a distance of twenty miles.[25] The absence of any accident during these excursions, such as those experienced by Allen and Wise early in the war, indicates that such methods were practical and that the crews could be proficient in such maneuvers.

The altitudes at which observations were taken varied with the nature of the positions and terrain under inspection, with the distance from observer to the enemy positions, and with the prevailing weather. The maximum elevation recorded for a captive ascension was 5,000 feet.[26] Lower altitudes, including 2,000, 1,500, 1,200, 1,000, 900, 800, and even down to 450

[21] Beaumont, p. 96; "Three Months with the Balloons," p. 102.

[22] James Allen to Lowe, March 23, 1863, MS, Lowe Papers.

[23] Ezra S. Allen to Lowe, April 29, 1863, 1 O. R., XXV ii, 289.

[24] Ezra S. Allen to Brigadier General Gouverneur K. Warren, June 11, 1863, MS, Warren Papers [Letters, Chancellorsville, May 1–June 22, 1863, Vol. I].

[25] Lowe to Steiner, January 14, 1862, MS, Lowe Papers.

[26] Deming to Stanton, April 12, 1862, MS Telegram, Bound Records of the Army of the Potomac, Vol. XXXIII [Misc. Telegrams]; New York Herald, April 13, 1862; New York Tribune, April 14, 1862; Philadelphia Inquirer, April 14, 1862; Philadelphia Press, April 12, 1862; Washington National Republican, April 14, 1862; Boston Transcript, May 22, 1862; Washington Sunday Morning Chronicle, April 13, 1862.

feet all appear in the reports and dispatches.[27] At these varying elevations, the aeronauts were able to observe a wide variety of objects and movements, and reported a highly diverse body of intelligence to the officers with whose commands they operated. At short ranges with favorable terrain features, troops, troop movements, and materiel were constantly observed and identified. Cavalry scouts, videttes, artillery batteries, emplaced guns, and infantry units in varying sizes were easily picked up in the observers' glasses when the enemy was reasonably close.[28] Earthworks and field fortifications were readily detected unless effectively masked. In some instances embrasures and the number of guns they contained were counted and reported.[29]

Encampments of the enemy were often recognized by concentrations of tents, many of which, bleached by sun and rain, showed up distinctly against the terrain.[30] Fair estimates of the size of a force could be made from the visible tentage and the area it covered. In short range observations the tents of small units could be counted with relative accuracy.[31] At night, campfires were similarly observed, large concentrations indi-

[27] Richmond *Enquirer,* April 12, 1862; Steiner to Lowe, February 10, 1862, MS M289, enc. 3, LRAP; James Allen to Lowe, March 22, 1863, MS, Lowe Papers; James Allen to Lowe, March 23, 1863, MS, *ibid.*; James Allen to Lowe, March 26, 1863, MS, *ibid.*; Lowe to Porter, September 20, 1861, MS, *ibid.;* New York *Times,* September 7, 1861; Edge, *McClellan and the Campaign on the Yorktown Peninsula,* journal entry for May 8, 1862, p. 194; Captain C. B. Fairchild, *History of the 27th Regiment, New York Volunteers* (Binghampton, 1888), journal entry for April 13, 1863, p. 155; Lieutenant S. M. Weld to his father, October 16, 1862, in *War Letters and Diary of Stephen Minot Weld, 1861-1865* (Privately printed, New York, 1912), p. 144; also numerous dispatches in 3 *O. R.,* III, 266-316, *passim.*

[28] Dispatches in the Lowe Papers, Heintzelman Papers, McClellan Papers, War Department Archives, and in 3 *O. R.,* III, 266-316, *passim.*

[29] Allen to Taylor, June 14, 1862, MS A85 [Supplementary], LRAP; Lowe to Butterfield, March 13, 1863, MS, Lowe Papers; Porter to Marcy, June 18, 1862, MS, McClellan Papers, Series I, Vol. LXV, MS 13264; Ezra S. Allen to Hooker, May 2, 1863, MS, Lowe Papers; Lowe to Marcy, June 14, 1862, 3 *O. R.,* III, 288; also other dispatches, *ibid.,* pp. 266-319, *passim.*

[30] Lowe to Colburn, December 16, 1861, 3 *O. R.,* III, 269; Lowe to Keyes, May 11, 1862, *ibid.,* 276-277; Lowe to Butterfield, March 13, 1863, *ibid.,* p. 296; Lowe to Sedgwick, April 29, 1863, *ibid.,* p. 310; " J. C." to editors of the Philadelphia *Press,* June 17, 1862, published June 23; Correspondence of the New York *Tribune,* June 24, 1862.

[31] For example, La Mountain's observations of Sewall's Point and its camp; see Chapter iv, *supra,* pp. 93, 101-102.

cating the enemy in force.[32] Scattered lines of small fires, close
to the friendly lines likewise disclosed the position and extent
of pickets and outer patrols.[33]

Smoke was also a valuable item of intelligence reported by
the aeronauts. When hostile camps were too far distant to permit the observation of tentage, smokes from campfires served to
mark the size and position of enemy forces.[34] On calm days the
smokes ascended vertically, and gave an excellent idea of the
positions and extent of the enemy.[35] Density and color also
provided bases of deduction for the observers. An unusually
large, thick cloud discerned in an area where a building had
been previously observed, was taken to be the destruction of a
supply depot. The total disappearance of the building, whether
it was a depot or not, confirmed that it had been burned.[36]
The dense smokes that attended the destruction of the Confederate base of supplies at Manassas when the Southern army
withdrew were reported similarly by the aeronauts, and their
conclusions were later confirmed.[37] Thin, curling wisps of
smoke at intervals along the front of the friendly lines disclosed
the enemy pickets almost as clearly as did the scattered campfires at night.[38] The smoke of battle was also a valuable source
of information. Dense clouds of light blue, mingled with
white, often accompanied by the sound of artillery and massed
musketry clearly heard from the aerial stations, indicated action

[32] Lowe to Captain Moses [AAG, Heintzelman's Division], March 5, 1862,
MS, Heintzelman Papers; Lowe to Butterfield, April 30, 1863, MS, Lowe Papers;
also numerous dispatches in War Department Archives.

[33] Lowe to Humphreys, June 7, 1862, MS, Lowe Papers.

[34] Lowe to Sedgwick, April 29, 1863, MS, Lowe Papers; Lowe to Butterfield,
February 7, 1863, MS L29 [Supplementary], LRAP; Candler to Sharp, February 27, 1863, MS [number omitted], Miscellaneous Papers, Army of the Potomac; Alexander to Barnard, May 30, 1862, 1 O.R., LI, i, 644; Steiner to
Lowe, February 10, 1862, MS M289, enc. 3, LRAP; also numerous dispatches
in Lowe Papers, McClellan Papers, War Department Archives, and in 3 O.R.,
III, 266-316 passim.

[35] Lowe to Marcy, May 6, 1862, MS, Lowe Papers.

[36] James Allen to Lowe, June 2, 1862, MS, Lowe Papers.

[37] Colonel Hiram G. Berry to Moses, March 12, 1862, MS, Heintzelman
Papers.

[38] Lowe to Porter, September 24, 1861, 3 O.R., III, 263; Lowe to Colburn,
November 21, 1861, ibid., p. 266.

joined at distant parts of the line.[39] Reports of such phenomena might be highly valuable to an army commander by acquainting him with events at distant sectors. In this way Sedgwick, in command of the extreme Union left at Falmouth, was kept informed of the action of Hooker's main force at Chancellorsville, over twelve miles away.[40]

Just as varying types of smoke indicated the presence of enemy camps, patrols, and pickets, clouds of dust observed on the terrain also disclosed the movement of troops when the distance was too great to permit direct observation of their columns.[41] In dry weather moving bodies of cavalry and infantry on the characteristic dirt roads of Virginia produced thick dust clouds visible for miles. Swifter moving cavalry could be distinguished by the rapid progress of the dust along a roadway; the clouds stirred by marching infantry moved at a noticeably slower rate.[42]

Wagon trains and their movements were noted frequently by the aerial observers.[43] Like the visible tentage their canvas covers, whitened by exposure to weather, afforded an unmistakable basis for identification.[44] Their movements in large numbers might be taken as indicative of many different tactical maneuvers, according to their number, the direction in which they moved, and other intelligence concerning the area involved.

[39] Lowe to Porter, June 1, 1862, 3 *O. R.,* III, 283; Lowe to Humphreys, June 27, 1862, *ibid.,* p. 290; Lowe to Marcy, June 30, 1862, MS, Lowe Papers.

[40] Lowe to Sedgwick [11 a. m.], May 1, 1863, 3 *O. R.,* III, 313; Lowe to Sedgwick [2:15 p. m.], May 1, 1863, *ibid.*; Lowe to Sedgwick [3:45 p. m], May 1, 1863, *ibid.*; Lowe to Butterfield [7:45 a. m.], May 2, 1863, *ibid.,* p. 314; Lowe to Sedgwick and Butterfield [7:15 a. m.], May 3, 1863, *ibid.,* p. 315.

[41] Porter to Marcy, October 1, 1862, MS P36 [Supplementary], LRAP; Porter to Lowe, September 25, 1861, MS, Lowe Papers; 3 *O. R.,* III, 263; New York *Times,* June 6, 1863; Washington *Sunday Morning Chronicle,* September 28, 1862; Philadelphia *Press,* September 29, 1862.

[42] Lowe to Porter, September 18, 1862, MS, Lowe Papers.

[43] Lowe to McClellan, May 31, 1862, 3 *O. R.,* III, 280; Lowe to Humphreys [11 a. m.], June 1, 1862, *ibid.,* p. 282; Marcy to Lowe, June 1, 1862, *ibid.*; Report of James Allen [8:20 a. m.], June 1, 1862, *ibid.,* p. 283; Lowe to Sedgwick [1:30 p. m.], April 29, 1863, *ibid.,* p. 310; Lowe to Sedgwick [10 a. m.] May 1, 1863, *ibid.,* p. 313; also numerous dispatches in the Lowe Papers, McClellan Papers, and the War Department Archives.

[44] Lowe to Marcy [4:30 p. m.], May 31, 1862, 3 *O. R.,* III, 280; also dispatches in the Lowe Papers.

When railway lines lay within range of observation, the movement of trains and locomotives became another important source of information furnished by the aeronauts.[45] During the Peninsular campaign the arrival and departure of trains to and from points within the Confederate lines, and also rail movements in and about Richmond were observed and reported.[46]

In addition to the direct discovery of enemy activity, the aerial reconnaissances performed the mission of " passive " observation. Enemy positions and encampments were continually watched and studied. When no changes were observed in their size and extent, it became apparent that no large scale movements were in progress.[47] For this reason roads were constantly watched and the reported absence of movement on them, even though night marches could not be detected, gave the Union commanders a greater measure of security than they would have had otherwise. On the other hand, the disappearance of a large camp previously observed meant that the troops occupying it had shifted position, thus warning the commanders to be on the alert.[48] When the army occupied positions along rivers, the aeronauts were bidden to focus their attention on all visible fords and roads leading to the crossings. The orders issued to the balloonists and the aerial reports submitted immediately before Hooker's flanking movement at Chancellorsville serve as good examples of this form of passive observation.[49]

When positions were occupied near navigable waters, the movement of enemy vessels also came under the eye of the aerial observer. In a message to General Wool, announcing that he

[45] Philadelphia Press, May 23, 1862; Lowe to Humphreys, May 26, 1862, MS, Lowe Papers.

[46] Washington Star, May 22, 1862; Washington National Republican, May 23, 1862; New York Times, May 23, 1862; New York Herald, May 23, 1862; Boston Transcript, May 22, 1862.

[47] Captain William L. Candler [ADC] to Colonel Sharpe (containing extract of Comstock's report of balloon observations), February 27, 1863, MS [number omitted], Miscellaneous Papers, Army of the Potomac; Philadelphia Press, October 10, 1862; Steiner to Lowe, February 10, 1862; MS M289, enc. 3, LRAP; Lieutenant G. C. Kaulback to Lowe, April 18, 1863, MS, Lowe Papers.

[48] Ezra S. Allen to Comstock, April 9, 1863, MS, Lowe Papers.

[49] Williams to Lowe, March 12, 1863, MS, Lowe Papers; Butterfield to Lowe, April 29, 1863, 3 O. R., III, 310; Butterfield to Hooker, April 28, 1863, MS LBAP, III, 269, letter 453.

was sending a balloon to Fortress Monroe, Colonel Macomb expressed the hope that the aerostat would be valuable in watching the movements of the much dreaded *Virginia*.[50] Hostilities adjacent to navigable waters also permitted the taking of observations from vessels which could sail to advantageous points for reconnaissance. Lowe's operations on the Potomac off Mattawoman Creek to observe the strength and extent of Confederate fortifications on the Virginia shore,[51] La Mountain's operations in Hampton Roads,[52] and the excursion made by James Allen on a vessel off Fort Powhatan illustrate this type of operation.[53] The use of the balloon-boat *G. W. Parke Custis* is another instance.[54]

The methods used for estimating the size of an enemy force are interesting. When columns were close enough for direct observation, the time required for them to pass a fixed point was taken.[55] This information, used in conjunction with average rates of march, permitted the reasonable calculation of the number of troops observed. The force contained in an encampment could be calculated by comparing the size and general extent of the tentage with those of the visible friendly encampment.[56] Since the observer knew the strength of the force to which he was attached, comparison with its extent and size as seen from the air, with modifications for the distance to the enemy and allowance for defilade, permitted fair estimates of the enemy's numbers. Lines of battle were similarly evaluated. " Their line

[50] Macomb to Wool, March 13, 1862, MS, Lowe Papers.

[51] Lowe to Colburn November 12, 1861, 3 *O. R.,* III, 265-266.

[52] See Chapter iv, *supra,* pp. 96-97, 103.

[53] Commodore Charles Wilkes to Commodore T. A. Jenkins, July 29, 1862, *Official Records of the Union and Confederate Navies* (Washington, 1894, *et seq.*), Series I, Vol. VII, 603 [cited hereinafter as *O. R. N.*].

[54] Washington *National Republican,* November 8, 1861; Lowe to Captain J. A. Dahlgren, USN [Comdt. WNY], November 16, 1861, MS, Dahlgren Papers; also MS [copy], Lowe Papers; Dahlgren to Secretary Gideon Welles, December 10, 1861, MS 195, Commandant's Letters, Washington Navy Yard, Vol. III, Naval Records and Library, Navy Department, Washington; Philadelphia *Inquirer,* November 25, 1861.

[55] Lowe to Sedgwick [10 a. m.], May 1, 1863, 1 *O. R.,* XXV, ii, 337; Lowe to Comstock, April 22, 1863, MS, Lowe Papers.

[56] Lowe to Marcy, March 6, 1862, MS, Lowe Papers; Lowe to Sedgwick, April 29, 1862, MS, *ibid.*

appears quite thin as compared with our force," Lowe reported after studying the Confederate position on the heights of Fredericksburg during the preliminaries to Chancellorsville.[57] " The force on the bank of the river is not very heavy as compared to ours," he again telegraphed General Sedgwick the same day.[58]

The range of observation depended on the altitude attained, the power of the glasses used, the state of atmosphere, and the nature of the objects observed. The longest range of observation reported was 30 miles, attention on this occasion being directed to vessels on the York River.[59] Heavy camp smokes were reported at distances calculated to be 25 miles,[60] and similar observations at ranges from 10 to 20 miles were not uncommon.[61] Camps with tentage visible were detected at varying ranges up to 5 and 6 miles from the observer.[62]

Ascensions were made at all hours, whenever the weather permitted. Routine observations were generally taken every morning just before sunrise to locate the positions of camp and cooking fires, as well as to discover immediately at daylight any changes in hostile dispositions that might have taken place during the night under cover of darkness.[63] Night ascensions were frequently made to observe and watch for changes in the position of bivouac and picket fires.[64] During Comstock's adminis-

[57] Lowe to Sedgwick [10 a. m.], April 29, 1863, MS, Lowe Papers; also printed in 3 O. R., III, 310.

[58] Lowe to Sedgwick, April 29, 1863, MS, Lowe Papers.

[59] Lowe's Report, 3 O. R., III, 276.

[60] Lowe to Butterfield, February 7, 1863, MS L29 [Supplementary], LRAP.

[61] James Allen to Lowe, June 3, 1862, MS, Lowe Papers; James Allen to Lowe, March 23, 1863, MS, ibid.; James Allen to Humphreys, July 16, 1862, MS, A25, LRAP; James Allen to Humphreys, July 16, 1862, MS H32 [Supplementary], LRAP.

[62] Ezra S. Allen to Lowe, March 26, 1863, MS, Lowe Papers; James Allen to Lowe, June 2, 1862, MS, ibid.; Ezra S. Allen to Hooker, May 1, 1863, MS, ibid.; also numerous dispatches in 3 O. R., III, 266-316, passim.

[63] Dispatches and orders in the Lowe Papers, McClellan Papers, War Department Archives, and in 3 O. R., III, 271-315, passim.

[64] Whipple to Lowe, August 21, 1861, 3 O. R., III, 260; Lowe to Colburn, November 12, 1861, ibid., p. 266; Oliver to Lowe, April 28, 1863, ibid., p. 310; Lowe to Moses, March 5, 1862, MS, Heintzelman Papers; also numerous other dispatches in the Lowe Papers, War Department Archives, and 3 O. R., III, 260-315, passim.

tration, ascensions were ordered to be made daily at dawn, at noon, and at nightfall, whenever the army was not moving or engaged in action.[65] Normally, throughout the day, the aeronauts went aloft whenever the commanding officers desired, and also at intervals without order, whenever the weather permitted.[66]

The duration of ascensions varied considerably depending upon enemy activity, the tactical operations in progress, and on the specific purpose of the observer. When General Barnard, McClellan's Chief Engineer, went aloft to study the Confederate fortifications at Yorktown, he remained up for nearly four hours.[67] Routine observations of the lines and encampments varied from a few minutes to two or three hours.[68] Comstock, when engaged in engineer surveys near Falmouth, was often reported to have remained aloft for an hour or an hour and a half.[69] When draftsmen went up to make maps and topographical surveys, the ascensions were generally of several hours' duration. During battle, the balloonists were ordered to remain aloft all day.[70]

Ascension for the purpose of making maps of the surrounding terrain and enemy positions became an important function of the balloon service. Engineer officers and civilian draftsmen often accompanied the aeronauts aloft, prepared maps and panoramic sketches, and made corrections on maps issued by the Engineer headquarters.[71] Occasionally the aeronauts prepared

[65] Comstock to Lowe, April 12, 1863, 3 *O. R.*, III, 303.

[66] Dispatches and orders in the source collections above cited.

[67] Cincinnati *Commercial,* May 5, 1862; St. Paul *Pioneer and Democrat,* May 4, 1862; Heintzelman's Diary, entry of April 16, 1862, MS, Heintzelman Papers.

[68] James Conner to his mother, May 13, 1862, *Letters of General James Conner* (Privately printed, Columbia, S. C., 1933), p. 102; New York *Tribune,* May 3, 1862; Ezra S. Allen to Lowe, April 18, 1863, MS, Lowe Papers; also numerous dispatches in the Lowe Papers, McClellan Papers, War Department Archives, and in 3 *O. R.*, III, 266-316, *passim.*

[69] James Allen to Lowe, March 23, 1863, MS, Lowe Papers; James Allen to Lowe, March 26, 1863, MS, *ibid.*

[70] Humphreys to Lowe, June 1, 1862, MS, Lowe Papers; also printed in 3 *O. R.*, III, 281.

[71] *Lowe's Report,* 3 *O. R.*, III, 255; Tyler to McDowell, June 24, 1861, MS 132, LRNV; Lowe to Hooker, November 30, 1861, 3 *O. R.*, III, 267-268;

rough explanatory sketches which they submitted with their observation reports.[72] Another important practice was the marking of enemy positions and the location of enemy activity on tactical maps carried aloft. Information thus graphically recorded was submitted to the headquarters of the army in the field.[73]

The accurate identification of objects observed and the correct interpretation of visible activity were of vital importance. Lowe and his civilian assistants had no prior experience or knowledge of military matters. Consequently they had to develop an ability to identify objects and movements on the terrain and to interpret them in terms of military intelligence. They were fully aware of their inexperience, and frequently requested that trained officers ascend with them for observations.[74] At times officers went aloft after the aeronauts had submitted reports, in order to confirm the information reported.[75]

For the first eighteen months the method of reporting observations consisted simply of a description of the object or movement, and its location with reference to some known or readily identified point, road, or map location.[76] To facilitate references, Lowe was furnished on one occasion with a descriptive

Lowe to Colburn, December 10, 1861, *ibid.*, p. 268; *Lowe's Report, ibid.*, p. 273; Marcy to Lowe, June 7, 1862, MS, Lowe Papers; James Allen to Lowe, July 12, 1862, MS, *ibid.*; Lowe to Colonel Eustis, April 22, 1863, MS, *ibid.*; Morrell to Lowe, June 19, 1862, MS, *ibid.*; New York *Illustrated News,* May 17, 1862; Philadelphia *Press,* May 30, 1862; Lieutenant Daniel G. Macnamara, *History of the Ninth Regiment, Massachusetts Volunteer Infantry* (Boston, 1899), p. 291.

[72] Ezra S. Allen to Hunt, May 3, 1863, MS, Lowe Papers.

[73] Lowe to Marcy [9:30 a. m.], June 14, 1862, 3 *O. R.,* III, 288; Lowe to Marcy, June 14, 1862, *ibid.,* p. 289; Lowe to Marcy, June 16, 1862, *ibid.*; Lowe to Marcy, May 6, 1862, MS, Lowe Papers; Butterfield to Lowe, February 7, 1863, 3 *O. R.,* III, 295.

[74] Lowe to Moses, March 5, 1862, MS, Heintzelman Papers; Lowe to Comstock, April 22, 1863, MS, Lowe Papers; Lowe to Sedgwick, April 29, 1863, MS, *ibid.*; Humphreys to Marcy, June 1, 1862, MS H453, LRAP; Lowe to Marcy, June 1, 1862, MS LBAP, Vol. XXXVI; Lowe to Butterfield, February 24, 1863, MS L56, LRAP; Lowe to Porter, September 20, 1861, 3 *O. R.,* III, 262; Lowe to Sedgwick, May 1, 1863, 1 *O. R.,* XXV, ii, 340.

[75] James Allen to Lowe, June 2, 1862, MS, Lowe Papers.

[76] Dispatches in the Lowe Papers, McClellan Papers, Heintzelman Papers, War Department Archives, and in 3 *O. R.,* III, 261-316, *passim.*

list of prominently visible houses in the area around Fredericksburg, which gave the name, general location, appearance, or color, and general shape of each house.[77] These buildings were evidently plotted on the headquarters maps, and from the list Lowe could select any convenient reference point for use in his reports. When Comstock took command of the corps, he inaugurated a new system of reporting enemy positions. Visible camps were numbered from right to left, and any new camp or observed activity was reported by compass bearing with reference to one of the numbered positions.[78] Thereafter, until the battle of Chancellorsville rendered this method ineffective, reports from the balloons followed the general form: " No. 2, west by south, three miles: wagon train, not in motion." Or, " No. 4, southwest by west, two miles: large camp." [79] Some of the reports in this form were also accompanied by exact compass bearings in degrees.[80]

Another important function of the balloon service was the direction of artillery fire by aerial observation. This " modern " method of artillery practice had been tried initially during Lowe's operations with the first balloon near Fort Corcoran.[81] The success of the first experiments led to its continued practice after the corps was created. At Island No. 10 the few ascensions made by Steiner were for the purpose of directing the fire from Commodore Foote's mortar batteries.[82] Similar technique was applied to fire with the field batteries in the Potomac lines in the fall of 1861.[83] And during the Peninsular campaign fire

[77] Captain W. H. Paine [ADC] to Lowe, with " Description and Location of Certain Buildings on the South Side of the Rappahannock," MS [n. d.], Lowe Papers.

[78] Comstock to Lowe, April 12, 1863, 3 O. R., III, 303.

[79] Lowe to Comstock, April 16, 1863, MS, Lowe Papers; Lowe to Comstock, April 14, 1863, 3 O. R., III, 306; Lowe to Comstock, April 17, 1863, ibid., p. 307.

[80] Kaulback to Lowe, April 21, 1863, MS, Lowe Papers.

[81] See Chapter vi, supra, pp. 212-213.

[82] Steiner to Lowe, March 27, 1862, MS, Lowe Papers; Commander A. M. Pennock [Fleet Captain, Western Flotilla] to Henry A. Wise [AIO, USN], March 27, 1862, 1 O. R. N., XXII, 702; St. Louis Daily Missouri Democrat, March 28, 1862; St. Paul Pioneer and Democrat, March 29, 1862; Cincinnati Commercial, March 29, 1862.

[83] Philadelphia Inquirer, October 9, 1861; Philadelphia Daily News, November 19, 1861; Comte de Paris, I, 511.

control by balloon observation was frequently conducted.[84] The system employed was in principle not unlike the procedure followed in similar practice by modern field artillery. From the aerial station shots that could not be observed by artillery officers were sensed for range and direction. Corrections were sent by telegraph or flag signal to the battery firing, which shortened or increased ranges or shifted its deflection according to the correction called for by the observers.[85] The gunners were thus able to register on targets hidden from their view, and at the same time they were permitted to occupy positions sheltered from enemy observation and fire.

Communication from the balloons was effected by three general methods, telegraph, dropped messages, and by visual signals. For general transmission of telegraphic messages, the usual process of spelling out words was followed, the same method used by commercial telegraphers. A code was developed for the transmission of standard information, which cut sending and receiving time to a minimum. This code consisted of a series of numbers, and some letters, each representing standard words, phrases, and sentences that were common to many of the messages sent. By proper combination of these numbers an observation could be reported in a minimum of signals.

For example, the number 51 designated the right of a friendly line or position; 8 designated the left; and 35 the center. The cardinal points of the compass were represented by the letters N, S, E, and W, with the usual combinations of these letters. The numeral 236 designated the sentence, " I can observe cavalry in the direction of." The signal for " advancing," " marching towards," or " moving," was the number 24.

---

[84] Beaumont, p. 99; New York *Herald*, May 25, 1862; New York *Times*, May 28 and June 24, 1862; Boston *Transcript*, May 25 and June 4, 1862; Cincinnati *Commercial*, June 6, 1862; Letter of Captain Charles A. Phillips, April 12, 1862, in *History of the 5th Massachusetts Battery* [Compiled by a Committee of Officers and Men (Boston, 1902)], pp. 212-213; second letter [n. d.], *ibid.*, pp. 303-304; Joel Cook, *The Siege of Richmond* (Philadelphia, 1862), pp. 118-119; Report of Captain John C. Tidball [2d US Artillery] to Lieutenant Colonel William Hays, May 25, 1862, 1 *O. R.*, XI, i, 656.

[85] Smith to Porter, September 23, 1861, 3 *O. R.*, III, 262; Smith to Porter, September 24, 1861, MS, LRPD; Porter to Lowe, September 24, 1861, MS, Lowe Papers; Smith to Porter, September 24, 1861, 3 *O. R.*, III, 263.

From these examples, a message given as " 236—Taylor's—24
—NW—8," would be read by the receiving operator as " I can
observe cavalry in the direction of Taylor's House advancing
northwest on our left." Numbers representing infantry, pickets,
artillery, wagon trains, smoke, and other similar intelligence
were adopted in the code, as well as procedure signals for the
convenience of the operators. A number of qualifying and
descriptive words and phrases were also reduced to signal numerals, including those which indicated the relative size of a force,
the speed at which it moved, and other tactical information.[86]

[86] Code Sheet, Telegraph Signals, MS, Lowe Papers. This document is in
poor condition, with a number of the code listings missing or obliterated. The
listings still readable are as follows:

1	" Use Dial "
2	" Nothing "
3	" Repeat "
4	" Yes "
5	" I don't know "
6	" Are You Ready? "
7	" No "
8	" Error—Not Proper Answer to my Signal "
23	" I understand "
34	" I am going to use signals "
52	" I am going to use alphabet "
86	" Revolve—I wish to regulate instrument "
25	" Wait "
26	" Make longer pause between signals "
31	" Higher "
35	" Lower "
32	" Quick " [" Fast "]
24	" Moving " [" Advancing "] [" Marching Towards "]
27	" Light"[" Small "]
29	" Heavy " [" Large "]
232	" I can observe EARTHWORKS in direction of "
235	" I can observe PICKETS in direction of "
236	" I can observe INFANTRY in direction of "
237	" I can observe CAVALRY in direction of "
238	" I can observe WAGON TRAINS in direction of "
239	" I can observe ENCAMPMENTS in direction of "
231	" I can observe FORTIFICATIONS in direction of "
241	" I can observe SMOKE RISING in direction of "
243	[blank in the MS]
244	[Blank in the MS]
421	" The enemy are advancing on our "
422	" Reinforcements are advancing on our "
423	" The enemy are retreating on our "

The code thus adopted naturally did not apply to all messages sent from the cars. Long, detailed reports of observations could not be reduced to such simple terms, but on many occasions, when information could be so compressed and speed was essential, the code system could be a valuable aid.

When there was no necessity for highly rapid transmission of information, or when instruments or operators were not available, communications from the air were effected by dropping messages. This procedure, sometimes referred to as the "paper express," consisted of weighting a written dispatch with a bullet, and fastening it to a ring atached to a mooring cable, which guided the message to the ground station below.[87] When time was not essential in the delivery of reports, the observers sometimes took notes during ascensions, and wrote detailed reports after descending, which they delivered by messenger or telegraph.[88]

During a major action, the telegraph lines from the balloons were often connected with lines running to general headquarters, in order that observations might be reported to the army commander as well as to the officer commanding the area where the balloon was operating.[89] At Seven Pines this procedure was carried to the extreme of connecting the balloon wires through to the War Department office in Washington, via Fortress Monroe.[90] Just prior to the battle of Chancellorsville, Hooker

427 " The enemy are in full retreat "
 51 " Right " [Friendly lines]
  8 " Left " [Friendly lines]
 25 " Center " [Friendly lines]
  N " North "
  S " South "
  E " East "
  W " West "

[87] Washington *Star*, August 30, 1861; Washington *National Republican*, August 30, 1861; New York *World*, August 29, 1861; New York *Commercial Advertiser*, August 28, 1861; Philadelphia *Inquirer*, August 28, 1861.

[88] The Lowe Papers contain several rough pencilled notes of observations which are incorporated into more detailed dispatches also found in this collection. A Brady photograph of Lowe writing a dispatch after descending appears in Miller's *Photographic History of the Civil War*, VIII, 380.

[89] Dispatches and orders in the Lowe Papers and War Department Archives.

[90] Philadelphia *Inquirer*, May 27, 1862; Parke Spring to Lowe, May 25, 1862, MS, Lowe Papers; Beaumont, p. 100; New York *Times*, June 5, 1862.

ordered communication established between both balloons and his headquarters, and during the action when the main lines got out of order, aerial dispatches were telegraphed to the adjacent ground station and were then relayed by visual flag signals to general headquarters.[91]

Visual signalling from the balloons was not practiced to any great extent. Some flag communication between Lowe's balloon at Pohick Church and another manned by one of his assistants at Budd's Ferry was established in March, 1862,[92] with the assistance of signal officers who took charge of the experiment.[93] Sometime later Colonel Myer, McClellan's signal officer, made recommendations and suggestions for visual communication during operations on the Peninsula.[94] He also suggested that similar communications might be opened from one of the balloons to the gunboats on James River by means of rockets fired from the car.[95] On several occasions balloons were sent up to observe rocket signals fired according to pre-arranged codes from distant parts of the lines.[96] From the air these pyrotechnic signals could be seen more readily and at greater distance than from ground signal stations. During the battle of Seven Pines visual signals supplemented the telegraph, and Myer instructed his signal officer at Heintzelman's headquarters to " look for signals from balloon nearest you and report to General McClellan, who is in front near Sumner. Balloon will reply to signal ' AF ' which [you will] make with 6-foot flag and 16-foot pole." [97]

The possibilities of long-distance signalling by balloon led Lowe to experiment with small " caloric " balloons for this pur-

[91] Butterfield to Hooker, May 1, 1863, MS LBAP, III, 297-299, letter 495; Lieutenant F. Wilson to Lowe, May 3, 1863, MS, Lowe Papers; Butterfield to Lowe, May 3, 1865, MS, *ibid.*

[92] Lowe to Marcy, March 6, 1862, MS, Lowe Papers; also printed in 3 O. R., III, 270.

[93] Lowe to Seaver, March 6, 1862, MS, Lowe Papers.

[94] Myer to Marcy, June 14, 1862, 1 O. R., XI, iii, 228.

[95] Reports of Colonel Myer, October 21 and 22, 1862, 1 O. R., XI, i, 244, 245.

[96] Oliver to Lowe, March 17, 1863, MS, Lowe Papers; printed in 3 O. R., III, 297.

[97] Myer to Lieutenant Johnson, June 1, 1862, MS LBAP, Vol. XXXVI.

pose in the fall of 1862. In November of that year he reported to General Burnside's chief of staff that his train was equipped with " small signal balloons that can be used day or night." Pyrotechnic flares used with these globes at night, he added, would be more effective than rockets.[98] Several weeks later he designed a complete signalling system based on the use of the hot-air balloons which he submitted to Burnside's headquarters.

The envelopes were constructed of strong linen paper, light and compact when folded, and required little space when carried in the field. Ascensive power was provided by a " fire-proof inflating apparatus," which furnished heated air to raise the envelopes. In size the signal balloons varied from 6 to 20 feet in diameter, depending on the distance over which information was to be conveyed. For day operations, white envelopes with distinctive stripes and markings of black were recommended. Elevated to a high altitude, these balloons and contrasting symbols could be observed with the aid of powerful glasses as far distant as thirty miles, Lowe asserted. In addition to the distinguishing marks on the envelopes, rigid panels or " flags " were to be suspended beneath the balloons, each bearing distinct and conspicuous devices and symbols. Combinations of these symbols and markings could be multiplied indefinitely with a small number of balloons, and thus any number of different signals could be sent. At night the same envelopes were to be equipped with powerful calcium flares of different colors and intense brilliance. Combinations of colors, arrangement, and of the number of flares sent up, like the symbols and panels for day use, also made possible an indefinite number of signals. The flares were designed to burn from five to thirty minutes, a feature that rendered them more useful than ordinary rockets. Such pyrotechnic signals, when sent up to a high elevation, Lowe declared, could be visible at ranges as great as forty miles.[99]

[98] Lowe to Parke, November 20, 1862, 3 *O. R.,* III, 293.

[99] Lowe to Burnside, December 22, 1862, Lowe Papers; Plans for long distance signalling system, with scale drawings and explanatory text [lithographed by P. S. Duval and Son, Philadelphia], File, " Balloon Corps," War Department Division, National Archives; also copy in the Lowe Collection, Aircraft Building, United States National Museum; Lowe to Parke, January 13, 1863, MS L12, LRAP; Miscellaneous MS Drawings, Lowe Papers. See Plate XLI, *infra.*

Burnside ordered the construction of several of these signal balloons for experimental use, but according to Lowe, they were prepared in a hurry from poor materials and were not capable of demonstrating the full merits of the system.[100] After Burnside had been relieved the new chief of staff, Major General Daniel Butterfield, conducted additional experiments with similar materiel.[101] The results of these tests are not disclosed in the records, but later developments indicate that the plan was not fully tried. In March, 1863, the proposed system came to the attention of Major General George Stoneman, commanding Hooker's cavalry corps, who was shortly to make his extensive cavalry raid in an effort to cut Lee's communications with Richmond. Stoneman, possibly anticipating the advantage of such a system of long-distance communication with Hooker during this excursion, notified Assistant Adjutant General Williams that

Professor Lowe has an arrangement for transmitting information from distant points by signal balloons, which I think might be made available and valuable with cavalry operating in the field. I have thought the subject over a good deal, and if the professor can get authority to secure the necessary apparatus, I will take measures to test, and if possible, put his plan into practice.[102]

Stoneman's suggestion was referred to Lowe for an estimate of cost and time necessary to provide the test materials.[103] Lowe replied that equipment for trying out the system from every angle and in all possible combinations, would require an expenditure of not more than $300, and the individual balloons complete would not exceed $6 each if ordered in quantity. A week would suffice for making all arrangements.[104] General Williams then inquired why the signal balloons used in Burnside's and Butterfield's experiments could not be used, and sug-

---

[100] Lowe to Williams, March 21, 1863, 3 *O. R.,* III, 298.

[101] Butterfield to Lowe, February 16, 1863, MS, Lowe Papers; Butterfield to Lowe, February 16, 1863 [second telegram], MS, *ibid.*; Butterfield to Lowe [n. d., pencil note on scrap], MS, *ibid.*

[102] Stoneman to Williams, March 19, 1863, 3 *O. R.,* III, 297.

[103] Williams' indorsement, March 19, 1863, *ibid.*

[104] Lowe to Williams, March 20, 1863, 3 *O. R.,* III, 297.

gested that if Lowe had any of these in his train a board would be convened to witness tests with them.[105] Lowe then explained that he had left only a few of the envelopes ordered by Burnside, and that because of their inferior quality they would not give a satisfactory demonstration of what could be accomplished. The seemingly high figure of cost quoted, he added, covered the cost of every variety of marking, panel, flare, and symbol, in order that the most effective might be selected for use.[106] The explanations did not satisfy Williams. The communication was returned. "Under the circumstances not favorably considered. Professor Lowe will notify General Stoneman," was the final indorsement from Hooker's headquarters.[107]

Despite this rejection, Lowe still hoped to secure the adoption of his plan, even after he left the army. In July, 1863, he offered the system to Colonel Myer, chief of the newly organized Signal Corps.[108] Although Myer was at first disposed to give the plan favorable consideration,[109] its adoption into the signal service was never effected.

Aerial photography, another possible and valuable function of the balloon service, was apparently not practiced by Lowe's corps, notwithstanding statements to the contrary that have appeared in a number of secondary works. Dollfus and Bouché, in their reputable survey of aeronautics have asserted that

in March 1862, during the War of Secession, the aeronaut Lowe took photographs from a captive balloon, before Richmond, and the prints, divided into squares, were used for strategical purposes.[110]

No source is given for this statement, and its accuracy is highly suspicious in view of the fact that Lowe was nowhere near Richmond at the time. He was still in Washington on March

---

[105] Williams' indorsement, March 20, 1863, *ibid.*

[106] Lowe to Williams, March 21, 1863, 3 *O. R.,* III, 298.

[107] Williams' indorsement, March 21, 1863, *ibid.*

[108] Lowe to Myer, July 13, 1863, MS, Lowe Papers.

[109] Myer to Lowe, July 21, 1863, MS Letterbooks, Signal Corps, USA, War Department Division, National Archives [cited hereinafter as LBSC]; Lowe to Myer, July 24, 1863, MS L25, Letters Received, Signal Corps, USA, War Department Division, National Archives [cited hereinafter as LRSC].

[110] Charles Dollfus and Henri Bouché, *Histoire de l'aérostatique* (Paris, 1932), p. 103.

23, and did not arrive before Yorktown until April 5.[111] Several other works on photography and aeronautics have also stated that aerial photographs were made from Lowe's balloons during the war. These assertions are also undocumented, and are no more convincing than the statement published in Dollfus.[112] In a recent article appearing in one of the most reputable historical journals the author declared that Lowe devised " A scheme for taking photographs from an elevation of thousands of feet for topographical study." [113] The source cited, however, does not support the statement specifically, and is so vague in its language as to preclude anything more than speculation on the subject.[114]

The value and possibilities of aerial photography, nevertheless, were brought to the attention of the military authorities on a number of occasions during the war. Less than a year before the opening of hostilities, William Black took the first successful aerial photographs made in this country from Samuel A. King's balloon over Boston on October 13, 1860.[115] Early in

---

[111] Williams to Lowe, March 23, 1863, 3 O. R., III, 273; Lowe's Report, ibid.

[112] Hippolyte Meyer Heine, La photographie en ballon (Paris, 1912), p. 2; Bernard E. Jones (ed.), Cassell's Cyclopaedia of Photography (London, 1911), p. 11; Martin Kiesling, Die Anwendung der Photographie zu militärischen Zwecken (Halle, a. s., 1896), p. 38; Hermann W. L. Moedebeck, Taschenbuch für Flugtechniker und Luftschiffer (Berlin, 1923), p. 224.

[113] J. Duane Squires, " Aeronautics in the Civil War," American Historical Review, XLII (July, 1937), 661.

[114] Lowe to Major General John G. Parke, Chief of Staff [not to Burnside, as the reference states], November 20, 1862, 3 O. R., III, 293.

[115] Boston Journal, October 15 and 16, 1860; Boston Herald, October 15, 1860; Boston Transcript, October 13 and 15, 1860. A copy of one of these photographs was recently published in Life, June 5, 1939. As early as the spring of 1859 the New York photographer W. J. Kuhns had arranged with John Wise to experiment with aerial photography from Wise's balloon Atlantic, but these trials were rendered abortive by inclement weather and other interruptions. American Journal of Photography and the Allied Arts and Sciences, New Series, III (New York, February, 1861), 265-266. Black had also made experiments from William H. Helme's balloon over Providence several months before his work with King in Boston. On this occasion he actually took photographs from the air, but these pictures can hardly be called successful, since lack of sunlight prevented good results, and the cracking of the collodion during the finishing process spoiled the plates. American Journal of Photography, III (New York, September, 1860), 105-106; Photographic News, IV (London, November, 1860), 347. For details of the first successful aerial photography by Black and King in October, 1860, see also King to editors of

June, 1861, an open letter to the editor of the New York *Tribune* appeared in the columns of that paper suggesting the advantage of photographs taken from the air. The writer had interviewed a number of prominent professional photographers, who agreed that the suggestion was entirely practical.[116] About the same time Black sent several of his prints to the editors of the same paper, who then published an article suggesting the value of such technique in the proposed balloon operations of the army.[117] Black had also sent prints to Professor Alexander D. Bache, of the Coast Survey, who had shown them to his uncle, Major Hartman Bache of the Topographical Engineers, under whose direction Lowe's balloon experiments were being conducted. Major Bache was evidently impressed with the possibilities of these pictures, and wrote to Black asking for details concerning the apparatus used in taking them.[118] Black's partner furnished the information requested,[119] but nothing more seems to have been done.

As early as May, 1861, more than a month before Lowe made his initial demonstrations of aerial telegraphy in Washington, members of the American Photographical Society had been discussing seriously the value of aerial photographs taken from balloons as an aid to military topographical operations. The question was formally presented for consideration by Charles Seely, a prominent member, and editor of the *American Journal of Photography,* at a meeting of the Society held on June 10. The subject received favorable discussion, and after some deliberation, it was agreed that the Society should tender

the *American Journal of Photography,* published in Volume III, p. 188; *Photographic Notes,* IV (London, January, 1861), 9-11; *Photographic News,* V (London, March, 1861), 144; Joseph Dixon to Charles Seely (Editor of *American Journal of Photography*), December 10, 1860, published in the *American Journal,* III, 222; *Photo-Miniature,* V (New York, July, 1903), 154-156.

[116] Payton Spence to the New York *Tribune,* June 10, 1861.

[117] New York *Tribune,* June 22, 1861.

[118] Bache to Messrs. Whipple and Black, June 25, 1861, MS LBTE, XII, 446.

[119] John A. Whipple to Bache, July [date missing], 1861, MS W853, LRTE. Black used, Whipple wrote, "what is called extra whole plate, Harrison tube, full aperture, and [the picture] was taken as quick as an aperture could be passed across the tube." Whipple suggested the use of a Voigtlander portrait instrument for Bache's purpose.

its joint services to the War Department in making available
the benefits of aerial photography to the army. A committee
was appointed to carry out this project,[119a] and on the following
day, the Society's president, Dr. John W. Draper, scholar, historian, scientist, and later author of one of the most original
interpretations of the Civil War,[120] addressed the Secretary of
War on behalf of the membership urging that aerial photography be included in the experiments and possible operations
with military balloons. The services of the members of the
Society in furthering the suggestion, he assured Secretary
Cameron, would be placed at the disposal of the Government.[121]
More than a month later a copy of this communication was read
before a meeting of the Society, at which it was announced that
no reply had been received.[122] This neglect by the military
authorities was attributed by Dr. Draper to the " overwhelming
mass of business which the Secretary of War had on his
hands "; Seely was rather of the opinion that " the Secretary
was not aware of the practical value of the aid tendered." It
was agreed, however, that though the failure of the military
officials to answer the communication was regrettable, the Society could do nothing more in the matter.[122a] The indifference of the Secretary of War to this offer of valuable service
by a body of respected and eminent men drew forth later an
editorial from the *Scientific American,* which reviewed the affair
and concluded:

Secretary Cameron, in the multiplicity of his public and private affairs,
never found time to reply to the communication, and the matter was
dropped. We see that Professor Lowe continues to make his ascensions, and we suggest to General McClellan, or any other officer who

---

[119a] Minutes of Meeting of the American Photographical Society, June 10,
1861, *Photographic Notes,* IV (London, August, 1861), 244; also in *American
Journal of Photography,* IV (New York, July, 1861), 67.

[120] Cf. John W. Draper, *History of American Civil War* (New York,
1867-1870), 3 vols.

[121] Draper to Secretary Cameron, June 11, 1861, MS D153, LRSW.

[122] *Scientific American,* September 21, 1861; *Photographic Notes* VI (London,
November, 1861), 311; *Photographic News,* V (London, October, 1861), 489.

[122a] *Scientific American,* September 21, 1861; *Photographic News,* V (October
11, 1861), 489.

may chance to see these remarks, the propriety of calling upon the Photographical Society for the services which they offered last year. Professor Draper, the President, is a man of European reputation. . . . Many of the leading members are persons of position in the world of science, and the high character of the Society is sufficient warrant that its suggestions are worthy of consideration. Let our military art accept the wonderful aid tendered to it by the most subtle department of science.[122b]

Officers of the army had also expressed their opinions on the subject. Captain Tracy's suggestion to General Butler has been mentioned in the narrative of La Mountain's operations,[123] and this was followed in December, 1861 by a similar proposal from a Captain E. B. Hunt, of the Corps of Engineers.[124] At the same time still another suggestion for aerial photography had been addressed to the military authorities and the letter was referred to Lowe, who was then in the midst of organizing the balloon corps. The chief aeronaut declared that the writer had " advanced no new ideas. As soon as other matters connected with the balloons are accomplished," he continued, " I shall give the photographic matter a thorough and practical test." [125] Another proposal was made directly to General McClellan in September, 1861, in which the writer, Robert B. Benson of New York, advocated the construction of large " bomber " dirigible balloons, which should also be used for making daguerreotypes of enemy positions and forces from the air.[125a] Five days later another New Yorker also called attention to the

[122b] *Scientific American*, May 3, 1862; *American Journal of Photography*, IV (May, 1862), 550.

[123] Tracy to Butler, June 16, 1861, MS, Butler Papers. See also Chapter iv, *supra*, p. 108. Announcement in the press that La Mountain had offered his services as a military aeronaut had also drawn editorial comment in the *American Journal of Photography* regarding the added usefulness of aerial photography to the proposed aeronautic operations. "Now we have to suggest," wrote the editor, Charles Seely, "that by all means a photographer should accompany the aeronaut. It has been demonstrated by Mr. Black, and Mr. Kuhns, of this city, that balloon photography is entirely practicable. A photographer and his apparatus will not at all incommode the operator of a large balloon. . . ." *American Journal of Photography*, IV (June, 1861), 40-41.

[124] New York *World*, December 25, 1861.

[125] Lowe to Colburn, December 16, 1861, 3 *O. R.*, III, 269. The author of the suggestion was a W. G. Fullerton.

[125a] Benson to McClellan, September 5, 1861, MS B154, LRAP.

value of aerial photography for military use in a communication to one of the large Philadelphia dailies.[125b]

There is no available evidence, either in Lowe's papers, in the voluminous archives of the War Department, or in the papers of McClellan, Macomb, or Warren, to show that Lowe ever carried on the proposed experiments, or took photographs from his balloons during the war. In November, 1862, he stated in a letter to Major General John G. Parke, Burnside's chief of staff, that he had with him " a set of powerful magnifying lenses with which a photograph three inches square can be magnified to the size twenty feet square." [126] But the letter does not mention any photographic operations, nor did Lowe, in enumerating the auxiliary equipment in his train, include mention of any cameras, or similar apparatus. It is only reasonable to assume that had such service been performed, some mention of it would have appeared in this letter, or in Lowe's long and detailed report of the corps' operations submitted to Secretary Stanton in 1863. The latter document's silence on the subject would seem to indicate conclusively that aerial photographs were not tried in the field, or, if attempted, were unsuccessful. This conclusion is further supported by absence of any photographic materials in the inventories and surviving property accounts of the corps. Cameras and other photographic equipment would not seem to be materiel that Lowe would buy with personal funds; and again, no such items are listed in his claims for reimbursement. Finally, none of the war correspondents whose accounts of the balloon service have been examined, mention photography from balloons. Surely such a new and unusual activity would not have escaped their inquisitive attention and search for dramatic material. Still more conclusive evidence lies in the fact that no aerial photographs have ever been found in the War Department Archives, or in the photograph collections in the archives of the Signal Corps. To assume that formerly existing prints of this kind have all been lost is not reasonable, since a number of maps made from the air have

[125b] Philadelphia *Ledger*, September 10, 1861.
[126] Lowe to Parke, November 20, 1863, 3 *O. R.*, III, 293.

survived. It is most unlikely, had aerial photography been practiced, that every print would have disappeared from the files.[127]

The assumption that photography was not practiced is further
strengthened by the letter of R. A. Maxwell on August 23,
1863, addressed to Secretary Stanton urging that ballooning in
the army be continued for the purpose of photographing enemy
positions from the air. Enlarged prints of the hostile works at
Manassas, Yorktown, and other theatres of action, Maxwell
stated, would have been invaluable to the generals in command.
Why not take advantage of this process hitherto neglected, and
use it in the future, he concluded.[128] Thus from all the evidence examined, the conclusion follows that aerial photography,
though recommended by laymen, scientists, professional photographers, and army officers, was not carried on by the balloon
corps. Whether it was tried and found ineffective; whether the
many duties that occupied Lowe's time prevented him from
adding this service to his other operations; or whether the supervising officers refused to provide the additional funds necessary for its prosecution are all questions that must remain
unanswered.

Several disadvantages were experienced in the operation of
the balloons in the field. The first of these was the complete
dependence on weather conditions. Wind interference with
ascension was an insurmountable problem. Frequently when
observations were of utmost importance, the aeronauts were
forced to report inability to go aloft because of prevailing
winds.[129] On several occasions interference of this kind took
place during critical moments preceding or during battles, as

---

[127] In reply to an inquiry on the subject from Dr. Albert F. Zahm, Chief of
the Division of Aeronautics, Library of Congress, the Acting Adjutant General
of the Army replied in May, 1933:

"No record has been found of such use of balloons during . . . [the Civil]
War, and inquiry of the Chief Signal Officer has elicited the information that
no record has been found among any of the collections of photographs in that
office to indicate that any were made from balloons during the war mentioned."
Brigadier General James F. McKinley to Dr. Zahm, May 10, 1933, Division of
Aeronautics, Library of Congress.

[128] R. A. Maxwell to Stanton, August 23, 1863, MS M725, LRTE.

[129] Dispatches and reports in the Lowe Papers, McClellan Papers, Heintzelman Papers, War Department Archives, and in 3 *O. R.*, III, 260-315, *passim*.

for example at the battle of Chancellorsville, when the chief of staff regretfully reported to Hooker that the balloons were grounded because of wind.[130] Similar examples of this kind are not uncommon throughout the records of the corps' operations. Even when the winds were not strong enough to ground the aeronauts entirely, they often prevented the gaining of altitudes sufficiently high to take the observations desired.[131] During the action at Chancellorsville Lowe was forced to report that " the wind continues so flawy that the balloon was blown from a thousand feet elevation to near the earth." [132] Again, even though the balloons could be sent up to the necessary elevation, there were times when winds rendered the cars so unsteady that observations were unsatisfactory.[132a] With the baskets thus affected, the aeronauts were unable to focus their glasses or hold them steady enough to take accurate observations.[133] On one occasion an officer reported that he could make out a camp, but the motion was such as to prevent his counting or estimating the number of tents it contained.[134] At Hilton Head, during nearly all of Starkweather's operations with Sherman, ocean and shore winds were particularly troublesome.[135] Damage to the envelopes and cordage was a further disadvantage produced by severe winds. Lowe's accident at Lewinsville, and the necessity of discharging gas during gales may also be cited as examples.

When the atmosphere was calm, ground mists and fog often oozed up from the earth and streams to plague the observers.

[130] Butterfield to Hooker [2 p. m.], May 2, 1863, 1 O. R., XXV, ii, 354.

[131] Lowe to Comstock, April 18, 1863, 3 O. R., III, 307; Lowe to Butterfield, April 30, 1863, ibid., p. 311; also numerous dispatches in the Lowe Papers.

[132] Lowe to Butterfield [1:05 p. m.], May 2, 1863, 3 O. R., III, 314.

[132a] The problem of wind interference was inherent in the spherical or semi-spherical shape of the balloons then used. This difficulty was not successfully overcome until the German officers, Major August von Parseval and Captain Hans Bartsch von Sigsfeld perfected their kite-balloon in 1896. Cf. Parseval and Sigsfeld, Der Drachen-Ballon (Berlin, 1897).

[133] Lowe to Butterfield [6:15 a. m.], May 2, 1863, 3 O. R., III, 314; also published in 1 O. R., XXV, ii, 353; Lowe to Butterfield [12:30 p. m.], May 2, 1863, 3 O. R., III, 314; New York Tribune, June 8, 1863.

[134] Starkweather to Lowe, January 13, 1862, MS, Lowe Papers.

[135] Starkweather to Lowe, April 12, 1862, MS, Lowe Papers; Starkweather to Lowe, April 15, 1862, MS, ibid.; Starkweather to Lowe, April 19, 1862, MS, ibid.; Starkweather to Lowe, January 13, 1862, MS, ibid.

Thick fog rolling up from the Rappahannock blinded the aeronauts at Fredericksburg, and later at times during the battle of
Chancellorsville.[136] "The balloon has not gone up on account of
fog," Sedgwick telegraphed to Hooker on the morning of the
critical first of May, 1863,[137] and on the same day the chief of
staff also reported the terrain and atmosphere " so foggy that the
balloons can see nothing." [138] General Heintzelman once
recorded fog so thick that the balloon at once became invisible
after it ascended. " We could see the ropes at the earth, but the
balloon was hid in the fog," the old veteran set down in his
diary.[139]

Even when the earth was not enveloped in this impenetrable
vapor, the atmosphere often became thick or hazy and thus
decreased visibility and prevented accurate observation.[140]
Steiner's operations at Island No. 10 were rendered useless for
several days because of the haze and smoky condition prevailing.[141] The same problem confronted the aeronauts in other
theatres of action throughout the war. During prolonged combat, powder smoke from sustained musketry and cannon fire
likewise reduced visibility and obscured the observers' view of
battle movements.[142]

It must be borne in mind that these several obstacles to successful observation—fog, ground mist, haze, and battle smoke
—were common to all methods of long distance reconnaissance.

[136] Dispatches in the Lowe Papers, and War Department Archives.
[137] Sedgwick to Butterfield [8:35 a. m.], May 1, 1863, 1 *O. R.,* XXV, ii, 336.
[138] Butterfield to Hooker, May 1, 1863, MS LBAP, III, 292, letter 487.
[139] Heintzelman's Diary, entry of May 1, 1862, MS, Heintzelman Papers.
[140] James Allen to Lowe, June 3, 1862, MS, Lowe Papers; Lowe to Humphreys, June 7, 1862, MS, *ibid.*; Ezra S. Allen to Comstock, April 9, 1863,
MS, *ibid.*; Lowe to Butterfield, February 22, 1863, MS [number omitted], Miscellaneous Papers, Army of the Potomac; also copy, MS, Lowe Papers. Also
numerous dispatches in the Lowe Papers, McClellan Papers, War Department
Archives, and in 3 *O. R.,* III, 266-315, *passim.*
[141] Steiner to Lowe, March 26, 1862, MS, Lowe Papers; St. Paul *Pioneer
and Democrat,* March 28, 1862; St. Louis *Daily Missouri Democrat,* April 3,
1862; Cincinnati *Commercial,* March 27 and 28, 1862.
[142] Lowe to Humphreys [n. d.], 1862, 3 *O. R.,* III, 284-285; Lowe to Humphreys, June 3, 1862, *ibid.,* p. 285; Lowe to Humphreys, June 13, 1863, *ibid.,*
p. 287; Lowe to Humphreys [?], June 27, 1862, *ibid.,* p. 290; Letters
of " W. D. B." in Cincinnati *Commercial,* June 14, 23, and 28, 1862.

Observers posted on signal towers, on hills, in trees, or in any observation posts at a distance from the terrain or objects under survey, were equally helpless when such atmospheric conditions prevailed. Consequently the failure of aerial observation under such conditions of visibility was not inherent to the use of balloons. Poor or zero visibility renders all observation ineffective, and when the balloonists failed because of such interference, any other observers would likewise have failed.

In winter, ice and snow combined with the other unfavorable weather conditions to interfere with the work of the balloonists and to damage their apparatus. " The balloon was loded down wis ice wich we coult not remove, and it was one inch thick all over the surfise," reported Steiner to his Chief in January, 1862. In characteristic spelling he described the damage to the fabric, resulting leaks in the envelope, and the serious loss of hydrogen.[143] The same day General Stone telegraphed Lowe repeating the substance of Steiner's report, and asked for another balloon to replace the ice-damaged *Intrepid*.[144] On another occasion Steiner again reported that freezing rain had encased the balloon and netting in ice, causing serious injury. " The cords of the nete was as thick as a thumm and waide a ton or more," he wrote.[145] Four days later he addressed Lowe that " I have still som gas in the balloon, but this snou and haile has plaid the Devl wis it." [146] Mason reported similar difficulties in March, 1862, and wrote that a coating of ice had covered his balloon " notwithstanding we turned out three times in the night to shake it off." [147] Snow was also troublesome. It was injurious to the envelopes and, according to General Hooker, it interfered with observation by obscuring the outlines of hostile camps and positions.[148] On a snow-covered landscape, however, any movement of troops must have been readily visible to the observers.

Unfavorable weather conditions were not the only obstacles

[143] Steiner to Lowe, January 25, 1862, MS, Lowe Papers.
[144] Stone to Lowe, January 25, 1862, 3 *O. R.*, III, 289.
[145] Steiner to Lowe, January 18, 1862, MS, Lowe Papers.
[146] Steiner to Lowe, January 22, 1862, MS, Lowe Papers.
[147] Mason to Lowe, March 4, 1862, MS, Lowe Papers.
[148] Hooker to Williams, February 18, 1862, 1 *O. R.*, V, 724.

to successful operations. In wooded territory, the cover afforded by foliage and patches of woods often rendered observations ineffective.[149] This disadvantage was recognized by General Porter, who welcomed the fall season for observations, and wrote to Colonel Colburn in November, 1861, that "loss of foliage will enable [Lowe] to discover the movements of the enemy, and perhaps his strength near Centreville and beyond."[150]

In addition to their being subject to weather and terrain conditions, the balloons presented another disadvantage by continually drawing artillery fire. From the first ascension with the *Union*, in August, 1861, to the close of the aeronautic operations of the Federal army, the Confederate artillery sought to bring down the balloons whenever guns could be brought to bear on them. Evidently apprehensive for the security of their own movements, and resenting the activities of the ever-present inquisitive observers, the Southern artillerists gave vent to their resentment in volleys, pot-shooting, and in constant gunnery practice at the globular targets whenever they rose into view.[151]

[149] Edge, *McClellan and the Campaign on the Yorktown Peninsula*, pp. 134, 167-168; Lowe to Humphreys, May 27, 1862, 3 *O. R.*, III, 279; Lieutenant Colonel William Allen, *The Army of Northern Virginia in 1862* (Boston, 1892), p. 470; Survivors Association, *History of the 118th Regiment, Pennsylvania Volunteers* (Philadelphia, 1906), p. 177.

[150] Porter to Colburn, November 6, 1861, MS 6604, McClellan Papers, 1st Series, Vol. XXX.

[151] Dispatches in the Lowe Papers, Heintzelman Papers, McClellan Papers, and in the various collections in the War Department Archives often mention artillery fire directed at the balloons.

Among the personal narratives of the war which also comment on the subject are: S. Millett Thompson, *13th Regiment, New Hampshire Volunteer Infantry, in the War of the Rebellion*, pp. 100, 106; Sergeant C. B. Fairchild, *History of the 27th Regiment New York Volunteers* (Binghampton, 1888), p. 47; Eugene A. Nash, *History of the 44th Regiment, New York Volunteer Infantry*, pp. 68, 79; Warren Cudworth, *History of the 1st Regiment, Massachusetts Infantry* (Boston, 1866), p. 155; Journal of Private David H. Grows, entry of June 22, 1862, in *History of the 5th Massachusetts Battery*, p. 308; Martin A. Haynes, *A History of the 2nd Regiment, New Hampshire Volunteer Infantry in the War of the Rebellion* (Lockport, N. H., 1896), pp. 58-59; Diary of Orderly Sergeant William P. Andrews, entry of January 2, 1863, in Lieutenant Charles A. Cuffel, *Durell's Battery in the Civil War* (Philadelphia, 1900), p. 111; Lieutenant Alfred Davenport to his mother, April 21, 1862, in Lydia Minturn Post (ed.), *Soldiers' Letters from Camp, Battlefield, and Prison*

So persistent was the Confederate artillery fire that a war correspondent later wrote:

It came at length to be our principal amusement in camp to watch the rebels fire at the balloon, as it sailed tranquilly above our picket line, and I have seen many a dollar staked by the " boys in blue " on the skill of the gray-coated artillerists. It was laughable to watch these bets, and I think I shall not go far astray from the truth, when I say that some very good patriots would have been glad to see the balloon struck, since it would have enabled them to win their wagers.[152]

Though the Confederates often scored very close shots, they never did register a direct hit. The cordage and baskets were occasionally struck by shell fragments, and at Seven Pines a large projectile passed between the mooring ropes close under the car. On another occasion registration of an enemy battery was within fifty feet of one balloon, and again another battery peppered the ground from which an ascension had just been made.[153] During the siege of Yorktown the Confederates fired

(New York, 1865), p. 81; Alfred Davenport, *5th New York Volunteer Infantry (Duryee Zouaves)*, New York, 1879, p. 162; T. S. C. Lowe, " The Balloons with the Army of the Potomac," *Photographic History of the Civil War*, VIII, 370; " Three Months with the Balloons," pp. 103-106; " A Balloon Adventure on the Potomac," p. 151; Cook, *The Siege of Richmond*, p. 32; *Lowe's Report*, 3 *O. R.*, III, 260, 275.

Further evidence that the balloons continually drew fire appears in numerous notices and dispatches in the contemporary newspapers for 1861, 1862, and 1863, from which the following may be quoted:

New York *Tribune*, August 31, 1861; Hartford *Daily Courant*, September 2, 1861; New York *Tribune*, September 1, 1861; Philadelphia *Inquirer*, December 2 and 3, 1861; Philadelphia *Daily News*, December 2, 1861; New York *World*, December 2, 1861; Philadelphia *Press*, April 28, 1862; Baltimore *American*, April 26, 1862; Cincinnati *Commercial*, May 12, 1862; Savannah *Republican*, May 28, 1862; Boston *Transcript*, June 5, 1862; New York *Tribune*, June 4 and 13, 1862; New York *Times*, June 24, 1862; Washington *Star*, June 13, 1863.

[152] " A Balloon Adventure on the Potomac," p. 151.

[153] Anonymous (" An English Combatant "), *Battlefields of the South, From Bull Run to Fredericksburg* (London, 1863), I, 242; Edge, *McClellan and the Campaign on the Yorktown Peninsula*, p. 134; New York *World*, December 2, 1861; Baltimore *American*, April 11, 1862; Boston *Transcript*, April 26, 1862; Cincinnati *Commercial*, May 12, 1862.

The war correspondent of the *Commercial* wrote to his editor: " Crossing the bridge over Wormley Creek we were just in time to see Professor Lowe's balloon *Intrepid* mounting into space with a reconnaissance officer. Soon after

incessantly at the balloons whenever they appeared,[154] and on one occasion, so Lowe relates, they elevated a large Armstrong to such a high angle and used such an excessive propelling charge in an attempt to reach one of the balloons, that the gun burst.[155] The war correspondent Cook also records at York-town that the continual presence of the Federal balloons " gave them [the Confederates] paroxysms of rage. They often brought out their artillery to shell the balloon, or shot musketry at it, or cried at it in derision." [156]

Several methods were worked out to destroy the aerostats with gunfire. The most feasible was the practice of laying batteries so that their planes of fire swept the areas into which the balloons were calculated to descend. In this way it was intended to lay down a curtain of fire through which the balloon would necessarily pass in descending to the earth. These attempts, however, were unsuccessful, though in one instance a battery of twelve rifled pieces was brought to bear and was fired simultaneously on the balloon, with the result that several shells passed through the rigging, and the effect of the volley was registered only 200 feet away.[157] One of the Southern papers reported another scheme for bringing down one of the balloons. Enterprising artillerists planned to run a gun by hand very close to the Union lines under cover of darkness, and to open with it on the envelope as soon as it rose at dawn.[158] The

the rebels discovered it and opened upon it furiously. Their first shot was admirably aimed and very little more powder would have settled both Professor Lowe and his balloon. The shell seemed to us to have exploded less than a hundred feet from the globe, producing a splendid effect, a white conical cloud, almost as large as the balloon itself, expanding and dispersing until it seemed to envelop the aerial machine. I suppose twenty or thirty shells were fired in rapid succession, but it [the balloon] was not drawn down until the reconnaissance was complete."

[154] Testimony of General Heintzelman before the Committee of Congress on the Conduct of the War, February 17, 1863, *Report of the Joint Committee,* Part I (Washington, 1863), p. 347; " An English Combatant," I, 242.

[155] Lieutenant Colonel W. J. Handy to Lowe, December 1, 1909, MS, Lowe Papers; also in Lowe's MS Memoirs, p. 131.

[156] Cook, *Siege of Richmond,* p. 32.

[157] T. S. C. Lowe, " Observation Balloons in the Battle of Fair Oaks," *American Review of Reviews,* XLIII (1911), 188.

[158] Savannah *Republican,* May 28, 1862; Philadelphia *Inquirer,* June 17, 1862.

Southern press later triumphantly announced the destruction of one of the Federal balloons by gunfire, under the charming headline of " ABE'S BALLOON PLUGGED." [159]   A Richmond daily described the alleged incident, and even identified the officer said to have directed the successful fire:

One of these balloons was sent up on Saturday morning [June 15, 1862] about nine o'clock, but had gone but a short distance in the air on its observing tour, when a well-directed shot from one of the pieces of the Purcell Battery, Captain Pegram, struck the balloon, tearing it all to pieces.[160]

Other Confederate newspapers took up the story, and spread the news of the proposed target shot in congratulatory notices.[161] A short while later another announcement appeared in the Richmond press that a Yankee balloon had been captured during McClellan's retreat to the James,[162] and an ardent Savannah editor relayed the news in an article gleefully headed " GOT THE BALLOON! " [163] These claims, however, appear to have had no foundation. All the balloons built for the Union service were accounted for intact when the corps was disbanded.[164]

Although the Confederate gunners were never successful in bringing down their aerial quarry, their incessant fire delivered at the balloons was often troublesome to the troops in position in the vicinity of the aeronautic operations. Shell and solid shot intended for the balloons frequently wrought severe damage and rendered the ground untenable near the observation stations. On one occasion Lowe had bombproofs dug and protected with sandbags to provide a safe place of operation for his ground crew while the balloon was aloft under fire.[165]   General

[159] Atlanta *Southern Confederacy*, June 20, 1862.

[160] Richmond *Enquirer*, June 17, 1862.

[161] Lynchburg *Republican*, June 18, 1862; Charleston *Mercury*, June 19, 1862; Savannah *Daily News*, June 20, 1862; Savannah *Republican*, June 20, 1862.

[162] Richmond *Whig*, July 1, 1862.

[163] Atlanta *Southern Confederacy*, July 3, 1862.

[164] " Report on the Condition of the Balloon Corps and the Balloons." June [date missing], 1863, MS, Warren Papers, Letters, Chancellorsville, May 1– June 28, 1863, Vol. I.

[165] Philadelphia *Inquirer*, June 25, 1862.

Stoneman, chief of McClellan's cavalry on the Peninsula, once complained to Lowe that the enemy fire directed at the balloons seriously endangered the extensive picket areas of his horses.[166] Shortly thereafter a 12-inch shell aimed at one of the observers greatly endangered Stoneman himself, and struck close by his tent-fly, showering him with dirt as he sat talking with Lowe.[167] Another projectile hurled at a balloon caused one of Stoneman's fellow generals equal inconvenience. "Occasionally the rebels train a gun on [the balloon]," wrote a company sergeant of a New York regiment, "and try to reach it with a shell."

They do not succeed, however; but one of these shells dropped down into the cook house at General Slocum's headquarters, scattering camp kettles and cooks, who were just then preparing breakfast.[168]

Before Yorktown General Heintzelman narrowly escaped death from the fire of a Hotchkiss gun that shelled one of the balloons as it descended. One projectile, he wrote" passed within five feet of me and passed through a crowd of men. . . . I did not think it possible for a stone to be thrown amongst them and not hit someone. . . . There were half a dozen generals around and several hundred men." [169] Under similar circumstances a 64-pounder struck close to where McClellan himself stood, and fearing that the elevated balloon might draw fire into an adjacent area thickly occupied by troops, Porter ordered the observer to descend.[170] Further orders directed that no more ascensions be made from the same position.[171]

Another disadvantage in the operation of balloons arose during periods of general inactivity. In times of major combat, troops are nearly always exposed to shellfire, whether it is directed at them or at other targets. But between battles when the armies occupied relatively close lines without engaging in action, the balloons also drew fire from batteries that would have otherwise

---

[166] Lowe's MS Memoirs, p. 134.
[167] T. S. C. Lowe, "The Balloons with the Army of the Potomac," *Photographic History of the Civil War*, VIII, 372. Lowe's MS Memoirs, p. 135.
[168] Fairchild, *History of the 27th Regiment, New York Volunteers*, p. 47.
[169] Heintzelman's Diary, entry of May 3, 1862, MS, Heintzelman Papers.
[170] *Lowe's Report*, 3 *O. R.*, III, 257.
[171] Monteith to Lowe, May 3, 1862, *ibid.*

remained silent. Gunners who did not bother to fire at the lines in quiet sectors during lull periods, invariably peppered away at the balloons, and thus produced damage and casualties that would ordinarily not have occurred.[172] It seems that target practice at the balloons, " hanging like oranges in the sky," as Jeb Stuart's chief of staff described them,[173] was too tempting for the Confederate artillerymen to resist.

One incident may be quoted to illustrate that this pot-shooting in quiet periods added to the woes of the common soldier as well as mussing up general officers' tent-flies and cook shacks. In February, 1863, a lieutenant of the 13th New Hampshire Infantry recorded the following entry in his diary while his regiment lay inactive at Falmouth:

Wednesday. Stinging cold. No drill. David Hogan of [Company] " E " has an experience that he can never forget. His round of [sentry] duty takes him near the regiment's sinks and cesspools. A large shell, intended for one of Professor Lowe's balloons, falls into one of them and bursts there, and scatters about two cartloads of the vile contents for rods around, nearly burying Hogan out of sight. Hogan is unhurt, beyond a scare, but his clothing and appetite are utterly ruined.[174]

From Private Hogan's point of view, the operation of balloons had serious disadvantages in its constant drawing of enemy fire.

---

[172] This paragraph is based on the general sources quoted in note 151, *supra.*
[173] Major Heros von Borcke, *Memoirs of the Confederate War for Independence* (London, 1866), I, 16; originally published in *Blackwood's Edinburgh Magazine* (Edinburgh, 1865), XCVIII, 275.
[174] Thompson, *Thirteenth Regiment, New Hampshire Volunteer Infantry,* journal entry of February 3, 1863, p. 106.

CHAPTER X

FIRST ORGANIZED SERVICE:

OPERATIONS ON THE POTOMAC LINES,

NOVEMBER, 1861–MARCH, 1862

By November 10, 1861, the four new balloons authorized by McClellan and the Secretary of War were completed and ready for service. At the same time Lowe reported that he had secured the services of competent aeronauts to operate the new equipment, and that the construction of field generators at the Washington Navy Yard was well under way. The Balloon Corps of the Army of the Potomac had emerged into being, and its chief now awaited orders for the assignment of his balloons and assistants to stations of observation.[1]

The operations of the corps for the next five months until the Grand Army embarked for the Peninsula, were for the most part a continuation of Lowe's activities of August and September on a much larger scale and in widely separated sectors of the extended Potomac lines. In the former period there had been but a single balloon managed by one aeronaut in the immediate vicinity of Washington. Now there were four, with a fifth in reserve, independently operated by aeronauts of the corps at various points of strategic importance, supervised and technically controlled by Lowe as Chief Aeronaut. At the same time a detached balloon train had been ordered to join the Port Royal Expedition for service in South Carolina, and a few months later a similar unit was sent west to the Department of the Missouri.

During the period under consideration, the tactical situation

---

[1] *Lowe's Report,* in *Official Records of the Union and Confederate Armies* (Washington [Government Printing Office], 1880, *et seq.*), Series 3, Vol. III, 265 [cited hereinafter as *O. R.*, with series number in arabic preceding abbreviated title]; Lowe to Colonel A. V. Colburn [AAG, AP], November 12, 1861, 3 *O. R.*, III, 265-266; Lowe to Lieutenant Colonel John N. Macomb, November 12, 1861, MS, Lowe Papers.

occupied by the two armies in the Potomac region remained much the same as before. Contact was maintained throughout the lines, and the opposing forces directed most of their activities to organization and to watching their opponents. A few minor changes did take place. On October the 19th the Confederate advance posts were withdrawn from the line of Mason's and Munson's Hills, and two days later the only serious engagement of the period, the battle of Ball's Bluff, took place on the upper Potomac.[2] Other than this one action, a number of skirmishes, reconnaissances in force, and collisions between outposts and foraging parties, inactivity prevailed until the first week of March, when the main Confederate army evacuated Centreville and Manassas and withdrew toward Richmond.

The absence of major combat operations limited the function of the balloon corps to routine inspection of hostile positions and activity to ascertain if there were in progress any large scale movements of a threatening nature. Consequently the observations taken during the fall and winter period revealed a similar type of information to that obtained by Lowe in August and September. In general the duties and accomplishments of the balloon corps were simply an expansion of the service performed by the one aeronaut before. Nothing spectacular or of startling significance was disclosed by the numerous aerial reconnaissances throughout the length of the lines. The most that the prevailing situation permitted was continual vigilance as a safeguard against surprise movements, and constant surveillance of enemy dispositions. This mission the newly established balloon corps accomplished with creditable success. At the same time, the aeronauts were able during the quiet period

[2] Lowe was still in process of building the balloons when the battle of Ball's Bluff was fought. Hence aerial observation was not available to the Union forces in the preliminaries to this engagement. A Baltimore paper expressed an opinion that proper reconnaissance from the air would have revealed the Confederate movements and the superior numbers that confronted Stone's units, and would have prevented the Federal rout. Baltimore *American,* November 14, 1861. This statement is somewhat strengthened by Stone's report, in which he declared that previous to the battle, he was " extremely desirous of ascertaining the exact position and force of the enemy in our front." Stone to Brigadier General Seth Williams [AAG, AP], October 29, 1861, 1 *O. R.,* V, 295.

to gain valuable training and experience in operation, which would increase the efficiency and value of the service during combat later.

At the same time that Lowe reported the new balloons ready for assignment to stations, the main body of the Army of the Potomac was concentrated in the immediate vicinity of Washington, with detached units occupying the north bank of the river from Liverpool Point in southern Maryland to Williamsport above Harper's Ferry. The left flank was held by Brigadier General Joseph Hooker's division at Budd's Ferry on the lower Potomac; the right, by Brigadier General Charles P. Stone's Corps of Observation with headquarters at Poolesville, near Edwards Ferry opposite Leesburg; and nine other divisions occupied positions in front of Washington and Alexandria, and at intermediate sectors of importance.[3] The distribution of balloons throughout this extensive theater was consequently designed to provide for observation at critical points on the Union lines. As rapidly as the generators were completed, Hooker and Stone on either flank were each assigned a balloon train. A third unit was attached to Porter's division at Hall's Hill in front of the Capital, and still another was held in reserve for general utility in adjacent commands around Washington.[4] A fifth was detached and sent to General Thomas W. Sherman at Port Royal.[5]

Hooker's command at Budd's Ferry was the first to receive its complement of aeronautic equipment. Since his position occupied the extreme left of the entire Union line, situated on the route of communications with Fort Monroe, and also on the water approach to Washington, observation of Confederate activity across the river in this region was of special importance. Late in October Hooker had notified Adjutant General Williams of his entire ignorance of the force opposite his command:

[3] George B. McClellan, *Report on the Organization and Campaigns of the Army of the Potomac* (New York, 1864), pp. 76-77; McClellan's report of operations, 1 *O. R.*, V, 32.

[4] Lowe to Macomb, January 16, 1862, MS, Lowe Papers; Lowe to Colburn, December 3, 1861, 3 *O. R.*, III, 268.

[5] Colburn to Lowe, November 16, 1861 [two dispatches], 3 *O. R.*, III, 266; Lowe to Sherman, November 27, 1861, *ibid.*, p. 267.

Masses of infantry . . . were descried from the glittering bayonets—in what numbers I have no means of determining. In this regard the opposite side of the river remains a sealed book to me. I have not been able to find anyone able or willing to furnish me with any satisfactory information as to the number of the rebel force opposed to me.[6]

A week later he recommended the sending of a civilian spy into the hostile territory across the river to reconnoitre the position and numbers of troops confronting his division.[7] Probably Hooker's declaration and suggestion induced McClellan to order a balloon sent at once to this sector of his lines. On November 10 Lowe arrived to establish the first balloon station operated under the expanded aeronautic service.[8]

The operations on the lower Potomac followed the same general pattern of activity and results that attended the work of the corps in other sectors of the Potomac lines. One innovation in this sector was the frequent use of the balloon-boat *G. W. Parke Custis,* an old vessel that had been specifically fitted out as a balloon carrier for operations on the water. Another step in the progress of the aeronautic service was the winning of General Hooker's interest, and that of his subordinate, General Daniel E. Sickles, officers who had not witnessed balloon operations in the field before. In Hooker's case, this accomplishment was no small task, for after several instances when ascensions and successful operations were rendered impossible by the weather, he was prone to regard the service as impractical. His unfavorable opinion was later reversed after a number of successful ascensions were made and valuable information obtained. The making of an excellent aerial map of the enemy dispositions by an officer of his division was also effective in convincing the general of the balloon's utility.

On November 10, on orders from General McClellan, Lowe placed the balloon *Constitution* and one of the new generators,

[6] Hooker to Williams, October 28, 1861, 1 *O. R.,* V, 384.

[7] Hooker to Williams, November 2, 1861, MS H298, Letters Received, Army of the Potomac, War Department Division, National Archives [cited hereinafter as LRAP].

[8] Lowe to Hooker, November 10, 1861, 3 *O. R.,* III, 265.

along with a supply of gas materials on the balloon-boat under tow of the naval tug *Coeur de Lion,* and proceeded down the Potomac to Hooker's command.[9] He was accompanied by William Paullin, assistant aeronaut who had been selected to take charge of the lower Potomac station.[10] The same afternoon anchor was dropped off Stump Neck, at the mouth of the Matta-woman Creek, opposite Cockpit Point.[10a] Lowe reported to Hooker and requested the detail of an officer and ground crew to assist in his operations. The following morning a detachment from the 72nd New York Infantry under Captain J. S. Austin reported for duty and preparations for observation were begun.[11] During the proceedings Hooker and Sickles visited the balloon-boat to inspect the apparatus and witness the inflation. By evening the balloon was filled. Sickles remained on

[9] *Ibid.*; Lowe to Colburn, November 12, 1861, 3 *O. R.,* III, 265-266; Lowe to Macomb, January 16, 1862, MS, Lowe Papers.

[10] Paullin to Lowe, November 12, 1861, MS, Lowe Papers.

[10a] A special correspondent of the New York *Herald* attached to Hooker's Division witnessed the arrival of Lowe and the balloon-boat, and also accompanied Lowe's messenger who reported to Hooker. He described the events of the day in a special letter to his paper:

"This afternoon, about four o'clock, a steamer was seen coming down the river towing a nondescript sort of craft. As they came near the steamer was found to be the *Coeur de Lion,* Captain Whittlesey, and the craft in tow the balloon vessel of Professor Lowe. . . . On the deck was a quantity of apparatus for the generating of hydrogen gas to inflate the balloon, that was protected by a voluminous fold of canvas. On going aboard the *Coeur de Lion* I saw Professor Lowe, with whom I had a previous acquaintance, and he promised me an ascent. A consultation was held and it was determined that the *Coeur de Lion* should anchor in the shelter of Mattawoman Creek for the night, and that a special messenger should be sent to appraise General Hooker of the arrival of Professor Lowe. Among the party was . . . a detachment of the 4th Michigan Regiment. A sergeant of this regiment was selected as messenger, and I volunteered to act as guide, having gone near Budd's Ferry before with Captain Eastman of the *Yankee.* We . . . found the headquarters of General Hooker half a mile from Posey's House, near Budd's Ferry. We had travelled three miles through a dense forest, with mud half-knee deep in some places, and had gone half a mile out of our way in addition. When we reached our destination the Sergeant delivered his dispatches, which were immediately acted upon, and we started on our cheerless way back, arriving on board the *Coeur de Lion* at eleven o'clock, having tramped through mud since half-past six o'clock. . . ." Special Letter to the New York *Herald,* November 10, 1861, published November 13.

[11] Lowe to Hooker, November 10, 1861, 3 *O. R.,* III, 265; Lowe to Hooker, November 11, 1861, MS, Lowe Papers.

board and accompanied Lowe and Paullin in the first ascension made on the lower Potomac.[12]

The information obtained during this ascent, though not of momentous importance, revealed with reasonable accuracy the position of the hostile forces across the river. The atmosphere was sharp and clear, and massed campfires visible at Dumfries and Occoquan indicated the enemy located at these points in considerable force. Lines of bivouac fires were also discerned extending from below Occoquan to Dumfries, and also inland from the river for some ten or twelve miles as far as Brentsville, in the direction of Manassas. Working parties were also observed building new batteries at Freestone Point.[13] The results of the ascension gave a fair idea of the extent and position of the forces opposite, and also set at rest a rumor that the Confederates had withdrawn nearly all their troops from this vicinity.[14]

Hooker was also anxious to obtain more detailed information of the enemy's works and water batteries opposite his main command. Since the station near Stump Neck was too far distant to permit observation of these positions in detail, the balloon was ordered the following day to Budd's Ferry, three miles down the river.[15] Lowe did not accompany the *Constitution* to the new position. After putting it and the attending equipment ashore under the charge of Paullin, he proceeded back to Washington to supervise the placing of the other balloons in service. The *Constitution* was moved overland to the vicinity of the

[12] Hooker to Williams, November 11, 1861; 1 *O. R.*, V, 649; Baltimore *American,* November 14, 1861; Providence *Press,* November 13, 1861; New York *Herald,* November 13, 1861 (Special Letter of November 12). See also other dispatches in the *Herald,* November 14, 15, and 24, 1861.

[13] Boston *Transcript,* November 13 and 14, 1861; Boston *Journal,* November 14, 1861; Providence *Press,* November 14, 1861; Providence *Post,* November 14, 1861; New York *Times,* November 13, 1861; New York *Tribune,* November 14, 1861; New York *World,* November 14, 1861; Philadelphia *Press,* November 14, 1861; Philadelphia *Ledger,* November 14, 1861; Detroit *Free Press,* November 14, 1861; Lowe to Colburn, November 12, 1861, 3 *O. R.,* III, 265-266. For reference to geographical locations of places in the general area under consideration, see *Atlas to Accompany Official Records* (Washington [Government Printing Office], 1891-1895), Vol. I, Plate viii, Sheet 2.

[14] Baltimore *American,* November 14, 1861.

[15] Hooker to Williams, November 12, 1861, 1 *O. R.,* V, 408.

Posey house near Hooker's headquarters,[16] where observation was resumed. Reconnaissance from this point disclosed clearly the heavy batteries at Shipping Point and at intervals for four miles down the river, established to command the channel. The number of guns mounted was not ascertained, and it appeared that the Confederates had carefully masked and camouflaged many of their embrasures and gun positions. Visibility was excellent, however, and several correspondents reported that the observers were able to identify one gun as one of the large Parrott rifles captured at Manassas. A short distance to the rear of these batteries a continuous line of camps extended for several miles parallel to the river. Troops estimated at a regiment were noted at drill on a hillside behind Shipping Point.[17]

These first observations with Hooker's command yielded accurate information. General W. H. C. Whiting's brigade, with three additional regiments attached, occupied Dumfries at this time,[18] and troops from the Aquia District under General T. H. Holmes were also located to the number of nearly two thousand at Evansport and its vicinity.[19] Early's Brigade had established headquarters at Mount Carmel Church near Occoquan in September, and had doubtless extended its positions to include the latter place by the time Paullin made his ascensions.[20] The information relating to the water batteries only confirmed ground and water reconnaissances made previously. A number of these batteries were in plain view from the river, and had been observed during boat reconnaissance for some time

[16] Paullin to Lowe, November 12, 1861, MS, Lowe Papers.

[17] New York *Tribune*, November 15, 1861; New York *Times*, November 15, 1861; New York *Evening Post*, November 15, 1861; Baltimore *American*, November 16, 1861; Washington *Star*, November 15, 1861; Washington *Sunday Morning Chronicle*, November 17, 1861.

[18] General Joseph E. Johnston to Acting Secretary of War Judah P. Benjamin, November 16, 1861, 1 *O. R.*, V, 955.

[19] Abstract, Return of Aquia District, October, 1861, 1 *O. R.*, V, 933.

[20] Special Orders, No. 326, HDQ, 1st Corps, Army of the Potomac [CS], September 15, 1861, MS Orderly Book, 1st Corps, Papers of Pierre Gustave Toutant Beauregard, Division of MSS, Library of Congress [cited hereinafter as Beauregard Papers].

In October Holmes' Brigade had also occupied Dumfries to guard against a rumored attack from adjacent Union forces. Benjamin to Johnston, October 4, 1861, 1 *O. R.*, V, 888-889.

before the balloon was sent to Hooker.[21] Observation from the air, though not adding new intelligence, nevertheless gave a fuller picture of the positions. The report that the enemy embrasures and gun pits seemed to be masked from aerial observation fits well with Beauregard's instructions to one of his engineer officers to use the utmost care in concealing his works.[22] Even the report that one unconcealed gun had been identified as a Parrott captured at Manassas seems to have been correct, since five of these captured pieces had been sent by General Johnston to Aquia Creek, not far below the positions under observation,[23] and one of these had been actually transferred in September to one of the Evansport batteries, directly opposite the balloon station.[24]

Thus at the outset, the balloon operations with Hooker's command were successful in making available accurate information of the force and disposition across the river, information Hooker had been anxious to secure. That the General now regarded the balloon as a means of securing valuable information is apparent in his letter to Adjutant General Williams on November 16, a few days after the first ascensions had been made. After discussing the general situation across the Potomac, Hooker declared that

on a reconnaissance from the balloon no doubt I shall be able to furnish you with more specific and satisfactory information.[25]

Despite this auspicious beginning, the aeronautic service on the lower Potomac shortly received a series of set-backs which for the time being caused Hooker to reverse his opinion of the balloon and its use. The favorable weather that attended the initial operations changed for the worse after the first few

[21] See Reports of Colonel Nelson Taylor [72d N. Y. Inf.], and Captain R. S. Williamson [Topl. Engrs.], 1 O. R., V, 372-377.
[22] Beauregard to Captain W. H. Stevens, September 1, 1861, MS Letterbook [Private and Official Correspondence], p. 27, Beauregard Papers; Colonel Thomas Jordan [AG, 1st Corps] to Stevens, September 1, 1861, MS Letterbook, 1st Corps, p. 121, ibid.
[23] Captain E. P. Alexander to A. L. Alexander, September 8, 1861, MS, Mackall Collection, Library of the Johns Hopkins University.
[24] General Samuel Cooper [A and IG, CSA] to Johnston, September 9, 1861, 1 O. R., V, 835.
[25] Hooker to Williams, November 16, 1861, ibid., p. 654.

days,[26] and for almost two weeks intermittent high winds and river mists prevented successful operations. Hooker became impatient at the delays and, to make matters worse, Lowe, who had intended to return immediately with a fresh supply of iron and acid, was delayed in Washington making arrangements to take a balloon to Fort Monroe, and was further detained to conduct ascensions at Porter's position at Minor's Hill.[27] During his absence Hooker complained that the Chief Aeronaut had not returned, and that Paullin was making no effort to reinflate his balloon.[28] The General was apparently forgetful of the fact that Paullin's supply of gas materials had been expended.

Lowe arrived back at Budd's Ferry on November 24, and having anchored the balloon-boat with its supply of gas materials at a point designated by Hooker,[29] proceeded on to Fort Monroe to leave a second balloon there for shipment to General Sherman at Port Royal.[30] Bad weather continued for the next few days, but on the 29th Paullin was able to inflate the *Constitution* and prepare for ascension. During the process one of the Confederate batteries across the river opened fire at the balloon at a range of two miles.[31] Paullin finished his preparations and ascended, and the battery loosed two more shells at him, registering excellent line shots which were reported to have burst within fifty yards of the balloon.[32] One correspondent later announced that a fragment had struck the car.[33] That evening a severe storm came up, and with it a gale of such intensity that Paullin was forced to discharge gas from the en-

[26] Hooker to Williams, November 14, 1861, *ibid.*, p. 422. See also Special Correspondence of the New York *Herald*, November 19, 1861, published November 24.

[27] Paullin to Lowe, November 19, 1861, MS, Lowe Papers; Lowe to Colburn, November 12, 1861, 3 *O. R.*, III, 265-266; Lowe to Colburn, November 21, 1861, *ibid.*, p. 266; Colburn to Lowe, November 16, 1861, *ibid.*; Colburn to Lowe, November 22, 1861, *ibid.*, p. 267.

[28] Hooker to Williams, November 22, 1861, 1 *O. R.*, V, 663.

[29] Hooker to Lowe, November 24, 1861, 3 *O. R.*, III, 267.

[30] New York *Times*, November 26 and 27, 1861.

[31] Hooker to Williams, November 29, 1861, MS H435, LRAP; also MS Letterbooks, Army of the Potomac, X, 403 [cited hereinafter as LBAP].

[32] Philadelphia *Inquirer*, December 2 and 3, 1861; Philadelphia *Daily News*, December 2, 1861.

[33] New York *World*, December 2, 1861.

velope to save the balloon from serious damage. Even with this precaution, some of the cordage was injured.[34] Then it was discovered that the second supply of gas materials had been sufficient for only one inflation, and the balloon was now grounded for want of hydrogen. At this juncture Hooker's impatience flared, and he denounced the balloon in general and Paullin in particular.

The history of the balloon while here [he wrote] is one of accidents and failures. I almost despair of being able to turn it to any account at this season of the year. Hitherto we have been prevented from taking satisfactory observations from the wind or smoky state of the atmosphere, and now we have no gas. In the hands of its present manager I apprehend that we will find it of little or no service.[35]

In view of what had been accomplished during the first few days, this condemnation appears hasty and unjustified. Paullin was not responsible for the weather, and the discharge of hydrogen was only a proper and sensible precaution to protect his equipment. The only apparent blunder was the shortage of iron and acid. Why Lowe did not bring a larger supply on his second trip, to meet such emergencies, cannot be answered; but in any case, the fault was not Paullin's. A new supply of gas materials was immediately dispatched to Budd's Ferry,[36] and within a few days the *Constitution* was again ready for ascension. But the storm of the 29th was unfortunately followed by a week of thick, foggy weather which continued to interfere with successful operations.[36a] "It is of no use to make ascensions while it is so smoky. We cannot see across the river," Hooker telegraphed Lowe in Washington on December 7th.[37]

[34] Hooker to Lowe [telegram], November 30, 1861, MS, Lowe Papers.

[35] Hooker to Williams, November 30, 1861, MS H439, LRAP.

[36] Lowe to Paullin, December 1, 1861, MS [telegram], Lowe Papers; also printed in 3 *O. R.*, III, 268.

[36a] " Budd's Ferry Correspondence," New York *Herald*, December 7, 1861, published December 10. The correspondent also recorded more pot-shooting at the balloon by the Confederate artillerymen across the river, and related how a 64-pounder shell landed in the yard of Posey's house, where it buried itself without exploding. A soldier retrieved this dud from the ground and sold it to a civilian visitor from Boston who wanted it for exhibition in the Massachusetts capital.

[37] Hooker to Lowe, December 7, 1861, MS [telegram], Lowe Papers.

The following day the mists dispersed and visibility was decidedly improved. Ascensions were resumed, and on this occasion the positions of the enemy and surrounding terrain were carefully mapped, an accomplishment that served to restore Hooker's confidence in the balloon and aeronaut. Some time before, General McClellan had personally directed Lowe to have prepared an aerial map of the hostile positions in the lower Potomac area, and in accordance with Lowe's request for a competent draftsman,[38] Hooker had assigned Colonel William F. Small, of the 26th Pennsylvania Infantry.[39] On December 8, Small ascended with Paullin to an elevation of 700 feet, and carefully inspected the Confederate batteries and encampments from Chopawamsic Creek to Freestone Point, a front of some seven miles opposite the Union positions. The various camps were observed to be within ready supporting distance of one another and the water batteries commanding the river. By comparing the visible extent of tentage, huts, and smoke with the camps of friendly troops below him, Small estimated that the force opposite Hooker did not exceed 12,000 men.[40] During the ascent the Confederate gunners again shelled the balloon with their long-range battery, and succeeded in exploding a number of projectiles uncomfortably close to their target. Several shells were reported to have burst directly under the balloon, and others registered within two hundred feet.[41]

Colonel Small, who proved to be an able draftsman, prepared

[38] Lowe to Hooker, November 30, 1861, MS, Lowe Papers; also printed in 3 O. R., III, 267-268.

[39] Hooker to Small, December 7, 1861[telegram], cited in MS H479, enc. 2, LRAP.

[40] Small to Hooker, December 9, 1861, MS H479, enc. 2, LRAP. Small's estimate appears to have been reasonably accurate. According to returns quoted by General Johnston, the Confederate force between Dumfries and Occoquan numbered 6,700 by November 7. Joseph E. Johnston, Narrative of Military Operations During the War Between the States (New York, 1874), p. 83. This strength was increased by the middle of November by three additional regiments, which, with troops from the Aquia District, brought the total to about 10,000. Johnston to Benjamin, November 16, 1861, 1 O. R., V, 955; Johnston to Whiting, November 16, 1861, ibid., p. 958.

[41] New York Times, December 10, 1861; Baltimore American, December 10, 1861; Alexandria Local News, December 11, 1861; Detroit Free Press, December 12, 1861.

a detailed panoramic map of the terrain under inspection from the balloon, on which the disposition of the Confederate forces, the topographical features of the country, and other information of value were carefully and accurately recorded. The map was drawn to scale, and displayed excellent perspective and panoramic depth. Important positions were labelled with letters and numerals, which referred to specific explanations and detailed information contained in an accompanying report of observations.[42]

As a result of these operations Paullin fully regained the favorable opinion of General Hooker and his officers. In two days of fair weather, he had conducted ten ascensions [43] and elicited the approval of all officers concerned. Colonel Small cited his efficiency in an official report to Hooker, in which he declared that

Mr. Paullin, the aeronaut in charge of the balloon, is entitled to much credit for the skill and zeal displayed in conducting the ascensions and in seconding my efforts to comply with your directions.[44]

Colonel Robert Cowdin, of the 1st Massachusetts, also ascended with Paullin,[45] and praised the aeronaut for his ability in managing the apparatus.[46] Small's map also drew well merited praise from Hooker and Adjutant General Williams, and its accuracy was also established by the testimony of a Confederate deserter named Peterson, a former resident of the vicinity.[47] Hooker's renewed interest and confidence is also manifest in the fact that he personally ascended with Paullin during the period of renewed operations.[48]

[42] " Sketch of Virginia and Rebel Camps and Batteries . . . December 8, 1861 . . . by Colonel William F. Small, 20th Reg. Pa. Vols.," MS H479, enc. 3, LRAP. See Plate XLIV, *infra.*

[43] Small to Hooker, December 9, 1861, MS H479, enc. 2, LRAP; Philadelphia *Inquirer,* December 9, 1861.

[44] Small to Hooker, December 9, 1861, MS H479, enc. 2, LRAP.

[45] Lowe to Colburn, December 10, 1861, 3 *O. R.,* III, 268.

[46] Paullin to Lowe, December 9, 1861, MS, Lowe Papers.

[47] Philadelphia *Daily News,* December 12, 1861.

[48] Statement of Private John Hare, 1st Regiment, Sickles' Brigade [prisoner held by Confederates], December 10, 1861, 1 *O. R.,* LI, ii, 410. Similar testimony of Hooker's ascensions was also given by Charles Smith, twelve-year-old

For the next two months the aeronautic service on the lower Potomac continued with no new developments. At times storms and fog interrupted operations,[49] and occasionally caused fresh indications of impatience from Hooker.[50] After heavy snowfalls, the observers also experienced difficulty in identifying the enemy dispositions, since the snow was found to obscure the outlines of camps.[51] When the weather continued fair, observations were conducted with routine frequency. On Christmas Day Paullin reported that a new battery was being erected at the outlet of Timber Branch, to bear on Stump Neck across the channel.[52] In February Hooker devised plans for crossing the river in force and carrying the Confederate works by assault. Information furnished by balloon reconnaissance had a full share in convincing the general that these plans were feasible. On the 20th he notified the Adjutant General that

My observations from the balloon satisfy me that the batteries in my front can be stormed and carried in the manner in which I have already communicated. . . .[53]

The same belief was shared by Captain Blake of the 11th Massachusetts, who later recorded that Hooker supposed from the information secured by balloon, that the high command would grant his request and permit the intended attack.[54] McClellan, however, vetoed the plan on advice from his Chief Engineer [55] although, unknown to them, General Whiting, in command of the Confederate positions in question, had admitted in Novem-

son of a lieutenant in the 5th Regiment of the same brigade, also captured by the Confederates. *Ibid.,* p. 409.

[49] Hooker to Williams, January 21, 1862, MS 7927, Papers of George Brinton McClellan, 1st Series, Vol. XXXVII, Division of MSS, Library of Congress [cited as McClellan Papers]; Paullin to Lowe, January [n. d.], 1862, MS, Lowe Papers.

[50] Hooker to Williams, January 27, 1862, 1 *O. R.,* V, 711; Hooker to Lowe, February 6, 1862, MS, Lowe Papers.

[51] Hooker to Williams, February 18, 1861, 1 *O. R.,* V, 724.

[52] Paullin to Lowe, December 25, 1861, MS, Lowe Papers.

[53] Hooker to Williams, February 20, 1862, MS 8729, McClellan Papers, 1st Ser., Vol. XLI.

[54] Captain Henry N. Blake, *Three Years in the Army of the Potomac* (Boston, 1865), p. 42.

[55] McClellan's Report of Operations, 1 *O. R.,* V, 47.

ber that the batteries could not be held against a formidable attack.[56] General Holmes, commanding the Aquia District, likewise admitted in January that "if General McClellan advances, it will certainly be irrespective of our batteries." [57] Continual aerial reconnaissance by the Federals in this region also caused apprehension at Johnston's headquarters at Manassas, and on February 22 the Confederate commanding general notified President Davis that

The enemy may not allow us much time for changes of position. He has been more active than usual lately. . . . Reconnaissances on the Lower Occoquan and on the Potomac have been frequent, the latter in balloons as well as boats.[58]

The disadvantage of being so closely and constantly watched by the aerial observers was also recognized by the Confederate commanders in the region involved, and at every opportunity the batteries along the river shelled the balloon in fruitless efforts to bring it down. On one occasion every battery in the vicinity that could be brought to bear opened on the ascending *Constitution*,[59] and the heavy cannonading led General Morrel at Minor's Hill near Washington to believe that a general engagement had taken place on the lower Potomac.[60] Towards the end of January Paullin was dismissed by Lowe for attempting to carry on an ambrotype business in addition to his military duties,[61] and Ebenezer Seaver replaced him as assistant aeronaut in charge of the lower Potomac station.[62] Clovis

[56] Brigadier General W. H. C. Whiting to Johnston, November 15, 1861, 1 *O. R.*, V, 956-957; Whiting to Johnston, November 16, 1861, *ibid.*, p. 957.
[57] Brigadier General Thomas H. Holmes to Johnston, January 17, 1862, 1 *O. R.*, V, 1035.
[58] Johnston to President Davis, February 22, 1861, 1 *O. R.*, V, 1079.
[59] Budd's Ferry telegrapher to Brigadier General George W. Morrell, February 10, 1862, MS 8457, McClellan Papers, 1st Series, Vol. XL; Telegrapher to Williams, February 10, 1862, MS 8461, *ibid.*
[60] Morrell to Williams, February 10, 1862, MS 8457, McClellan Papers, 1st Series, Vol. XL.
[61] Lowe to Captain John B. Howard [AQM, AP], January 31, 1863, MS L100, enc., Letters Received, Secretary of War, War Department Division, National Archives [cited hereinafter as LRSW]; Paullin to Lowe [n. d.], 1862, MS, Lowe Papers.
[62] Lowe to Hooker, January 23, 1862, MS, Lowe Papers.

Lowe had been assigned as an assistant to Paullin since January 17 [63] and was retained at the station with Seaver. The new aeronaut in charge, according to the elder Lowe, took himself and his authority very seriously, and caused much amusement among Hooker's officers by his first appearance in a gaudy and unofficial uniform. His self-importance is best revealed in his formal announcement " of assuming command," a document which must have been surprising as well as amusing to all concerned.[64] The elder Lowe took Seaver's pompous affectations in good humor and cooperated with him fully, though he did occasionally poke fun at him in private letters to his son.[65] Despite Seaver's personal peculiarities, he conducted the aeronautic operations with Hooker's division with skill and ability, and remained with the station until the Union forces began the movement to the Peninsula.

A final example may be cited to show the effective operations in this sector and to indicate Hooker's reliance on aerial observation for information. Early in March, General Sickles, one of his brigade commanders, reported that the Confederate force opposite had been strengthened by four additional regiments. Hooker, in sending this report on for the information of McClellan, advised Adjutant General Williams that he would confirm or disprove the presence of the reinforcements by observations from his balloon.[66]

While one unit of the balloon service was functioning under Paullin with Hooker's division at the extreme left of the lines, a second station was established on the right, at Edwards Ferry on

[63] Clovis Lowe to T. S. C. Lowe, January 17, 1862, MS, Lowe Papers.

[64] Seaver's declaration, penned a few hours after his arrival, reads like a commanding general's admonition to an insubordinate staff:

" Having received and accepted my appointment to take charge of the balloon and apparatus now here, I hereby announce my determination to permit no one to interfere with or dictate to me in the performance of the duties appertaining to the proper performance of the charge entrusted to me. Hereafter all orders in any way relating to the management of the balloon will be given by myself in person, or by someone especially authorized by me." Declaration of Ebenezer Seaver, January 25, 1862, MS, Lowe Papers.

[65] Clovis Lowe to T. S. C. Lowe, January 25, 1862, MS, Lowe Papers.

[66] Hooker to Williams, March 5, 1862, MS 9125, McClellan Papers, 1st Series, Vol. XLIII. Also printed in 1 *O. R.,* V, 735.

the upper Potomac. Early in November an outpost commander in this sector had reported a Confederate balloon hovering over Leesburg,[67] a fact that probably led the Federal commander in the region to call for similar equipment. A week later General Stone, commanding the Corps of Observation holding this portion of the lines, requested the assignment of a balloon for the use of his command.[68] Delays at the Washington Navy Yard in the construction of the field generators, and General Porter's requisition of the first generator available after Hooker's station had been established, prevented the arrival of the balloon train at the upper Potomac station for several weeks after Stone's request.[69] On December 14, Lowe, accompanied by Seaver as assistant aeronaut, reached Edwards Ferry with the balloon *Intrepid* and all necessary auxiliary equipment and established a second observation station.[70] Ascensions were conducted the same day, during which General Stone personally ascended to reconnoiter the Confederate positions and activity about Leesburg and its vicinity.[71]

The operations of the upper Potomac station were generally similar to those of the corps at other points along the extensive Federal lines. Routine observations as a security measure; vigilance for the detection of change in the hostile positions and force, such as the appearance of new camps and the disappearance of old; and the watching of important roads for troop movements: these in general were the function of the aeronautic service with Stone's command. That few changes of

[67] Brigadier General W. A. Gorman to Captain Charles Stewart [AAG, Corps of Obs.], November 5, 1861, MS, Letters Received, Corps of Observation, Army of the Potomac, War Department Division, National Archives [cited hereinafter as LRCO].

[68] Stone to McClellan, December 10, 1861, MS 7020, McClellan Papers, 1st Ser., Vol. XXXII.

[69] Lowe to Colburn, December 3, 1861, 3 *O.R.*, III, 268; Lowe to Porter, December 3, 1861, MS, Lowe Papers.

[70] Stone to McClellan, December 14, 1861, MS 7076, McClellan Papers, 1st Ser., Vol. XXXIII; Captain Richard Derby to [addressee omitted], December 15, 1861, Lydia Minturn Post (ed.), *Soldiers' Letters from Camp, Battlefield, and Prison* (New York, 1865), p. 50; Seaver to Lowe, December 17, 1861, MS, Lowe Papers.

[71] Stone to McClellan, December 14, 1861, MS 7076, McClellan Papers, 1st Ser., Vol. XXXIII; Lowe to Colburn, December 16, 1861, 3 *O.R.*, III, 269.

importance in the enemy dispositions occurred or were reported by the observers did not lessen the practical value of the service performed. Stone's position on the Union flank was one that naturally required safeguard against surprise movements, and the presence of the balloon made possible a greater measure of security against possible sudden attack. Also, reliable information as to the strength of the forces opposite was of prime importance. On the same day Lowe arrived with the *Intrepid*, Beauregard instructed General D. H. Hill, commanding at Leesburg to spread the rumor that his force was to be reinforced by 10,000 men.[72] Such news would naturally have caused Stone grave concern had he not been provided with the means of checking its accuracy.

In this sector, too, unfavorable weather conditions proved to be a serious hindrance to operations, and at intervals prevented effective observation. At times heavy winds made ascension impossible,[73] and occasional prolonged rainstorms not only stopped all observations, but also proved injurious to the fabric of the envelope. " The balloon has now been in the rain for nine days, and is completely soaked. . . . The varnish is decomposing," the aeronaut in charge reported in January.[74] Worst of all were the severe winter storms that brought heavy snow, sleet, and ice, which worked hardship in operation and maintenance and damaged the apparatus. On several occasions blinding snowstorms, in addition to their damage to the balloon, made attempts at observation useless even though the balloon was ordered aloft.[75] Again, freezing rain and hail completely encased the *Intrepid* and its netting in ice, causing strained segments in the envelope, holes in the silk, and loss of hydrogen.[76] Toward the middle of January the balloon was so badly injured that General Stone requested that it be replaced.[77] The *Union*,

[72] Beauregard to Hill, December 14, 1861, 1 *O. R.*, V, 995.
[73] Steiner to Lowe, January 11, 1862, MS, Lowe Papers.
[74] Steiner to Lowe, January 20, 1861, MS, Lowe Papers.
[75] Seaver to Lowe, December 26, 1861, MS, Lowe Papers; Steiner to Lowe, February 15, 1862, MS, *ibid.*; Steiner to Lowe, February 18, 1862, MS, *ibid.*
[76] Steiner to Lowe, January 18, 1862, MS, Lowe Papers; Steiner to Lowe, January 22, 1862, MS, *ibid.*; Steiner to Lowe, January 25, 1862, MS, *ibid.*
[77] Stone to Lowe, January 20, 1862, MS, Lowe Papers; Stone to Lowe, January 25, 1862, MS, *ibid.*; also printed in 3 *O. R.*, III, 269.

held in reserve at Washington, was sent to the upper Potomac station, and the damaged *Intrepid* was recalled for repairs.[78] In general, winter operation at Edwards Ferry was fraught with hardship, difficulties, and unavoidable interference.

Notwithstanding the unfavorable weather conditions that often prevailed, the upper Potomac station performed satisfactory service, and maintained effective, though interrupted, observation of the hostile forces across the river. The initial ascensions made by Lowe and his assistants upon their arrival revealed with satisfactory accuracy the strength and disposition of the Confederate units in the Leesburg area. After ascending with Lowe, Stone notified McClellan that the force opposite was relatively small. He observed troops estimated at four regiments encamped in rear of the town along the road leading toward Manassas, with three companies of infantry supported by an artillery battery in position near Edwards Ferry on the Virginia side. An outpost was also noted opposite Conrad's Ferry four miles up the river.[79] Similar information was also reported by Lowe, who noted with interest that tentage visible within the earthworks between Leesburg and Edwards Ferry during the first ascensions, had been struck later in the day, indicating that the enemy might be trying to conceal his numbers from aerial observation.[80] General Stone also took aloft with him a map of the region, on which he plotted information obtained during his reconnaissance.[81]

Stone's estimate of the enemy force, based on his observations, was essentially correct. The Confederate units in the vicinity of Leesburg had originally consisted of four regiments, comprising Evans' Seventh Brigade of Beauregard's First Corps, with a six-gun battery of artillery attached.[82] At the time the

[78] Steiner to Lowe, February 10, 1862, MS M289, enc. 3, LRAP; Lowe to Macomb, February 18, 1862, MS L83, LRAP.

[79] Stone to McClellan, December 14, 1861, MS, 7076, McClellan Papers, 1st Ser., Vol. XXXIII.

[80] Lowe to Colburn, December 16, 1861, 3 *O. R.*, III, 269.

[81] *Ibid.*; Philadelphia *Inquirer*, December 16, 1861; Baltimore *American*, December 17, 1861.

[82] General James Longstreet, *From Manassas to Appomattox* (Philadelphia, 1896), p. 61.

observations were taken, Evans had been replaced by General D. H. Hill, but no increase in troops had accompanied the change of command.[83] The strength of the brigade in December was slightly less than 3,000 effectives,[84] a number that agrees well with Stone's calculation of four regiments. If the intended rumor that Hill was directed to circulate concerning his reinforcement by ten thousand men had reached Stone and caused him any apprehension, such misgivings must surely have been put at rest by the nature of Stone's observations at this time, and later.

Continued reconnaissance disclosed but little significant change in the dispositions opposite the Union lines. A slight diminution in January led Stone to conclude that one regiment had gone into winter quarters, since the hostile camp area had decreased in size.[85] A month later, the enemy appeared to have been slightly reinforced. Two new encampments were observed on a road leading west out of Leesburg, and the existing camps about the town appeared to be somewhat larger. Additional troops were also visible occupying the south bank of Goose Creek, with fresh earthworks in progress in this area. The Confederate fortifications and earthworks around Edwards Ferry and Leesburg were reported unchanged, save for some improvements, and the apparent mounting of several heavy guns.[86] The information obtained seemed to indicate that the enemy intended defensive rather than offensive operations. The balloon was in constant operation whenever weather conditions permitted. Stone instructed the aeronaut to keep the hostile camps and positions under continual surveillance, and to report any movements and changes.[87]

In January, Seaver, who had been detailed to operate the station with Stone's command, had been transferred to Hooker's

[83] Johnston, p. 83.

[84] Abstract Return of the Department of Northern Virginia, December, 1861, 1 O. R., V, 1015.

[85] Stone to McClellan, January 12, 1862, MS 7725, McClellan Papers, 1st Ser., Vol. XXXVI.

[86] Steiner to Lowe, February 10, 1862, MS M289, enc. 3, LRAP; also printed in 3 O. R., III, 270.

[87] Steiner to Lowe, January 23, 1862, MS, Lowe Papers.

division, and John H. Steiner was sent to replace him at the upper Potomac station.[88] Neither of these aeronauts had encountered any difficulty in arousing the interest and confidence of General Stone. This officer had early manifested interest in the advantages of balloon observation, and while waiting for the aeronautic train to arrive, had complained to McClellan that " no balloon has arrived here, although one was ordered nearly four weeks since." [89] The *Intrepid* had been in service with his division only a month before he asked for a second balloon. " General Stone thinks it of great importance to have two balloons here. . . ."[90] He wishes to use them at once and at different places," Steiner notified Lowe. One of these balloons Stone wanted particularly for observations farther up the river, and at the same time he wished the Edwards Ferry station to continue operations.[91] The general's recognition of the aerial service is also apparent from his statements when the aeronaut's operations were temporarily suspended. On one occasion when the balloon was deflated and Seaver was waiting for a new supply of iron and acid, Stone complained to Lowe at the loss of the calm weather without ascensions. " I wanted the balloon greatly yesterday . . . and I should like to use it early tomorrow should the weather be calm. Were it inflated it would now be used," he telegraphed the Chief Aeronaut.[92] Again, he urged all possible speed in shipping a replacement balloon when the *Intrepid* had become seriously damaged by ice. " Please send up the small balloon immediately," he telegraphed,[93] and when it had not arrived in five days, he again wired Lowe to " hurry up " the replacement aerostat.[94]

Stone was not alone in his interest in aerial observation. General John Sedgwick, commanding an adjacent division at Harper's Ferry in February, telegraphed Lowe and requested

[88] Lowe to Macomb, January 16, 1862, MS, Lowe Papers.
[89] Stone to McClellan, December 10, 1861, MS 7020, McClellan Papers, 1st Ser., Vol. XXXII.
[90] Steiner to Lowe, January 20, 1862, MS, Lowe Papers.
[91] Steiner to Lowe, January 23, 1862, MS, Lowe Papers.
[92] Stone to Lowe, December 30, 1861, MS, Lowe Papers.
[93] Stone to Lowe, January 20, 1862, 3 *O. R.*, III, 269.
[94] Stone to Lowe, January 25, 1862, *ibid.*

that the balloon operating at Edwards Ferry also be made available for his use.[95] This request was followed in March by a similar suggestion from General Nathan P. Banks, whose division then occupied Charlestown [West] Virginia, who asked that the same balloon be assigned to his command. " The General," Banks' adjutant wired McClellan's headquarters, " thinks it would be more serviceable here, and requests an order from the Major General Commanding to transfer it to this division." [96] Here, then, was no lack of interest among the general officers, with three brigadiers—full division commanders in the field—asking for the use of the same balloon.

Ascensions at the upper Potomac station, and at points selected by General Stone in the vicinity often brought the balloon and its crew within rifle range of the Confederate pickets. On one occasion the balloon party was fired on by sharpshooters, with the result that a soldier had his hat knocked off by a bullet. Steiner wrote Lowe after the incident and naively asked for a revolver with which to defend his balloon.[97] During previous ascensions in positions exposed to attack, Seaver had asked the detail of a party of riflemen, supported by one or two field guns and their crews, for security.[98] In approving this request, Stone inaugurated the only balloon train with artillery attached that appears in the records of the corps.[99] Troops assigned to balloon duty as guards and ground crew were furnished by the 84th New York, 4th Michigan, and the 9th New York Infantry.[100] One of the details included a competent draftsman, who was assigned to special duty in making aerial maps and sketches.[101]

Although Seaver and Steiner were fully responsible for the

[95] Sedgwick to Lowe, February 21, 1862, MS, Lowe Papers.
[96] Major R. Morris Copeland [AAG, Banks' Division] to Assistant Adjutant General Williams, March 9, 1862, MS 9286, McClellan Papers, 1st Ser., XLIV.
[97] Steiner to Lowe, January 20, 1862, MS, Lowe Papers.
[98] Seaver to Stone, December 18, 1861, MS S32, LRCO.
[99] Stone's indorsement, ibid.
[100] Brigadier General Fitz John Porter to Williams, January 21, 1862, MS 7938, McClellan Papers, 1st Ser., Vol. XXXVII; Copeland to Williams, February 4, 1862, MS 8340, ibid., XXXIX; Seaver to Stone, December 28, 1861, MS S42, LRCO.
[101] Seaver to Stone, December 23, 1861, MS [number omitted], LRCO.

operation of the station, Lowe exercised a vigilant supervision. He kept in constant communication, attended to the supply of iron and acid, exacted reports of duty performed, and sent instructions at intervals to his assistants.[102] As chief of the corps he felt the responsibility for the success of all stations under his command, and urged both Seaver and Steiner to take advantage of every opportunity to make themselves useful.[103] His assistants responded loyally and frequently reported the progress of their activities. In his zeal to keep the Chief Aeronaut constantly informed of the operations of his station, Steiner occasionally sent Lowe reports of observations even before written reports were submitted to General Stone. This irregularity in the channels of command drew sharp reprimands from Stone, and the stodgy German balloonist once ruefully reported that " I am giting my sealf in to hotte water." [104] At the same time, Lowe was in the midst of his difficulties with La Mountain, and Steiner loyally informed his chief that

I saw by the papers that John La Mountain was going to take your place. If so let me know. I will *quit* . . . I not only saw it in the paper but was told so at Headquarters. I will not serve around J. La Mountain.[105]

About the middle of February, Steiner was relieved of duty with the upper Potomac station and detailed to take charge of a balloon train ordered west to the Department of the Missouri. Grapevine rumors of the proposed aeronautic service in the west had reached Edwards Ferry in January, and Steiner had asked Lowe for this assignment.[106] The request was granted. About February 21 Jacob C. Freno, recently enrolled as an assistant aeronaut, arrived to relieve Steiner at Edwards Ferry. He was accompanied by Corporal James Starkweather, brother of the

---

[102] Lowe to Steiner, January 14, 1862, MS, Lowe Papers; Lowe to Seaver, December 29, 1861, MS, *ibid.*; Lowe to Brigadier General W. A. Gorman, February 21, 1862, MS, *ibid.*; Lowe to Macomb, February 13, 1862, MS, Lowe Papers.

[103] *Ibid.*; the statement is also supported by the letters from Seaver and Steiner to Lowe, quoted in notes above, *passim.*

[104] Steiner to Lowe, February 15, 1862, MS, Lowe Papers.

[105] Steiner to Lowe, February 12, 1862, MS, Lowe Papers.

[106] Steiner to Lowe, January 11, 1862, MS, Lowe Papers.

aeronaut at Port Royal, as general assistant.[107] Freno remained in charge of the station until March, when the apparatus was brought in to Washington, preparatory to the movement to the Peninsula.[108]

While the flank stations were being operated at either end of the lines with the divisions of Hooker and Stone, a third balloon unit was in constant operation with Fitz John Porter's command, in front of the Capital. At this station, Lowe, occasionally assisted by Seaver and Ebenezer Mason, personally conducted the ascensions and kept the Confederate forces under observation. The operations of this unit were simply a continuation of the Chief Aeronaut's previous service from the time of his original appointment. The Confederate withdrawal from the line of hills just across the river from Washington removed these positions as critical points for constant observation, and as the former hostile strong points were occupied by Federal troops, the balloon was moved out for ascensions from this high ground. Observations were conducted from Hall's, Minor's, and Munson's Hills, with the observers' attention being largely focused on the Confederate encampments at Fairfax Court House, Centreville, and Manassas.[109]

The nature of the information obtained from the air at this station was practically identical with that previously reported. The main Confederate force was still observed at Manassas and Centreville, with advance guard troops in the vicinity of Fairfax Court House. Picket smoke revealed the outpost line regularly along the Leesburg-Alexandria Turnpike. Such was the general situation when Lowe took observations on Novem-

---

[107] Lowe to Freno, March [n. d.], 1862, MS [number omitted], LRCO.

[108] Lowe to Clovis Lowe, March [n. d.], MS, Lowe Papers.

[109] Alexandria *Local News*, November 14, 1861; Philadelphia *Inquirer*, November 16, 1861; *ibid.*, November 20, 1861; *ibid.*, November 22, 1861; Philadelphia *Press*, November 22, 1861; Philadelphia *Daily News*, November 23, 1861; Philadelphia *Ledger*, November 22, 1861; Lancaster *Daily Evening Express*, November 22, 1861; Baltimore *American*, November 22, 1861; New York *Evening Post*, November 22, 1861; New York *World*, November 22, 1861; New York *Tribune*, November 22, 1861; Boston *Transcript*, November 22, 1861; Providence *Post*, November 22, 1861; Alexandria *Local News*, November 23, 1861; Seaver to Porter, November 21, 1861, MS, Lowe Papers; Captain Eugene A. Nash, *History of the 44th New York Volunteer Infantry in the Civil War, 1861-1865* (Chicago, 1911), p. 55.

ber 21.[110] A week later he reported that Fairfax Court House seemed to be entirely deserted, a fact soon verified by a reconnaissance by part of General Wadsworth's brigade.[111] Observation was continued throughout December and January, with little or no change revealed in the disposition of the enemy forces. Lowe's mission continued to be that of routine surveillance and watchful vigilance.

As at the other stations, unfavorable weather conditions intermittently prevented operations and interfered with routine observation;[112] but in general, the service was steady and reliable. The hardship and difficulties experienced on the upper Potomac as a result of severe winter inclemency do not seem to have prevailed in the same degree at the station with Porter's division. Being close to Washington and the source of facilities for shelter, this unit probably escaped to a large extent the severities experienced by Seaver and Steiner at Edwards Ferry.

The marked interest displayed by General Porter in the earlier period with the first balloon did not wane. Even while Lowe was in the midst of constructing the new balloons, Porter wrote Assistant Adjutant General Colburn and suggested that the aeronaut be temporarily recalled to the field for observations.[113] He was also annoyed when shortage of gas caused the loss of a fine clear day without ascensions. " I shall be disappointed," he telegraphed Lowe, " the first calm day if the balloon is not inflated so as to be serviceable." [114] At the time, the Navy Yard had not turned out enough generators to provide Porter with a source of hydrogen in the field, and while the envelope was being reinflated in Washington,[115] Porter telegraphed Lowe that

[110] Lowe to Colburn, November 21, 1861, MS L73, Letters Received, Headquarters of the Army, War Department Division, National Archives [cited hereinafter as LRHQA]. Also printed in 3 O. R., III, 266.

[111] Philadelphia *Inquirer,* November 28, 1861.

[112] Philadelphia *Inquirer,* November 20, 1861; Lowe to Porter, December 3, 1861, MS, Lowe Papers.

[113] Porter to Colburn, November 6, 1861, MS 6604, McClellan Papers, 1st Ser., Vol. XXX.

[114] Porter to Lowe, November 27, 1861, MS, Lowe Papers.

[115] Since it was possible to inflate the balloons used around the Washington defenses at the gas works in the city, the generators were assigned to distant

Your balloon is wanted, and it is of the highest importance that it should be here to take advantage of the first calm. . . . I will send in men now if you will send it.[116]

When high wind prevented the safe towing of the inflated envelope out to Hall's Hill the next day, the General again wired Lowe that he was "very desirous" of having the balloon in service as early as possible.[117]

The aerostat originally assigned to Porter's division was the *Intrepid*. On December 14, it was transferred to the upper Potomac station, and the *United States* was placed in service with Porter.[118] The latter balloon was operated from positions near the Capital until the latter part of February, when it was taken to Heintzelman's division near Fort Lyon.[119] Early in March, during preparations for the Richmond campaign, Porter requisitioned the balloon recalled from the upper Potomac, and ordered ascensions from Fairfax Court House, which had recently been evacuated by the Confederates.[120]

Through Lowe's initiative and desire to make the benefits of his observations available to as many units of the army as possible, the balloon service was extended to include operations with the division of General S. P. Heintzelman. This unit occupied the area south and southwest of Alexandria, as far as Accotink and Pohick Church. Below its outpost lines lay Occo-

stations first, as it was not possible to obtain gas for their operation except by the use of field generators. Thus in November and early December, the balloon with Porter's division was still inflated in Washington and towed by hand to the lines.

[116] Porter to Lowe, November 30, 1861, MS, Lowe Papers; also printed in 3 *O. R.*, III, 267.

[117] Porter to Lowe, December 2, 1861, MS, Lowe Papers.

[118] Lowe to Colburn, December 3, 1861, 3 *O. R.*, III, 268; Stone to McClellan, December 14, 1861, MS 7076, McClellan Papers, 1st Ser., Vol. XXXIII; Seaver to Lowe, December 17, 1861, MS, Lowe Papers; Lowe to Macomb, January 16, 1861, MS, *ibid.*

[119] Diary of General S. P. Heintzelman, entry of February 27, 1862, MS, Papers of Samuel Peter Heintzelman, Division of MSS, Library of Congress [cited hereinafter as Heintzelman Papers].

[120] Porter to Lowe, March 8, 1862, MS, Lowe Papers; also printed in 3 *O. R.*, III, 271; Porter to Williams, March 9, 1862, MS 9304, McClellan Papers, 1st Ser., Vol. XLIV; Colonel Fred T. Locke [AAG, Porter's Division] to Lowe, March 9, 1862, MS, Lowe Papers.

quan, with its encampments of Confederates and their field works and batteries on the south bank of Occoquan Creek. The Confederate forces in this area had been under observation from the balloon with Hooker's division farther down the river on the Maryland side, and though a fair understanding of their positions had been available from the Budd's Ferry station, it was apparent that frontal observation from Heintzelman's outposts would prove more effective. Lowe consequently resolved to carry his operations into this area.

On February 21 he called on General Heintzelman at his headquarters at Fort Lyon and suggested that ascensions be made from the outer lines of his division. Heintzelman received the proposal with interest, offered all necessary assistance and facilities, and recommended that the balloon be operated from Pohick Church.[121] Within a week arrangements for transportation, the assignment of an outpost detachment to guard the balloon and its crew, and other details were completed: the sixth balloon train to be sent into the field arrived at Pohick Church on February 27.[122] The following day Lowe completed inflation and began a series of ascensions that were continued with regularity for the next two weeks. The balloon and its equipment were then recalled for the general movement to the Peninsula.

During the brief period at Pohick Church, Lowe performed effective and satisfactory service. Assisted by Mason and Sergeant Charles F. Eaton, a soldier permanently assigned to balloon duty because of his proficiency, he maintained a vigilant watch over the forces opposed to Heintzelman's command, and also reported information of value regarding the main Confederate army at Manassas, which lay some eighteen miles west of his position. Heintzelman and his subordinate officers were duly impressed with the advantages of aerial observation and

[121] Diary of General Heintzelman, entry of February 21, 1862, MS, Heintzelman Papers.

[122] Lowe to Lieutenant C. B. Ferguson [AQM], February 26, 1862, MS, Lowe Papers; Ferguson to Captain Isaac Moses [AAG, Heintzelman's Division], February 27, 1862, MS, Heintzelman Papers; Heintzelman's Diary, entries of February 27 and 28, 1862, MS, *ibid.*; Emil Moons [Adj., 2nd Michigan Inf.] to Lowe, March 1, 1862, MS, Lowe Papers.

displayed interest in Lowe's work. The division adjutant general, Captain Isaac Moses, made frequent ascensions, and Heintzelman likewise often went aloft, sometimes several times in one day.[123] On one occasion the General declared that as soon as his troops should occupy the north bank of the Occoquan, he expected to ascertain from the balloon the position and strength of all forces opposite that might oppose his crossing of the stream.[124] Colonel Hiram G. Berry, of the 4th Maine, also ascended on several occasions.[125]

The information obtained was similar in nature to that reported by the other stations, and the observers' chief function was again to keep the hostile forces under careful surveillance. Offensive movements, however, were not expected, and the aeronauts and the officers who ascended with them directed their attention to the discovery of any withdrawals that might take place. In this mission they were successful on two occasions of importance. On March 5 Lowe reported fires along the Occoquan and at Fairfax Station. Heavy smokes were visible across the stream extending from Wolf Run Shoals to the Potomac. The hostile force present appeared inferior in strength to that of Hooker, across the Potomac on the Maryland shore.[126] The following day Heintzelman made three ascensions and reported similar information. Smokes and encampments were still clearly visible along the Occoquan, as far back as Bacon Race Church.[127] Lowe, ascending later in the evening, obtained more significant information. The smokes at Fairfax Station had largely increased, and there were heavy clouds visible at Dumfries,

---

[123] Mason to Lowe, March 4, 1862, MS, Lowe Papers; Eaton to Lowe, March 11, 1862, MS, *ibid.*; Moses to Lowe, March 5, 1862, MS, *ibid.*; Heintzelman's Diary, entries of February 28, March 4, 5, and 6, MS, Heintzelman Papers; Moses to Heintzelman, March 3, 1862, MS, *ibid.*; Moses to Heintzelman, March 9 [two telegrams], MSS, *ibid.*; Lowe to Brigadier General R. B. Marcy [C of S, AP], March 6, 1862, MS, Lowe Papers,; also printed in 3 *O. R.*, III, 271; *Lowe's Report, ibid.*

[124] Heintzelman's Diary, entry of March 5, 1862, MS, Heintzelman Papers.

[125] *Lowe's Report*, 3 *O. R.*, III, 271; Berry to Lowe, March 10, 1862, MS, Lowe Papers.

[126] Lowe to Moses, March 5, 1862, MS, Heintzelman Papers; Lowe to Marcy, March 6, 1862, MS, Lowe Papers; see also 3 *O. R.*, III, 270.

[127] Heintzelman's Diary, entry of March 6, 1862, MS, Heintzelman Papers.

Brentsville, and Manassas.[128] Diminishing visible strength, coupled with increased volumes of smoke, as though from burning abandoned property, indicated preparations for withdrawal. On the 7th of March Colonel Berry reported from the balloon that the camps and bivouacs along the Occoquan had disappeared and the area appeared to be deserted.[129] Subsequent ascensions made by Eaton corroborated Berry's conclusions.[130] Berry and Eaton were right. By the 9th the Confederates had effected a general evacuation of their positions and had abandoned their batteries on the Potomac and Occoquan. The Union flag was planted on the works at Cockpit Point, from which the Confederate batteries had recently commanded the river.[131]

In the meanwhile rumors and suspicions of the impending Confederate evacuation of Manassas and Centreville had reached the Federal high command. The heavy smoke at Manassas, reported on the 6th, strengthened these suspicions. As a result, Lowe had been ordered to prepare another balloon for use at Fairfax Court House to watch for the expected evacuation, and was absent from Pohick Church when the movement occurred. Delays in transportation prevented his getting the additional balloon aloft in time to note the movement,[132] but the observers at Pohick reported on March 10 that huge fires were visible in the vicinity of Manassas, indicating that stores and other property were being destroyed and that evacuation had taken place.[133] Contraband negroes entered Berry's camp next day and announced the withdrawal and the destruction of large quantities of munitions and supplies, as well as the bridge over Bull Run. This information, Berry notified Moses, " confirms the observations made last evening from the balloon,

[128] Lowe to Marcy, March 6, 1862, MS, Lowe Papers; also printed in 3 O. R., III, 271.

[129] *Lowe's Report*, 3 O. R., III, 271; Baltimore *Republican*, March 14, 1862.

[130] Eaton to Lowe, March 11, 1862, MS, Lowe Papers; Eaton to Heintzelman, March 11, 1862, MS, Heintzelman Papers.

[131] Heintzelman to Lieutenant S. Hunt [ADC], March 9, 1862, MS, Heintzelman Papers.

[132] *Lowe's Report*, 3 O. R., III, 271.

[133] Eaton to Heintzelman, March 10, 1862, MS, Lowe Papers.

which discovered numbers of fires in that direction." [134] The observations taken and the conclusions drawn from them were entirely correct. Stuart's cavalry, left as a rear guard after the main body had left, had fired the storehouses and burnable remaining property on March 10, and then withdrew to Warrenton Junction, leaving Manassas entirely abandoned. [135] Three days later, the balloon at Pohick Church was packed and preparations were made to move back to Washington. [136]

While arrangements for the movement of the Army of the Potomac to the Peninsula were in progress, McClellan ordered a balloon and aeronaut sent to General Wool at Fort Monroe. [137] The mission of this unit, the last assigned before the opening of the Peninsular Campaign, was to afford means of ascertaining the movements of the dreaded Confederate ironclad *Virginia,* which had recently wrought havoc among the Union warships at Hampton Roads. Also, the aeronaut was expected to take observations and secure other information of value " not otherwise obtainable," particularly with respect to the region into which the Grand Army's advance was to be pushed towards Richmond. [138] Seaver, still attached to Hooker's division, though comparatively inactive since the Confederate withdrawal from the lower Potomac, was selected for this mission. Having received his orders and credentials, he embarked with the *Constitution* from Budd's Ferry on March 15, and arrived the following day at Old Point, where he reported for duty at Wool's headquarters. [139]

After much difficulty in securing men and teams with which to move his apparatus, Seaver took up a position in the vicinity of Newport News, and began operations where La Mountain

[134] Berry to Moses, March 12, 1862, MS, Heintzelman Papers.

[135] Johnston, pp. 103-104.

[136] Mason to Moses, March 13, 1862, MS, Heintzelman Papers.

[137] Macomb to Lowe, March 12, 1862, MS, Lowe Papers; also printed in 3 *O. R.,* III, 271; Macomb to Lowe, March 13, 1862, MS, Lowe Papers.

[138] Macomb to Wool, March 13, 1862, MS, Lowe Papers.

[139] Lowe to Seaver, March 13, 1862, MS, Lowe Papers; Lowe to Wool, March 13, 1862, MS, *ibid.;* also printed in 3 *O. R.,* III, 272; Lowe to Macomb, March 15, 1862, MS, Lowe Papers; also printed in 3 *O. R.,* III, 272; New York *Tribune,* March 17 and 18, 1862; Boston *Transcript,* March 17, 1862.

had formerly made ascensions under Wool's predecessor, Butler.[140] Several ascensions from this vicinity revealed all quiet in the direction up the Peninsula, and no trace of the *Virginia* appeared on the waters visible at the time.[141] A short while later, Lowe received orders to embark with his entire balloon equipment and proceed to Fort Monroe, where he was to await further orders.[142] The arrival of the first divisions of the Army of the Potomac, and Lowe with the remainder of his corps, cut short Seaver's separate operations. The Peninsular Campaign was about to begin, and the balloon service was soon to function not on fixed lines of comparative inactivity, but on the battlefields of its first campaign.

In general, the service performed by the balloon corps on the Potomac lines during the fall and winter of 1861-1862 was all that could be expected under the prevailing conditions and circumstances. Despite poor administrative organization as a result of its hybrid military-civilian status, and despite the rigors of winter weather, the aeronauts had accomplished their mission of vigilance and watchful observation from one end of the Potomac lines to the other. The information reported, though not of vital importance, had been accurate and reliable, and had provided the various commanders with a knowledge of the strength and position of the hostile forces confronting them that they would not have otherwise been able to obtain. Lowe had succeeded in organizing and maintaining continual and simultaneous observation at several stations of tactical importance. He had supervised the work of his assistants as well and as efficiently as circumstances permitted. And in every case, the general officers under whom the stations were assigned, were interested and satisfied with the results obtained. They had learned to regard the balloons as a valuable part of their intel-

[140] Baltimore *Republican*, April 1, 1862; Seaver to Lowe, March 17, 1862, MS, Lowe Papers. " ' Red Tape,' ' Red Tape,' ' Red Tape,' I am surfeited with ' Red Tape ' since my arrival here yesterday p. m. at 3 o'clock," Seaver complained at the delays in securing assistance from Wool.

[141] Seaver to Lowe, March 19, 1862, MS, Lowe Papers; Seaver to Lowe, March 22, 1862, MS, Lowe Papers.

[142] Williams to Lowe, March 22, 1862, 3 *O. R.*, III, 272-273; Williams to Lowe, March 23, 1862, *ibid.*, p. 273.

ligence service. The expansion of the service from one balloon to a corps, loosely organized as it was, had proved justified and successful. The support of Porter and McClellan had been fully rewarded.

Another aspect of the success attained was present in the reaction that the balloons produced at the various Confederate headquarters. The continual presence of the Union aerostats caused apprehension and misgivings more than once in the minds of the Southern commanders. In November Beauregard had written to General Whiting that he believed the numerous balloon ascensions to be indicative of impending offensive operations on the part of the Federal army.[143] At Leesburg, General D. H. Hill commented on the fact that four Yankee balloons were constantly hovering over the lines.[144] And a deserter from his brigade declared that General Stone's balloon at Edwards Ferry had been regarded as a serious menace. " Its occupants, the man asserted, must have had a full view of [Fort Evans] and the men there at work." [145] At Centreville Longstreet directed his brigadiers to conceal carefully the dummy guns mounted in the outer works in order to prevent the enemy balloonists from discovering their real nature.[146] Thus the aeronautic operations not only provided material advantage for the forces with which the corps functioned, but also served to affect the morale of the enemy. The balloon service had proved itself on the inactive lines. Now it was about to receive the test of the battlefield.

---

[143] Beauregard to Whiting, November 13, 1861, 1 *O. R.,* V, 950.

[144] Hill to Beauregard, December 14, 1861, Catalogue of MSS offered by Stan V. Henkels, Philadelphia, October, 1915, Beauregard Papers.

[145] Testimony of William H. Demorey, before E. J. Allen [Allan Pinkerton], in Allen's Report of February 19, 1862, MS 8683, McClellan Papers, 1st Ser., Vol. XLI.

[146] General James Longstreet to Brigadier General Philip St. George Cooke, December 5, 1861, MS, courtesy of Mr. Forrest H. Sweet, Battle Creek, Michigan.

CHAPTER XI

OPERATIONS IN THE DEPARTMENT OF THE SOUTH AND
THE WESTERN DEPARTMENT, 1862

The need of cooperation with land forces to maintain an effective blockade of the south Atlantic coast was early recognized by the Federal high command. One measure adopted was the dispatch of an expeditionary force under Brigadier General Thomas W. Sherman in the fall of 1861 to gain a foothold near Port Royal, South Carolina. On November 7, after a bombardment of Forts Walker and Beauregard at the mouth of Port Royal Entrance, Sherman seized these forts and the adjacent ground on Hilton Head and Phillips Islands commanding the inlet to Port Royal and Beaufort. From this base he and succeeding commanders carried on offensive operations to threaten Savannah and other adjacent points of strategic importance.[1]

The possible value of balloon observation in the operations of Sherman's expeditionary force seems to have occurred to General McClellan shortly after the command sailed for South Carolina. At the time, Lowe was engaged in building the additional balloons authorized in September, and a week after Sherman's troops had landed on the coast, the General-in-Chief instructed Lowe to reserve the first balloon completed for assignment to the Port Royal Expedition.[2] McClellan evidently considered the matter of special importance and sent the Chief Aeronaut a second message on the same day, repeating his instructions.[3] On November 22 Lowe received further orders to send the balloon at once to Fort Monroe whence it would be

---

[1] The details of the Port Royal Expedition appear in official reports, orders, and correspondence published in 1 O. R., VI. A good summary of the mission and objectives of the expedition, with a detailed narrative of Sherman's early operations can be found in Rear Admiral Daniel Ammen, "Du Pont and the Port Royal Expedition," Battles and Leaders of the Civil War (New York, 1884-1887), I, 671-691.

[2] Colonel A. V. Colburn [AAG, AP] to Lowe, November 16, 1861, 3 O. R., III, 266.

[3] Colburn to Lowe, November 16, 1861 [second dispatch], ibid.

forwarded to Port Royal,[4] and arrangements were made with the Navy Department to release the steam tug *Coeur de Lion* to carry the apparatus to Old Point.[5] Rumors of the intended aeronautic operations in the Southern department appeared in the press almost at once, and several correspondents announced that Lowe himself would go to Port Royal for duty.[6] The exacting duties of organizing and directing the work of the newly expanded balloon service required Lowe's presence on the Potomac lines, and the assignment was given to John B. Starkweather, the Boston aeronaut who had recently joined the corps.[7]

Starkweather sailed from Washington on November 24, and arrived three days later at Old Point,[8] where he disembarked with his apparatus and waited for further transportation to Port Royal.[9] His equipment consisted of the balloon *Washington*,[10] two generators without running gear, tentage, tools, and a supply of materials for generating gas.[11] He was also provided with a letter of introduction to General Sherman, who was requested to furnish the aeronaut with troops for a ground crew, wagons for the generators, and any other assistance that might be required.[12] Arrangements were also made to place the

---

[4] Colburn to Lowe, November 22, 1861, 3 *O. R.*, III, 267.

[5] Indorsement of Assistant Secretary of the Navy G. V. Fox, November 22, 1861, *ibid.*; Philadelphia *Inquirer*, November 25, 1861; Boston *Journal*, November 26, 1861.

[6] New York *Times*, November 22, 1861; New York *World*, November 26, 1861; Boston *Journal*, November 27, 1861; Philadelphia *Inquirer*, November 28, 1861; Detroit *Free Press*, November 28, 1861; New York *Herald*, November 30, 1861.

[7] Lowe to Sherman, November 27, 1861, MS, Lowe Papers; also printed in 3 *O. R.*, III, 267; Boston *Journal*, November 30, 1861; Philadelphia *Inquirer*, December 2, 1861; Baltimore *American*, November 30, 1861.

[8] Baltimore *American*, November 30, 1861.

[9] Philadelphia *Inquirer*, December 2, 1861.

[10] The Boston *Journal* [November 26] and the Philadelphia *Inquirer* [November 26] erroneously reported that the balloon *Constitution* had been assigned for service with Sherman. A similar erroneous announcement appeared in the Providence *Post*, December 5, 1861.

[11] Lowe to Macomb, January 17, 1861, MS M106, enc. 2, Letters Received, Army of the Potomac, War Department Division, National Archives [cited hereinafter as LRAP]. Also MS [copy], Lowe Papers.

[12] Lowe to Sherman, November 27, 1861, 3 *O. R.*, III, 267; MS, Misc. File, "Balloon Corps," dated January 9, 1861, War Department Division, National Archives.

aeronaut in proper relations with Sherman's staff and quarter-master in order that administrative details, such as pay, supplies, and subsistence might be effected without difficulty.[13]

Although Starkweather left Washington within two days after Lowe received orders to send the balloon to Sherman, continual delays occurred before the aeronaut was able to secure passage from Fort Monroe to Port Royal. He was promised transportation on several steamers that successively cleared Old Point, but in every case these vessels sailed for other destinations. Despite his annoyance and disappointment at the repeated delays, Starkweather made good use of the time by giving the *Washington* and its attending generators a complete overhauling, in order that the equipment might be ready for immediate service when he should eventually arrive at Sherman's headquarters.[14] Finally after waiting for more than five weeks, he was given passage on the brig *Empire,* which sailed from Old Point on December 28. Starkweather's departure for balloon duty with the Port Royal expedition aroused widespread comment, and notices of the event appeared throughout the Northern press.[15] He arrived at Hilton Head on January 3, and immediately began preparations for active service.[16]

In contrast to the success and official recognition that attended the early period of Lowe's services on the Potomac lines, Stark-

[13] Macomb to Williams, December 16, 1861, MS M526, LRAP; McClellan's indorsement, December 17, *ibid.;* also copy, MS Indorsement Book, Records of the Army of the Potomac, Vol. XXIII, p. 143, War Department Division, National Archives; Brigadier General Lorenzo Thomas [AG, USA] to Sherman, December 30, 1861, MS. Misc. File, "Balloon Corps."

[14] Philadelphia *Inquirer,* December 24, 1861; Lowe to Starkweather, November 28, 1861, MS, Lowe Papers; Lowe to Starkweather, December 1, 1861, MS, *ibid.;* Starkweather to Lowe, December 16, 1861, MS, *ibid.;* Starkweather to Lowe, December 17, 1861, MS, *ibid.;* Starkweather to Lowe, December 19, 1861, MS, *ibid.*

[15] Boston *Transcript,* December 30, 1861; Boston *Journal,* December 30, 1861; Hartford *Daily Courant,* December 31, 1861; New York *Evening Post,* December 31, 1861; New York *Tribune,* December 31, 1861; New York *World,* December 31, 1861; Philadelphia *Inquirer,* December 31, 1861; Philadelphia *Ledger,* December 31, 1861; Philadelphia *Daily News,* December 31, 1861; Philadelphia *Press,* December 31, 1861; Lancaster *Daily Evening Express,* December 30, 1861; Baltimore *American,* December 31, 1861; Detroit *Free Press,* December 31, 1861; New York *Herald,* December 31, 1861.

[16] Starkweather to Lowe, January 3, 1862, MS, Lowe Papers.

weather's operations with the Department of the South were disappointing. A number of factors combined to render his services ineffective, though none of them reflect discredit on Starkweather as an aeronaut, or on the general value of balloon service. The absence of success in this theatre of operations, in contrast to Lowe's accomplishments in the Potomac region, serves to demonstrate some of the difficulties confronting the new branch of the service and, at the same time, some of its admitted limitations.

The first major problem faced by Starkweather upon his arrival at Hilton Head was the complete indifference with which his proposed services were received by General Sherman. The expedition commander was simply not interested in balloon observation. There is no evidence that he had asked for the services of an aeronaut prior to Starkweather's arrival, and possibly he felt that the balloonist had simply been wished on his command. Regardless of reasons, Sherman maintained a frigid indifference towards Starkweather and his suggestions for observations. Immediately after his arrival, Starkweather put his apparatus in order and was ready to inflate the *Washington* and conduct ascensions. He was instructed to wait for orders.[17] Months went by without any further orders being issued. Starkweather remained idle and totally useless to the forces to which he was assigned. On March 10 he explained his position in reply to a letter from Lowe, in which the Chief Aeronaut had expressed displeasure at his inactivity:

You say you are sorry that I have not had a chance to operate with the balloon. You are not more sorry than I am, but if I could have a chance, it would be better than doing nothing here. . . . I have not had any orders from the General. . . . The balloon is all in good order and ready for use at any moment, if I should be called upon. You say that you are anxious that every member of the Aeronautic Department should render all the valuable service possible. I am anxious to do the same. . . . All I want is the chance.[18]

Still another month passed, and Starkweather reported the same

[17] Starkweather to Lowe, January 7, 1862, MS, Lowe Papers.
[18] Starkweather to Lowe, March 10, 1862, MS, Lowe Papers.

state of affairs. " I am still at this place [Hilton Head] with the balloon you gave me in charge, waiting orders to use and work the balloon for the Army," he wrote early in April.[19] Three months to a day had elapsed, and the officer commanding the forces to which the balloonist was attached had not seen fit to order a single ascension. Very naturally no results could be accomplished under these circumstances.

At the end of March, the Department of the South was officially created, comprising the states of South Carolina, Georgia, and Florida, under the command of Major General David Hunter.[20] Within the Department two districts were created, of which the Northern, with headquarters at Port Royal, was assigned to the command of Brigadier General Henry W. Benham, who superseded Sherman.[21] Starkweather notified Lowe of the change of command with little enthusiasm. " I understand General Benham is to take General Sherman's place," he wrote, " but I suppose I shall have to wait orders from the generals now in command." [22] At first it appeared that the new commander would effect no change of policy with respect to balloon observation, and a week after his arrival, Starkweather reported that

I have been trying to see General Benham ever since he has been here, but have been unable to see him yet.[23]

Benham, however, was at the moment busily engaged in preparations for the bombardment of Fort Pulaski and could not spare attention for others matters.[24] The fort fell on April 11,[25] and three days later Starkweather secured an interview with the General, who expressed interest in the balloon and requested

[19] Starkweather to Lowe, April 3, 1862, MS, Lowe Papers.
[20] General Orders, No. 1, HDQ Department of the South, March 31, 1862, 1 O. R., VI, 257-258.
[21] Ibid., p. 258; General Orders No. 2, HDQ Department of the South, 1 O. R., VI, 258; Benham to Brigadier General E. L. Viele, April 1, 1862, ibid., p. 259.
[22] Starkweather to Lowe, April 3, 1862, MS, Lowe Papers.
[23] Starkweather to Lowe, April 10, 1862, MS, Lowe Papers.
[24] See Benham's Report, Bombardment and Capture of Fort Pulaski, Ga., 1 O. R., VI, 136-139.
[25] Terms of Capitulation, 1 O. R., VI, 139-140.

that it be taken to the captured fort for observations of Savannah and its river approaches.[26] Thus after three months of idleness Starkweather was finally permitted to perform the service for which he had been assigned.

Another serious obstacle in the way of successful balloon observation during Starkweather's service with Sherman and Benham was the constant interference by the winds that prevailed in the theatre of operations. The low, flat islands and adjacent mainland of the coast were continually swept with stiff ocean breezes and off-shore winds. In brief, the geographic location and physical features of the region were unsuited to dependable operations or continuous ascensions such as were being carried on along the Potomac.

In his first interview with Benham, the aeronaut had pointed out this difficulty, and advised the general that " the balloon could not be used to any advantage on this island [Hilton Head] as it was too near the seacoast and the wind has been blowing [both] from the sea and off shore." [27] He later wrote Lowe after Benham had ordered him to make ascensions from Fort Pulaski, that he would be glad when his operations would be made on the mainland, since the inland regions were better suited to successful ascensions.[28] When he had first arrived to report for duty, Starkweather had inflated the balloon in expectation of orders from Sherman, but a severe gale blowing in from the sea had compelled him to discharge most of his hydrogen three days after the envelope had been filled.[29] This loss of gas fully confirmed his report to Lowe a week before that " this place is too near the shore and is a very bad place for a balloon, as it blows off shore all the time." [30] During the bombardment of Fort Pulaski, when aerial observation of the effect of the fire would have been valuable, the balloonist again recorded that the wind was too high and " a balloon could not have been used to advantage." [31] And finally, severe damage to the *Washington*

[26] Starkweather to Lowe, April 15, 1862, MS, Lowe Papers.
[27] *Ibid.*
[28] Starkweather to Lowe, April 19, 1862, MS, Lowe Papers.
[29] Starkweather to Lowe, January 18, 1862, MS, Lowe Papers.
[30] Starkweather to Lowe, January 13, 1862, MS, Lowe Papers.
[31] Starkweather to Lowe, April 12, 1862, MS, Lowe Papers.

resulting from a heavy wind was to terminate the aeronautic operations in the Department.

The assumption may be made that this obstacle of weather had been apparent to Sherman, and thus caused him to issue no orders for ascensions. Although such a hypothesis may contain some element of truth, it is not fully convincing. In the period of Starkweather's inactivity, there must have been many days when the atmosphere was sufficiently calm to permit successful ascension. The likelihood of continuous wind prevailing every day for three months is too remote for serious consideration. Also, the numerous reconnoitering operations made by the subordinate units of Sherman's force were frequently pushed into the rivers and inlets that extended in from the coast,[32] and ascensions with the balloon would have been feasible and of advantage. Then too, since Sherman was totally unfamiliar with the technique of ballooning, the question of whether an ascension could or could not be made should have been a matter for the aeronaut to decide. By and large, the period of idleness imposed on Starkweather from January to April, 1862, appears to have been caused by indifference rather than weather conditions.

A final difficulty that detracted from Starkweather's success lay in the problem of obtaining supplies of iron turnings and sulphuric acid for the manufacture of gas. He had carried in his initial supply eighty carboys of acid and thirteen barrels of iron, or enough of the latter for only two inflations.[33] Since the gas generated for the first inflation had to be released because of a gale, he was immediately left with only enough iron for one replenishing. Communications between Port Royal and sources of supply were slow, since requisitions for additional materials had to go through the chain of command to Lowe and thence on for approval by the Chief Aeronaut's superiors.[33a] Almost a

[32] Cf. Reports, Correspondence, &c., January 7–March 31, 1862, 1 *O. R.*, VI, 217-256, *passim*.

[33] Starkweather to Lowe, January 7, 1862, MS, Lowe Papers.

[33a] When the iron turnings were obtained from gun factories in the various navy yards, the channels of communication through which requisitions of the Balloon Corps had to pass included those of the several Navy Department bureaus as well as the normal army channels. Cf. Macomb to Commodore Joseph Smith, Chief of the Bureau of Yards and Docks, USN, February 1, 1862,

month elapsed before an additional supply of twenty barrels of iron reached Starkweather.[34] On one occasion when the stores of these materials were low, lack of acid prevented the making of important ascensions during a series of skirmishes on James Island. Starkweather had been ordered to move his balloon from Fort Pulaski to join General Stevens' forces on James Island. He asked permission to tow the aerostat inflated, but was ordered to discharge the gas in order to move the balloon without delay. Shortage of acid prevented replenishing the envelope sufficiently to make effective ascensions after he reached Stevens.[35] To add to the difficulties of supply, a shipment of acid and iron sent on by Lowe a short time later was lost through shipwreck on the way to Hilton Head.[36]

Despite the combination of unfavorable factors that prevented Starkweather from attaining the degree of success enjoyed by his fellow aeronauts in Virginia, he was nevertheless able to conduct a number of ascensions and secure valuable information. On April 19 General Benham issued instructions for the balloon to be taken to Fort Pulaski for observations of Savannah and the river.[37] A ground crew was detailed from the 7th Connecticut Infantry, of which six men were assigned to Starkweather for the duration of his duty in the Department. The balloon was inflated and placed on board the steamer *Mayflower*, which steamed up the Savannah River to within three miles of the city. A number of observations of the city and its surrounding approaches were taken, and the number of Confederate gunboats and armed river steamers was noted. An engineer officer, Lieutenant P. H. O'Rorke, accompanied Starkweather in the ascension.[38]

requesting a supply of iron turnings from the Brooklyn Navy Yard for the use of Starkweather at Port Royal, MS 15, Misc. Letters, Navy Department, February, 1862, Vol. I, Naval Records and Library, Washington, D. C.

[34] Lowe to Colonel Daniel H. Rucker [CAQM], January 27, 1862, MS, Lowe Papers.

[35] Starkweather to Lowe, June [n. d.], 1862, MS, Lowe Papers.

[36] Starkweather to Lowe, June 5, 1862, MS, Lowe Papers.

[37] Starkweather to Lowe, April 19, 1862, MS, Lowe Papers; Baltimore *Republican*, May 4, 1862.

[38] O'Rorke to Brigadier General Quincy A. Gillmore, May 7, 1862, MS, Misc. File, Balloon Corps, War Department Division, National Archives; Stark-

Other observations revealed the burning of large quantities of cotton along the docks of Savannah, and the presence of strong earthworks southeast of the city. An ironclad gunboat under construction was also observed and reported,[39] and detailed inspection of the shores of the Savannah River revealed the absence of batteries mounted to command the channel.[40] An ascension on May 20 disclosed the presence of two Confederate camps, one on each side of the Savannah River. The one on the south bank was identified near Four Mile Point by Lieutenant O'Rorke, who estimated that it contained four or five companies.[41] The balloon was also used on the Stono River, for observations of the vicinity of Charleston,[42] and Brigadier General Alfred H. Terry, in command of the Union garrison at Fort Pulaski, recommended its use in a sham movement against Savannah in order to create an alarm in this area while a real attack was to be pushed against Charleston.[43] On James Island, during General Stevens' occupation prior to the action at Secessionville, observations from the air permitted the Union commander to extend his lines with security.[44] In general, after Benham superseded Sherman, Starkweather was able to perform creditable service, even though his observations were not frequent or productive of highly significant information. In one of his letters he stated that he had had " the balloon inflated for over a month, and in use most of the time." [45] On other occasions he reported that the commanding officer and his subordinates were well pleased with the balloon and its service.[46]

weather to Lowe, May 9, 1862, MS, Lowe Papers. In the last cited letter, Starkweather enclosed a fragment of the Confederate colors that had been flown from the staff over Fort Pulaski. This scrap is still preserved in the Lowe Papers.

[39] Starkweather to Lowe, September 14, 1862, MS, Lowe Papers.

[40] Starkweather to Lowe, June 5, 1862, MS, Lowe Papers.

[41] Terry to Benham, May 20, 1862, 1 O. R., XIV, 343.

[42] Starkweather to Lowe, June 5, 1862, MS, Lowe Papers.

[43] Terry to Benham, May 20, 1862, 1 O. R., XIV, 343-344; Starkweather to Lowe, June 25, 1862, MS, Lowe Papers.

[44] William Todd [Co. " B," 79th N. Y. Inf.], The Seventy-Ninth Highlanders, New York Volunteers, In the War of The Rebellion (Albany, 1886), p. 145.

[45] Starkweather to Lowe, June 5, 1862, MS, Lowe Papers.

[46] Starkweather to Lowe, May 9, 1862, MS, Lowe Papers; Starkweather to

Other witnesses, however, have asserted that the aeronautic operations were of very little value. One officer recorded that the balloon " was not made available to any very useful extent in the Department," [47] and another declared that although the balloon was intended to be a great factor in the operations of Benham's force, it was found to be impracticable.[48] Comments in the Confederate newspapers in the vicinity of its operations, however, indicated that the balloon was taken seriously by the Southerners.[49]

In June Starkweather was ordered to take the *Washington* from Fort Pulaski to James Island for observations. The balloon was accordingly taken in tow by the steamer *Mayflower,* which proceeded up the coast to the position ordered. The movement was attempted during a heavy wind which developed into a gale. The *Mayflower* ran aground and the balloon, moored to her deck, got out of control and was dashed against her super-structure. A large hole was torn in the envelope, which could not be repaired for want of materials.[50] With this unfortunate maneuver, Starkweather's activities with the Department of the South came to an end. He packed up the damaged aerostat and other equipment and took passage for New York, where he turned the equipment over to the Assistant Quartermaster General's Office. The apparatus was then sent on to the Smith-sonian Institution to await reissue to Lowe's corps.[51]

Starkweather's six months of service, admittedly unsuccessful, reveal one or two points of importance. From his experience it is obvious that the attitude of the general officer in command became an essential factor in the success of aeronautic opera-tions. Without the support and full cooperation of the com-

Lowe, June [n. d.], 1862, MS, *ibid.*; Starkweather to Lowe, September 14, 1862, MS, *ibid.*

[47] Captain Isaiah Price, *History of the 97th Regiment, Pennsylvania Volun-teer Infantry, During the War of the Rebellion* (Philadelphia, 1875), p. 114.

[48] Captain Daniel Eldridge, *The Third New Hampshire* (Boston, 1893), p. 99.

[49] Savannah *Daily News,* May 9, 1862; *ibid.,* May 29, 1862; Savannah *Re-publican,* May 9, 1862; Savannah *Southern Confederacy,* May 11, 1862.

[50] Starkweather to Lowe, June [n. d.], 1862, MS, Lowe Papers; Starkweather to Lowe, September 14, 1862, MS, *ibid.*

[51] Lowe to Colonel Daniel D. Tompkins [AQMG, NY], July 22, 1862, MS, Lowe Papers; Tompkins to Lowe, July 23, 1862, MS, *ibid.*

mander, little or nothing could be accomplished. The success that attended Lowe's first months of service could not have been possible had he received the indifference that Starkweather experienced at the hands of Sherman. Also, the Port Royal operations demonstrated that geographic location could impose definite limitations on the effectiveness of the balloon service. Because of its physical features and locality, the region of the South Carolina and Georgia coast was not adapted to successful aerial observation. Finally, a third lesson was apparent. In operations at long distance from bases of supply, with slow communications, better planning for the supply of materials became essential to effective operation. This much might have been learned from the period of service performed in the Department of the South.

*　　*　　*

While Starkweather was idly waiting for instructions from Sherman, General McClellan again decided to extend the operations of the balloon service to still another theater of the war. On February 4 he ordered Colonel Macomb to send a balloon train and aeronaut to Cairo, Illinois, for service with the Army of the Mississippi.[52] Major General John Pope, commanding this army, was at the time preparing to move on New Madrid, a strategic point on the river above the hairpin bend at the Kentucky-Tennessee boundary. The apex of this bend was commanded by the formidable Confederate stronghold on Island No. 10, the capture of which was an ultimate objective in Pope's plan of campaign.

Macomb referred McClellan's order to Lowe, with instructions to assign one of his assistant aeronauts and appropriate equipment for service in the West.[53] Accordingly Lowe selected his experienced German assistant, John H. Steiner, who had been on duty for several months with General Stone's Corps of Observation on the upper Potomac.[54] The balloon *Eagle* was

[52] Lieutenant Colonel N. B. Sweitzer [ADC] to Macomb, February 4, 1862, MS M192, LRAP.

[53] Macomb to Lowe, February 5, 1862, MS, Lowe Papers.

[54] Lowe to Macomb, February 13, 1862, MS M289, enc. 1, LRAP; Macomb to Steiner, February 17, 1862, MS M251, LRAP.

designated to accompany Steiner's train.[55] By February 13, all
the appendages, generator, cordage, and auxiliary equipment
had been packed for shipment and turned over to the Quartermaster Department for delivery.[56] The Chief of Ordnance also
arranged for the delivery of an initial supply of iron and acid,[57]
and Macomb requested that the Adjutant General issue appropriate orders that would place the aeronaut " in a proper relation to some staff officer at or near Cairo, to whom he shall look
for assistance in men and means . . . for managing the balloon
and apparatus . . . and also for his daily pay." [58] By the 21st,
all details for Steiner's replacement at his station with Stone
were completed, and Lowe ordered him to proceed at once to
Cairo.[59] Taking the *Eagle's* envelope with him in his personal
baggage to insure proper care and handling, Steiner left for the
West on the following day.[60] He arrived at Cairo three days
later, and reported for duty to Brigadier General George W.
Cullum, chief of staff of the Department of the Missouri, the
geographical organization under which Pope's army was
operating.[61]

The history of Steiner's service with the western command is
simply a repetition of the same indifference of commanding
officers experienced by Starkweather in the South, with the
added difficulty of much worse administered supplies and pay.
These obstacles were even aggravated by ridicule and contempt
from the commanders to whom the aeronaut had been accred-
ited by the eastern authorities. In the five months that he spent
in the western department, Steiner was permitted to conduct

[55] Abstract Log, Mortar Division, Western Flotilla, entry of March 25, 1862,
Official Records of the Union and Confederate Navies (Washington [Govern-
ment Printing Office], 1894-1908), Series 1, Vol. XXII, 771 [cited hereinafter
as *O. R. N.*].

[56] Lowe to Macomb, February 13, 1862, MS M289, enc. 1, LRAP.

[57] Macomb to Williams, February 22, 1862, MS M292, LRAP.

[58] Macomb to Williams, February 19, 1862, MS M251, LRAP; also MS Letterbooks, Army of the Potomac, X, 646; MS Indorsement Book, Army of the
Potomac, XXIII, 278.

[59] Lowe to Steiner, February 21, 1862, MS, Lowe Papers.

[60] Lowe to Macomb, February 13, 1862, MS, Lowe Papers; Lowe to Macomb,
February 22, 1862, MS, *ibid.*

[61] Steiner to Lowe, March 1, 1862, MS, Lowe Papers.

ascensions for barely a week, and these operations were made possible only through the independent initiative of a naval officer to whom the balloonist had not been officially assigned to duty. Although the result of his observations in this one instance was successful and proved the value of his services, he was not allowed to perform further duty of any kind, but was compelled to spend months in idleness waiting for orders that were never issued.

The indifference of the general officers encountered by Steiner was far more pronounced than that which Starkweather had received from Sherman. When the aeronaut first arrived and reported to General Cullum, he was bluntly informed that no one at departmental headquarters knew anything about the intended balloon operations or the equipment that had been shipped there for Steiner's use. The officers to whom he appealed for assistance and recognition merely laughed at him.[62] After the aeronaut personally located his apparatus at the Cairo quartermaster depot and put it in order, he again reported to Cullum, who reiterated that he knew nothing of the matter, and would have nothing to do with the proposed aeronautic service. When asked to bring the matter to the attention of Major General Henry W. Halleck, commanding officer of the Department, Cullum refused.[63]

Resentful of this treatment, Steiner wrote several complaints to Lowe, in which his broken English did not conceal his opinion of the western commanders and their staffs. " I can not git eny ascistence here. Thay say thay know nothing about my balloon business [and] thay even laugh ad me. . . . Let me hear from you as soon as possible and give me a paper [order] from Headvuarters to show theas blockheads hoo I am," he wrote soon after his arrival.[64] After being ignored and refused assistance for two more weeks, he again urged Lowe to get some action taken in his behalf, and declared that " all the officers hear are as dum as a set of *asses*." [65] Lowe promptly for-

[62] *Ibid.*
[63] Steiner to Lowe, March 8, 1862, MS, Lowe Papers.
[64] Steiner to Lowe, March 1, 1862, MS, Lowe Papers.
[65] Steiner to Lowe, March 13, 1862, MS, Lowe Papers.

warded the complaints to Macomb,[66] who in turn took up the matter with the Adjutant General of the Army.[67] The original orders drafted to introduce Steiner to the western command and place him in proper relation with its staff had been in some way delayed in Washington, but were immediately sent on to Halleck when the aeronaut's difficulties had been made known.[68] Even then, Steiner's situation was no better. In the middle of April, nearly two months after he had first reported for duty, he advised Lowe that the general commanding at Cairo had denied receiving any orders regarding his services, and referred him to the headquarters of the Department of the Mississippi, at St. Louis.[69] Proceeding there, Steiner likewise obtained no satisfaction, and wrote on April 15:

I came here but I finde that no such papers arrivid here. I told the Gen. here how I was situated and ounder wot sircumstands I had bin send out here. He loock ad me a momment and then sed I can doo nothing for you and walk off, and such has bin my treedment every where.[70]

Eventually Steiner wrote directly to the Chief of the Topographical Bureau for copies of all the orders that had been issued in his behalf, and armed with these he again stated his case to Halleck. Although he was treated with more consideration and was furnished with the necessary materials for his balloon, he was ordered to remain at Cairo until further orders should call him for active duty.[71] The orders were never issued. A month later he reported that he had appealed again to

[66] Lowe to Macomb, March 13, 1862, MS, Lowe Papers.

[67] Macomb to Adjutant General Lorenzo Thomas, March 16, 1862, MS, Letters Received, Headquarters of the Army, War Department Division, National Archives [cited hereinafter as LRHQA].

[68] Lowe to Steiner, March 30, 1862, MS, Lowe Papers; Macomb to Steiner, April 26, 1862, MS, LBAP.

[69] The Department of the Missouri, The Department of Kansas, and the southwestern portion of the Department of the Ohio were consolidated into the Department of the Mississippi by the President's War Order of March 11. Halleck, with Headquarters at St. Louis, assumed command of the new department on March 13. President's War Order, No. 3, March 11, 1862, 1 O. R., VIII, 605; General Orders, No. 1, HDQ, Dept. Miss., March 13, 1862, ibid., p. 611.

[70] Steiner to Lowe, April 15, 1862, MS, Lowe Papers.

[71] Steiner to Lowe, May 7, 1862, MS, Lowe Papers.

General Halleck, now at Pittsburg Landing, but had been instructed to remain at Cairo until he should be sent for.[72] Finally, towards the middle of June, Steiner sent copies of all his credentials to General Pope, and begged that he be given an opportunity for field service. " I cannot see why I am kept out of active duty so long," he declared. " I am anxious to be placed in proper relation with your command." [73] After more than ten days elapsed without a reply, he decided to return east, even though he had no authority to do so. On June 28 he wrote Lowe and asked for transportation to rejoin the Army of the Potomac.[74]

In addition to rendering the balloon service in the West practically useless, the refusal to recognize Steiner and assign him to active duty imposed severe personal hardship on the aeronaut. The arrangement for salary and subsistence made by Macomb prior to Steiner's departure from Washington called for payment by the authorities of the western department. As a result, when these officers denied all responsibility in the matter, and made no effort to clarify the aeronaut's official status, Steiner found himself in a most unfortunate position. " I am here like a dog wisout a tail and I dond know ware I will be abel to draw my pay for no one seams to know eny thing abought this thing," he complained to Lowe.[75] A week later he wrote that unless the authorities in Washington sent direct orders, the officers at Cairo " will not pay me a cent." [76] In the meanwhile he was forced to meet all expenses from his own slender means, and by the middle of March he had exhausted his personal funds.[77] More than a month later he reported the same state of affairs. " I am treeded wis contempt and if I had the means to return to Washington I wold start today . . . now that I can git no pay

[72] Steiner to Lowe, June 9, 1862, MS, Lowe Papers.
[73] Steiner to Pope, June 16, 1862, MS, Letters Received, Army of the Mississippi, War Department Division, National Archives [cited hereinafter as LRAM].
[74] Steiner to Lowe, June 28, 1862, MS, Lowe Papers.
[75] Steiner to Lowe, March 8, 1862, MS, Lowe Papers.
[76] Steiner to Lowe, March 13, 1862, MS, Lowe Papers.
[77] Steiner to Lowe, March 8, 1862, MS, Lowe Papers; Steiner to Lowe, March 16, 1862, MS, *ibid.*

out here," he wrote bitterly.[78] How he managed to live under these circumstances, after his own money ran out, is difficult to imagine. Not until June, more than three months after his arrival, was he able to draw any compensation, and this was forthcoming only after he had sent to Washington for copies of the original orders sent out shortly after his departure.[79] Even then the paymasters granted him only half the amount due, and the remainder was still unpaid a year later, despite Lowe's efforts to secure a settlement of his assistant's account.[80] Such delays might possibly be understandable in the case of line troops on active campaign, but when the aeronaut was present at headquarters practically all the time, neglect of this sort, particularly since Steiner as a civilian could not even draw rations, appears utterly inexcusable.

The matter of supplies, though Steiner was given opportunity to use them but once, was equally mismanaged. When he arrived at Cairo, he found that the iron turnings shipped by the Chief of Ordnance were missing.[81] This material was not located for some two weeks.[82] After he had used most of his initial supply in the few ascensions he was permitted to make, Steiner was bluntly refused any assistance in procuring more. In April he wrote:

I have the balloon ready ad New Madrid wis a small quanty of material, not a nuff for one inflation. I can not git eny more as my requisision was refust by every quartermaster ad St. Louis, Cairo, and New Madrid. . . . I am now out of monny and creded . . .[83]

Although Halleck did eventually instruct his quartermasters to furnish the aeronaut with the supplies he requested, this action was not taken until two months after Steiner first arrived.[84]

[78] Steiner to Lowe, April 15, 1862, MS, Lowe Papers.

[79] Steiner to Lowe, June 9, 1862, MS, Lowe Papers.

[80] Certificate dated April 1, 1863, signed by T. S. C. Lowe, certifying to two months unpaid salary due John H. Steiner from the Western Department, MS, Lowe Papers.

[81] Steiner to Lowe, March 1, 1862, MS, Lowe Papers; Steiner to Lowe, March 8, 1862, MS, *ibid.*

[82] Steiner to Lowe, March 13, 1862, MS, Lowe Papers.

[83] Steiner to Lowe, April 15, 1862, MS, Lowe Papers.

[84] Steiner to Macomb, April 30, 1862, MS S228, Letters Received, Bureau of Topographical Engineers, War Department Division, National Archives.

This experience with pay and materials made Ingalls' administration of Lowe's corps in the East seem a model of efficiency.

The reasons for the indifference of Halleck, Pope, and their subordinates towards the aeronaut and his service are difficult to understand. The mere fact that Steiner's original credentials were delayed in arriving should not have been any legitimate excuse for the treatment that the balloonist received for months on end. The drafting of one or two letters or the sending of a telegram by some member of Halleck's staff would have cleared up Steiner's status at once and obviated all difficulties with regard to pay and supplies. But the delay in the initial orders from Washington can present no answer whatever to the question of the aeronaut's continued forced idleness. Even after supplementary orders and credentials were received and presented several times, neither Halleck nor Pope would allow the balloonist to take the field. It is clearly apparent that they had no interest in balloon observation and were not disposed to bother about its possibilities, even though advantages might be secured. Perhaps they were convinced in advance that balloon reconnaissance was humbug; but such a conviction does not explain why they did not allow Steiner to make at least a trial. Failure in such a test would have fully justified their adverse opinion and attitude; success would have gained them tactical benefits. In any case Halleck and his subordinates had nothing to lose in giving the aeronaut a chance. Obstinate, uncompromising indifference appears to be the only answer to their action in the matter.

After three months of vain application for service, Steiner arrived at the same conclusion. " I am satisfyde that Gen. Halleck is no friend to the Aeronautick Core," he wrote in June. " I could have bin of grade servis at Corince [Corinth], and explained it to General Halleck at Pittsburgh Landing, but he told me to stay ad Cairo ontill he wold sende for me." [85] The refusal of the generals in the western department also drew fire from one of the eastern papers, which seized this opportunity to criticize these officers in no uncertain terms:

[85] Steiner to Lowe, June 9, 1862, MS, Lowe Papers.

Captain Steiner, of Prof. Lowe's Corps of Observation, is now at Cairo . . . with all necessary appurtenances, for making reconnaissances, but very strangely, he has had no orders to proceed with his work. . . . But then, what can we expect when the matter is left to the discrimination of one or two fossils who have no time for anything else but to prepare official dispatches wherein " I " largely figures.[86]

Possibly Halleck resented the action of the authorities in the East in thrusting the aeronaut on his command. True enough, McClellan had not inquired whether the services of a balloonist were wanted, but resentment over such an issue is scarcely worthy of a resourceful and efficient general officer. It would have cost Halleck nothing in prestige or money to have tried out the possibilities of balloon observation, and opportunities were by no means lacking.

The forces in Halleck's department were active almost continuously throughout Steiner's period of forced idleness. While the aeronaut was vainly seeking to perform duty in the field, Pope's army moved on New Madrid, laid siege, and forced its evacuation. In April the bloody campaign of Shiloh was fought, and the siege and occupation of Corinth took place soon afterward. In all these operations, Steiner might have been useful had he been assigned to duty. Possibly observations from his balloon might have prevented the surprise accompanying the Confederate onslaught that fell upon the unsuspecting Union commander at Pittsburg Landing. In the few observations he was permitted to make at Island No. 10, he demonstrated that his services could be of definite value; but this success did not alter the attitude of the Department commander and his subordinate generals.

The suggestion might be made that Steiner was snubbed and refused proper recognition because of his uncouth, broken English and his illiterate, curiously spelled letters. Such an assumption can scarcely be taken seriously. The language difficulty might have caused him to make a less favorable impression when he first arrived, but this can hardly be credited as the cause of the contempt and indifference he received after two sets of credentials were sent to the western commanders in his

[86] Philadelphia *Inquirer*, March 14, 1862.

behalf. Steiner was not sent west to display a command of English; his function was to conduct balloon ascensions for the benefit of the Union forces, and in this he was well qualified despite his linguistic shortcomings. And this fact he demonstrated in his brief period of service during the investment of Island No. 10.

The few ascensions that Steiner was permitted to make at this time are of interest, first as the only active balloon operations that were conducted in the West throughout the War, and second as an example of the practical value of the aeronaut's services. The opportunity for performing this brief tour of duty came through the interest and cooperation of Commodore Andrew H. Foote, of the Navy. After the capture of New Madrid, Pope's next objective was the reduction of the Confederate fortifications on Island No. 10, which commanded the hairpin bend in the river some ten miles below the captured town. In this mission he was aided by the gunboat flotilla and mortar division under Foote, which had arrived above the island fortress on March 15.[87] The commodore had been at Cairo two days before, and Steiner, who had vainly sought active duty with the military forces, offered his services to Foote, who accepted them.[88]

A large flatboat was placed at the aeronaut's disposal, and on March 20 the *Eagle* and its generator were loaded on board and taken down the river to the gunboat fleet.[89] The preparations for the first aerial operations in this region drew widespread notice from the war correspondents of the western armies, and comments on Steiner's movement to join Foote's flotilla appeared in newspapers from Minnesota to the east coast.[90] Foote's naval batteries, supported by fire from the accompanying mortar-boats, had begun a bombardment of the

[87] Foote to Secretary Welles, March 17, 1862, 1 *O. R. N.,* XXII, 693.

[88] Steiner to Lowe, March 13, 1862, MS, Lowe Papers; Baltimore *Republican,* March 25, 1862.

[89] Steiner to Lowe, March 24, 1862, MS, Lowe Papers; Steiner to Lowe, March 27, 1862, MS, *ibid.*

[90] St. Paul *Pioneer and Democrat,* March 23, 1862; Cincinnati *Commercial,* March 22, 1862 [two dispatches]; *ibid.,* March 24, 1862 [two dispatches]; New York *Herald,* March 23, 1862; New York *Tribune,* March 24, 1862; *ibid.,* March 17, 1862; Philadelphia *Inquirer,* March 24, 1862; Baltimore *American,*

island fortress and its adjacent shore batteries on March 17, but for some days little progress in reducing the positions was effected. Commander Henry Walke of the gunboat *Carondelet* later recorded that the bombardment was " attended with such poor results that it became a subject of ridicule among officers of Pope's army." [91] Part of this failure seems to have been caused by inaccuracy of fire from the mortar batteries, for Steiner's activities were at once directed toward observation of this fire.[92]

His mission with the flotilla was twofold: first to inspect the enemy position and ascertain the exact location of his batteries; and second, to observe the Union mortar fire and report any error in range and deflection.[93] Since the mortar boats were anchored above the Island beyond the bend in the river, their officers were unable to observe accurately the effect of their high-angle fire which was delivered diagonally across the protruding neck around which the meandering river flowed.

Steiner's flatboat was anchored at a point just above Foote's flagship *Benton,* and on March 23 inflation was completed and preparations were made for ascension.[94] High winds, unfortunately, forestalled operations for the next two days.[95] By the

March 28, 1862; Baltimore *Republican*, March 29, 1862; Washington *Star*, March 24, 1862; Washington *National Republican*, March 24, 1862; Washington *Sunday Morning Chronicle*, March 23, 1862.

[91] Rear Admiral Henry Walke, " The Western Flotilla at Fort Donelson, Island No. 10, Fort Pillow and Memphis," *Battles and Leaders*, I, 441.

[92] Steiner to Lowe, March 25, 1862, MS, Lowe Papers.

[93] *Ibid.*; St. Louis *Daily Missouri Democrat*, March 28, 1862.

The correspondent described Steiner's mission in enthusiastic terms: " Information which can be obtained in this way [from the balloon] will be of great service to Commodore Foote. The enemy's forts can be closely inspected, the number of guns and their range obtained. It can be learned whether or not they have an infantry force in the rear, which for some time has been a matter of speculation. The attempt to evacuate can be immediately discovered, and the effect of our own shooting can be learned and corrected if in any respect at fault."

[94] Cincinnati *Commercial*, March 26, 1862; *ibid.*, March 27, 1862 [letter of " C. D. M." dated Flagship *Benton*, March 25, 1862]; Baltimore *American*, March 29, 1862 (reprint).

[95] Steiner to Lowe, March 27, 1862, MS, Lowe Papers; St. Louis *Daily Missouri Democrat*, March 24, 1862; *ibid.*, March 25, 1862; Cincinnati *Commercial*, March 25, 1862; Boston *Transcript*, March 25 and 26, 1862; New York *Herald*, March 25 and 26, 1862; New York *Tribune*, March 25, 1862; New York *Times*, March 26, 1862; Philadelphia *Press*, March 25 and 26, 1862; Philadelphia *Inquirer*, March 25, 1862; Baltimore *American*, March 25 and

25th the winds had abated, and in the calm that followed, Steiner went aloft at 3 p. m., to conduct the first ascension to be made in the western department. Again, extraneous circumstances conspired to prevent full success. The atmosphere was clouded with smoke from the continuous bombardment, and was also thick with haze from the river. Steiner was able to observe seven steamers anchored at the lower end of the Island, and reported that no enemy gunboats were in the vicinity, but poor visibility prevented his observing the enemy positions accurately or noting the effect of the Federal mortar fire.[96] The log of the mortar division, however, recorded that the " experiment proved satisfactory." [97]

The atmosphere cleared on the following day, and Steiner, accompanied by Colonel N. B. Buford and Captain H. E. Maynadier, officers commanding the mortar division, again ascended to observe the effect of the mortar fire. From the air, the position of the enemy batteries was clearly defined and observed. The mortar shells were seen to fall well beyond the hostile batteries, without noticeable effect on the enemy guns or the crews that manned them. The direction of fire was satisfactory, but the elevation and propelling charges were too great, thus causing the mortars to overshoot their targets. The defect was ordered corrected at once, with the result that the mortar fire became accurate and effective.[98] Commander Walke, though

27, 1862; Baltimore *Republican*, March 26, 1862; Washington *Star*, March 25, 1862; Washington *National Republican*, March 26, 1862.

[96] Steiner to Lowe, March 26, 1862, MS, Lowe Papers; St. Paul *Pioneer and Democrat*, March 28, 1862; St. Louis *Daily Missouri Democrat*, April 3, 1862; Cincinnati *Commercial*, March 27 and 28, 1862; Boston *Transcript*, March 28, 1862; New York *Herald*, March 28, 1862; New York *Tribune*, March 28, 1862; Philadelphia *Inquirer*, March 28 and April 3, 1862; Philadelphia *Press*, March 28, 1862; *ibid.*, March 31, 1862; Philadelphia *Inquirer*, March 31, 1862; Baltimore *American*, March 28, 1862; Washington *Star*, March 28, 1862; Washington *National Republican*, March 29, 1862.

[97] Abstract Log, Mortar Division, Western Flotilla, Entry of March 25, 1 *O. R. N.*, XXII, 771.

[98] Steiner to Lowe, March 27, 1862, MS, Lowe Papers; Commander A. M. Pennock (Fleet Captain, Western Flotilla) to Henry A. Wise [AIO, USN], March 27, 1862, 1 *O. R. N.*, XXII, 702; Frederick Milnes Edge, *Major General McClellan and the Campaign on the Yorktown Peninsula* (London, 1865), p. 32; St. Louis *Daily Missouri Democrat*, March 28, 1862 [second dispatch]; St. Paul *Pioneer and Democrat*, March 29, 1862; Cincinnati *Commercial*, March

he omits mention of the aerial observation, later wrote that these pieces threw their 13-inch shells so effectively that " the Confederates were driven from their batteries." [99] Within ten days, Pope's forces had effected a crossing of the river below the Island. With their communications thus cut off, and harassed with the fire from Foote's batteries, the Confederate garrison at Island No. 10 surrendered on April 7.[100] With this capitulation, Steiner's active service with the flotilla came to an end, and he was not again employed in the western department.

The failure of McClellan's attempt to extend the operations of the balloon corps to the western theatre of the war again illustrates the same lesson demonstrated by Starkweather's experience in the South. Without the cooperation and interest of the commanders to whom aeronauts were assigned, nothing could be accomplished. This was true in Steiner's case even after he showed that his work could be of practical value. The lack of success in the West also showed the fallacy of sending an aeronaut to a distant command without ascertaining the wishes of the general concerned, or making effective arrangements in advance. Regardless of any advance arrangements, however, the cardinal factor was the interest of the general in command, without which, no credentials were of value in furthering the newly adopted aeronautic service. Twenty-two years before, when Colonel Sherburne advocated the use of balloons in the Florida War, Secretary Poinsett had replied that the adoption of such a measure would be useless unless the commanding officers approved it, and added that such an innovation " can only succeed in willing hands." [101] The truth of the Secretary's statement cannot be better illustrated than by the experience of Steiner and Starkweather in 1862.

29, 1862; Boston *Transcript,* March 28, 1862 [second dispatch]; New York *Tribune,* March 29, 1862; New York *Herald,* March 29, 1862; New York *Times,* March 29, 1862; Philadelphia *Inquirer,* March 29, 1862; Philadelphia *Press,* March 29, 1862; Baltimore *American,* March 29, 1862; Washington *Star,* March 29, 1862.

[99] Walke, " The Western Flotilla," *Battles and Leaders,* I, 439.

[100] Foote to Welles, April 8, 1862, 1 *O. R. N., XXII,* 720; see also dispatches, correspondence, &c., April 1–April 8, 1862, 1 *O. R.,* VIII, 653-675.

[101] Poinsett to Sherburne, November 18, 1840, MS, Misc. File 284 [formerly AGO serial], War Department Division, National Archives. See also Chapter i, *supra,* p. 27.

INDEX

ABBOT, LIEUT. HENRY L., 53, 61, 72, 201, 298
Ascension with James Allen, 48-49
Describes failure of Allen's balloons, 51
Opinion on balloon observation, 49
Portrait, Plate VI
ABOUKIR, 13
Academie de Sciences,
Carnot's address before, on balloons, 5, n. 15
ACCOTINK, 369
ADAMS, JOHN QUINCY,
Diary quoted on French balloon service, 14
Adriatic (tug),
La Mountain's ascensions from, 103, 106
AERIAL BOMBARDMENTS,
Proposals for,
Cronstadt, 1854, 20
San Juan de Ulúa, 1846, 29-32
Sepoy Mutiny, 1857, 21
Southern cities, 1861, 104
Toulon, 1793, 5
Rogier's pamphlet on, 1819, 19
Venice, in siege of, 1849, 18
AERIAL BOMBING CHARTS,
Used in the siege of Venice, 1849, 18
AERIAL MAPS, 308, 320, 355-356
Suggested in 1783, 3
Facsimile of Col. Small's, Budd's Ferry, Plate XLIV
AERIAL PHOTOGRAPHY, 34, 171, 329-335
Balloon Corps, not practiced by, 329, 334-335
Editorial on military use of, 332-333
Errors in secondary works on use of, 329-330
First photographs from the air, 34, 330, and n. 115
Suggested for military operations by, *American Journal of Photography,* 333, n. 123
American Photographical Society, 331-333
Robert B. Benson, 333
Dr. John W. Draper, 332

W. G. Fullerton, 333, n. 125
Capt. E. B. Hunt, 333
R. A. Maxwell, 335
Charles Seely, 331, 333, n. 123
Payton Spence, 331
Capt. Albert Tracy, 108, 333
AERIAL SIGNALLING, (see also *Balloons, military*)
First suggested, 1783, 3
Lowe's long distance visual system, 326-329; drawings illustrating, Plate XLI
Tactical function of French balloon service, 9
Telegraphic, see *Balloons, military*
Visual, flag and pyrotechnics, 326
AERIAL TELEGRAPH,
First message sent by, 175
First practically demonstrated by Lowe, 171, 174-175
Press comment, on, 177-179
Use of, in military operations, 325-326
AERIAL TORPEDO BATTALIONS,
Organized by Austrian Army, 1849, 18
AERONAUTICS, MILITARY,
Austria, 1849, 18
Denmark, 1807, 15
England, 1803-1865, 19-21
Crimean War, 20
France, 1793-1802, 5-15
Algerian Campaign, 1830, 17
Second Republic, 21, n. 100, 260
Italian Campaign, 1859, 21-23
Italy, 1848, 17
Russia, 1812, 16-17
Siege of Sebastopol, 21
United States, prior to 1861, 23-32
Mexican War, 28-32
Seminole War, 24-28
Aérostiers, 1er Compagnie d',
Created, 1794, 8
Disbanded, 1802, 15
Operations of, 10 ff.
Organization of, 8-9
Tactical functions of, 9
Training of, 9
Uniform of, 8, and n. 32

Sends air photographs to Professor Bache, 331

BLAKE, CAPT. HENRY N., 357

BLANCHARD, JEAN PIERRE,
Makes first aerial crossing of English Channel, 23
Makes first successful ascension in America, 23

BLENKER, GEN. LOUIS, 125, 126, 127

BOATWRIGHT, W. H., 165

BOMBING, see *Aerial bombardments*

BORCETTE,
French balloon depot at, 11

BRENTSVILLE, 350, 352

BROOKS, S. M., 34

BROWN, JOHN, 159, n. 10.

BROWN, JUNIUS, 167, 168

BROWN, RICHARD,
Employed as barge-master for balloon-boat, 269

BRUSSELS,
French balloon operations at, 11

BRYAN, E. P., 217

BUDD'S FERRY (MD.), 136, 258, 261, 265, 347, 349, 350, 353, 354, 370, 373

BUFORD, COL. N. B.,
Ascends with Steiner at Island No. 10, 396

BUISSART, ANTOINE,
Correspondence with Carnot on balloons, 5, and n. 16

BULL RUN, BATTLE OF, 69, 71, 76, 91

BUREAU OF TOPOGRAPHICAL ENGINEERS, 60-61, 72, 74, 76, 77, 80, 182-184, 190, 197, 199, 200-201, 260, 280, 281, 285, 288, 294, 389

BURNS, GEORGE MCDOWELL,
Assists Lowe in first aerial telegraph experiment, 172, 174

BURNSIDE, GEN. AMBROSE E., 40, 44, 47, 50, 52, 261, 299, 309, 326-327, 328-329, 334
Estimate of James Allen, 55

BUTLER, GEN. BENJAMIN F., 87-90, 92, 95, 99, 103-105, 106-108, 111-112, 116, 140, 149-150, 174, 333, 374
Offers La Mountain post as military aeronaut, 85-86
Opinion on balloon reconnaissance, 86
Accompanies La Mountain on tug during ascensions, 103-106
His satisfaction and approval of La Mountain's work, 104-105

Interested in La Mountain's plan for aerial bombing, 106
Press comment on his interest in balloons, 106
His initiative, liberal views, and encouragement of balloon observation, 109
Portrait, Plate X

BUTLER, BENJAMIN F. (Secretary of War),
Praises Col. Sherburne's plan for balloon operations in Seminole War, 26

BUTT, CAPT. RICHARD, 202

BUTTERFIELD, GEN. DANIEL, 89, n. 22, 328

CAFARELLI, GENERAL,
Correspondence with Napoleon I on balloons, 14, n. 65

CAIRO (ILL.), 263, 297, 386, 387-389, 391-394

CALDWELL, JOHN, 164, n. 32 and n. 35

CAMERON, SIMON, 40, 63, 87-88, 113, 150, 170, 180-181, 226, 332
Favorable opinion on balloon reconnaissance, 87
Ignores Edward La Mountain's application for service, 109

CAMOUFLAGE,
To prevent balloon observation, 135, 136, 215 and n. 78, 216

CAMP SPRAGUE, 47

CAMP TYLER, 184

CAMP WILLIAMS, 122, 123

CARLYLE, JANE WELSH,
Mazzini discusses military balloons with, 17, n. 81
Opinion of Mazzini's scheme, 18, n. 81

CARNOT, LAZARE,
Urged to adopt balloons in French Army, 1793, 5
Reads paper on balloons before the *Academie de Sciences*, 5, n. 15
His *Mémoire sur les ballons*, 5, n.15
Commends balloon operations to Gen. Jourdan, 7

Carondelet (gunboat), 395

CARS, BALLOON, see *Balloons, military, construction.*
Construction of, for Wise's balloon, 66

CASE, PRIVATE JAMES F., 276